AF595025

Urban Planning and Sustainable Land Use

Urban Planning and Sustainable Land Use

Editors

Qingsong He
Jiayu Wu
Chen Zeng
Linzi Zheng

MDPI • Basel • Beijing • Wuhan • Barcelona • Belgrade • Manchester • Tokyo • Cluj • Tianjin

Editors
Qingsong He
College of Public
Administration
Huazhong University of
Science and Technology
Wuhan
China

Jiayu Wu
Insitute of Landscape
Architecture
Zhejiang University
Hangzhou
China

Chen Zeng
Department of Public
Management-Land
Management
Huazhong Agricultural
University
Wuhan
China

Linzi Zheng
College of Public
Administration
Huazhong University of
Science and Technology
Wuhan
China

Editorial Office
MDPI
St. Alban-Anlage 66
4052 Basel, Switzerland

This is a reprint of articles from the Special Issue published online in the open access journal *Sustainability* (ISSN 2071-1050) (available at: www.mdpi.com/journal/sustainability/special_issues/sus_urban_land_use).

For citation purposes, cite each article independently as indicated on the article page online and as indicated below:

LastName, A.A.; LastName, B.B.; LastName, C.C. Article Title. *Journal Name* **Year**, *Volume Number*, Page Range.

ISBN 978-3-0365-8047-0 (Hbk)
ISBN 978-3-0365-8046-3 (PDF)

Contents

Editorial

Urban Planning and Sustainable Land Use

Qingsong He

College of Public Administration, Huazhong University of Science and Technology, Wuhan 430074, China; heqingsong@hust.edu.cn

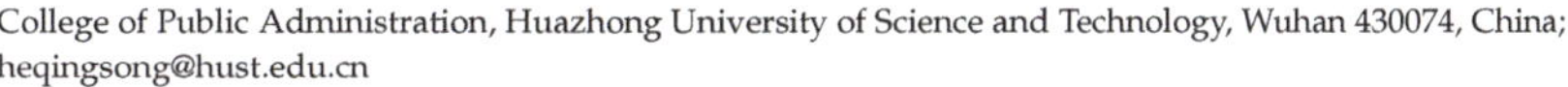

1. Introduction

The main purpose of this Special Issue is to gather the literature from diverse disciplines on contemporary urban planning and land use in different regions, in order to contribute to addressing the global challenges of sustainable urban development. We conduct research on sustainable urban land use by combining the perspectives of land resource sustainability and urban spatial structure, and by drawing upon disciplines such as urban planning, geographic information systems (GIS), ecology, and others. We aim to reveal the spatial patterns and dynamic changes in urban land use and summarize and propose strategies for implementing sustainable land utilization to meet the various requirements of urbanization, while minimizing harmful social and ecological impacts as much as possible. We are delighted to receive attention from scholars across different regions of the world. They have contributed new research cases, academic insights, and advanced technologies for sustainable land use studies. This has positive implications for us in terms of understanding spatial patterns, driving mechanisms, and predictive simulations of urban planning and land use with different backgrounds.

2. Overview of Published Papers

As of 5 June 2023, we have received a total of 23 submissions. After careful screening and professional review by the editorial team and expert reviewers, 13 papers have been accepted for publication. The author team consists of researchers from Australia, Turkey, South Korea, and China. The research topics can be categorized into three areas, which are spatial simulation/optimization models, indicator-based quantitative measurements/evaluations, and studies on planning practices and management.

2.1. The Development of Spatial Simulation/Optimization Models

The article 'Hybrid Economic–Environment–Ecology Land Planning Model under Uncertainty—A Case Study in Mekong Delta' [1] and the article 'A Two-Stage Fuzzy Optimization Model for Urban Land Use: A Case Study of Chongzhou City' [2] both propose land use optimization model tools based on uncertainty. The authors of the former article focus on three uncertainties in land use systems, which are interval uncertainty, fuzzy uncertainty, and random uncertainty, and propose an interval probabilistic fuzzy land use allocation (IPF-LUA) model to manage these uncertainties. They also apply their model to the Mekong Delta to help policy makers find a balance between economic and ecological efficiencies subject to different objectives in land use planning cycles. From a different perspective, the perspective of structural reform in land supply, the latter article utilizes Bayesian networks and fuzzy mathematical programming to investigate the impact mechanisms of land use under uncertain conditions in Chongzhou, China. It optimizes the land use structure, providing methodological references for the development of more sustainable and ecologically friendly land use plans.

The article 'Multi-Scenario Simulation of Urban Growth under Integrated Urban Spatial Planning: A Case Study of Wuhan, China' [3] and the article 'The Evolution of Ecological Space in an Urban Agglomeration Based on a Suitability Evaluation and

Citation: He, Q. Urban Planning and Sustainable Land Use. *Sustainability* **2023**, *15*, 9524. https://doi.org/10.3390/su15129524

Received: 12 June 2023
Accepted: 13 June 2023
Published: 14 June 2023

Cellular Automata Simulation' [4] both propose improved CA models. The former focuses on construction land as the main object, combining the trade-off between urban development and nature conservation with zone-based planning implementation mechanisms. By adjusting preference parameters in different planning zones, multiple planning constraint scenarios are generated. The authors use the case of the Wuhan Urban Development Area, China, to apply the model for simulating current and future urban scenarios. The results demonstrate that compared to the baseline model without planning constraints, the simulated model exhibits higher accuracy. The latter focuses on ecological land as the main object and simulates and predicts the ecological spatial evolution of the Changsha–Zhuzhou–Xiangtan urban agglomeration in China based on the suitability assessment of land use zoning. The results suggest that the total area of ecological space in this region will decrease by 2035.

The article 'An Innovative Framework on Spatial Boundary Optimization of Multiple International Designated Land Use' [5] focuses on addressing the issue of overlapping boundaries among different types of protected areas. It constructs a research framework called Candidate Area–Natural background–Heritage Resource–Construction (C-NHC) and takes the Jiangshan Nature Reserve in China as an example to optimize the boundaries of the protected area. This technical framework can provide a valuable reference for the boundary delineation of protected areas, not only in China but also worldwide.

The article 'Multi-Scenario Simulations of Land Use and Habitat Quality Based on a PLUS-InVEST Model: A Case Study of Baoding, China' [6] employs the PLUS-InVEST model and ESV model to simulate and analyze the land use and habitat quality changes in Baoding City, China, in four scenarios: natural development (ND), water protection (WP), forest rehabilitation (FR), and cultivated land protection (CP). The results demonstrate that the CP and FR scenarios will establish a land use pattern characterized by "high ecological quality and high value", which can effectively balance the economic development and ecological conservation in Baoding.

2.2. Index Measurement and Assessment

The article 'Does Agroforestry Correlate with the Sustainability of Agricultural Landscapes? Evidence from China's Nationally Important Agricultural Heritage Systems' [7] qualitatively evaluates the correlation between agroforestry and the comprehensive sustainability of the landscape. The research findings indicate that agriculture and forestry are closely associated with most sustainability indicators, such as biodiversity, income diversity, resource utilization, hydrogeological protection, and water resource management. Based on these findings, the article discusses the role of agroforestry in promoting sustainability.

The article 'Multi-Source Data-Based Evaluation of Suitability of Land for Elderly Care and Layout Optimization: A Case Study of Changsha, China' [8] utilizes multiple data sources and constructs suitability evaluation indicator systems for community-based residential land and institutional elderly care land development. The results reveal significant spatial variations in the suitability of elderly care facility land. The authors accordingly propose a coping strategy to establish a three-tiered senior care service structure of "district–county–community–street (township)" that effectively connects urban and rural areas.

The article 'Study on Green Utilization Efficiency of Urban Land in Yangtze River Delta' [9] contributes a set of indicators and evaluation methods to measure urban land green utilization efficiency. Building upon the calculation of urban land green utilization efficiency in the Yangtze River Delta region, China, the study employs the Dagum Gini coefficient, decomposition method, and exploratory spatial data analysis method to analyze the spatiotemporal evolution characteristics of urban land green utilization efficiency in the Yangtze River Delta region. The results indicate a fluctuating upward trend in urban land green utilization efficiency in the region, with the formation of two agglomeration patterns: high agglomeration in areas with the strongest comprehensive strength and low agglomeration in areas with the weakest comprehensive strength.

Based on the space syntax theory, the article 'Sustainable Planning and Design of Ocean City Spatial Forms Based on Space Syntax' [10] measures the sustainability of urban spatial morphology in Shenzhen Bay, China, and assesses whether the internal spatial configuration of the city can effectively support the sustainable and healthy operation of a marine city with different features. Based on the research findings, the authors summarize the strategies for the spatial design framework of a transitional marine city.

2.3. Urban Planning and Governance

The main contribution of the article 'A Low-Carbon Land Use Management Framework Based on Urban Carbon Metabolism: A Case of a Typical Coal Resource-Based City in China' [11] lies in its proposal for a land use management framework oriented towards low-carbon development in coal resource-based areas. It is based on the analysis of carbon metabolism networks in the urban three-life spaces of "ecology–production–living" and the identification of ecological relationships between different land use types. The application case in Yuling, China, demonstrates the effectiveness of this framework.

The article 'The Assessment of Greyfields in Relation to Urban Resilience within the Context of Transect Theory: Exemplar of Kyrenia–Arapkoy' [12] focuses on the buffer zone between urban and rural areas, known as "greyfields", and proposes a roadmap for redeveloping greyfields for public uses based on the framework of transect theory. This roadmap aims to guide future planning activities and enhance urban resilience.

The article 'Kerbside Parking Assessment Using a Simulation Modelling Approach for Infrastructure Planning—A Metropolitan City Case Study' [13] observes spatial views of parking through manually collected and video-captured camera data. It discusses the impacts of curbside parking demand and supply on short-term parking (STP) and freight activity space (FAS). Furthermore, it identifies the mismatched areas in time and space between parking demand and load distribution in Parramatta, Australia. The study explores the potential of reducing peak parking times by changing parking limits. The proposed benchmarking model provides recommendations for infrastructure planning and the formulation of travel demand management strategies in future predictive scenarios.

3. Discussion and Perspectives

Land is a finite resource. Effective land use research enables the optimal utilization of available space. It involves analyzing land suitability, determining appropriate land uses, and allocating land for residential, commercial, industrial, and recreational purposes to meet the needs of a growing population. Urban planning involves the systematic and strategic organization of land use within urban areas. It encompasses the allocation of land for various purposes such as residential, commercial, industrial, recreational, and public spaces. Urban planning and land use research play a crucial role in shaping sustainable, equitable, and livable cities, and provide the foundation for creating sustainable, inclusive, and resilient cities that improve the quality of life for their residents while balancing environmental and economic considerations.

The academic community welcomes the diverse development in this research field, but here I would like to emphasize the application of artificial intelligence (AI) in this domain. AI in urban planning and land use research holds significant potential for transforming how cities are planned, developed, and managed. For example, AI can analyze large and diverse datasets, including satellite imagery, sensor data, and social media feeds, to provide valuable insights for urban planners and land use researchers. It enables data-driven and evidence-based decision making by identifying patterns, trends, and correlations that were previously difficult to discern. AI algorithms can forecast future scenarios by integrating historical data with real-time information. This capability allows urban planners to anticipate population growth, traffic patterns, land use changes, and other factors, enabling more proactive and efficient planning processes. AI techniques, such as machine learning and computer vision, can extract valuable information from geospatial data. This includes land cover classification, urban form analysis, identifying vacant or underutilized land, and assessing the suitability of areas for specific purposes such as green spaces or affordable housing. AI also can facilitate public participation in the planning process by collecting and analyzing citizen input, sentiment analysis of social media data, and creating virtual environments for collaborative design. This allows for more inclusive and transparent decision-making processes.

However, while the application of AI in urban planning and land use research presents promising opportunities, it also raises concerns related to privacy, bias in algorithms, and ethical considerations. Therefore, careful consideration of these issues and robust governance frameworks are essential to ensure the responsible and inclusive implementation of AI in this domain.

Funding: This work was funded by the National Natural Science Foundation of China (ID. 42001334) and Independent Innovation Fund for Young Teachers of Huazhong University of Science and Technology (ID. 2021WKYXQN031 and 2022WKFZZX025).

Conflicts of Interest: The author declares no conflict of interest.

References

1. Ma, Y.; Zhou, M.; Ma, C.; Wang, M.; Tu, J. Hybrid Economic-Environment-Ecology Land Planning Model under Uncertainty—A Case Study in Mekong Delta. *Sustainability* **2021**, *13*, 10978. [CrossRef]
2. Yao, J.; Qiu, B.; Zhou, M.; Deng, A.; Li, S. A Two-Stage Fuzzy Optimization Model for Urban Land Use: A Case Study of Chongzhou City. *Sustainability* **2021**, *13*, 13961. [CrossRef]
3. Wang, H.; Liu, Y.; Zhang, G.; Wang, Y.; Zhao, J. Multi-scenario simulation of urban growth under integrated urban spatial planning: A case study of Wuhan, China. *Sustainability* **2021**, *13*, 11279. [CrossRef]
4. Chen, Y.; Zheng, B.; Liu, R. The Evolution of Ecological Space in an Urban Agglomeration Based on a Suitability Evaluation and Cellular Automata Simulation. *Sustainability* **2022**, *14*, 7455. [CrossRef]
5. Gao, H.; Weng, Y.; Lu, Y.; Du, Y. An Innovative Framework on Spatial Boundary Optimization of Multiple International Designated Land Use. *Sustainability* **2022**, *14*, 587. [CrossRef]
6. Hu, N.; Xu, D.; Zou, N.; Fan, S.; Wang, P.; Li, Y. Multi-Scenario Simulations of Land Use and Habitat Quality Based on a PLUS-InVEST Model: A Case Study of Baoding, China. *Sustainability* **2022**, *15*, 557. [CrossRef]
7. Zhang, M.; Liu, J. Does Agroforestry Correlate with the Sustainability of Agricultural Landscapes? Evidence from China's Nationally Important Agricultural Heritage Systems. *Sustainability* **2022**, *14*, 7239. [CrossRef]
8. Yang, J.; Lou, Z.p; Tang, X.p; Sun, Y. Multi-Source Data-Based Evaluation of Suitability of Land for Elderly Care and Layout Optimization: A Case Study of Changsha, China. *Sustainability* **2023**, *15*, 2034. [CrossRef]
9. Lin, Q.; Ling, H. Study on green utilization efficiency of urban land in Yangtze River Delta. *Sustainability* **2021**, *13*, 11907. [CrossRef]
10. Zhang, L.; Yuan, J.; Kim, C. Sustainable Planning and Design of Ocean City Spatial Forms Based on Space Syntax. *Sustainability* **2022**, *14*, 16620. [CrossRef]
11. Li, L.; Bai, Y.; Yang, X.; Gao, Z.; Qiao, F.; Liang, J.; Zhang, C. A Low-Carbon Land Use Management Framework Based on Urban Carbon Metabolism: A Case of a Typical Coal Resource-Based City in China. *Sustainability* **2022**, *14*, 13854. [CrossRef]
12. Akansu, V.; Karaman, A. The Assessment of Greyfields in Relation to Urban Resilience within the Context of Transect Theory: Exemplar of Kyrenia–Arapkoy. *Sustainability* **2023**, *15*, 1181. [CrossRef]
13. Samaranayake, P.; Gunawardana, U.; Stokoe, M. Kerbside Parking Assessment Using a Simulation Modelling Approach for Infrastructure Planning—A Metropolitan City Case Study. *Sustainability* **2023**, *15*, 3301. [CrossRef]

Article

Multi-Scenario Simulation of Urban Growth under Integrated Urban Spatial Planning: A Case Study of Wuhan, China

Haofeng Wang [1], Yaolin Liu [1,*], Guangxia Zhang [1], Yiheng Wang [1] and Jun Zhao [2]

[1] School of Resource and Environmental Science, Wuhan University, Wuhan 430079, China; hhfwang2009@whu.edu.cn (H.W.); zhangguangxia@whu.edu.cn (G.Z.); wangshulingxiao@whu.edu.cn (Y.W.)
[2] Qingdao Surveying & Mapping Institute, Qingdao 266000, China; whuzhaojun@163.com
* Correspondence: liuyaolin1999@126.com

Abstract: Although many publications have noted the impact of urban planning on urban development and land-use change, the incorporation of planning constraints into urban growth simulation has not been adequately addressed so far. This study aims to develop a planning-constrained cellular automata (CA) model by combining cell-based trade-off between urban growth and natural conservation with a zoning-based planning implementation mechanism. By adjusting the preference parameters of different planning zones, multiple planning-constrained scenarios can be generated. Taking the Wuhan Urban Development Area (WUDA), China as a case study, the planning-constrained CA model was applied to simulate current and future urban scenarios. The results show a higher simulation accuracy compared to the model without planning constraints. With the weakening of planning constraints, urban growth tends to occupy more ecological and agricultural land with high conservation priority. With the increase in preference on urban growth or natural conservation, the future urban land pattern will become more fragmented. Furthermore, new urban land beyond the planned urban development area can be captured in future urban scenarios, which will provide certain early warning. The simulation of the current urban spatial pattern should help planners and decisionmakers to evaluate the past implementation of urban planning, and scenarios simulation can provide effective support for future urban planning by evaluating the consequences.

Keywords: cellular automata; planning constraints; scenario simulation; urban growth

Citation: Wang, H.; Liu, Y.; Zhang, G.; Wang, Y.; Zhao, J. Multi-Scenario Simulation of Urban Growth under Integrated Urban Spatial Planning: A Case Study of Wuhan, China. *Sustainability* **2021**, *13*, 11279. https://doi.org/10.3390/su132011279

Academic Editors: Qingsong He, Jiayu Wu, Chen Zeng, Linzi Zheng and Tan Yigitcanlar

Received: 18 September 2021
Accepted: 12 October 2021
Published: 13 October 2021

Publisher's Note: MDPI stays neutral with regard to jurisdictional claims in published maps and institutional affiliations.

1. Introduction

Urbanization is one of the greatest social transformations in human history [1,2]. In order to meet the unprecedented demand of urban population growth and economic development, urban land presents a violent and disorderly expansion trend [3]. Nowadays, the average growth rate of urban land is twice that of their populations [4]. Although urban land is only a small part of the Earth's surface, its rapid growth has been identified as a major contributor to natural habitat loss, cropland reduction and ecosystem services degradation [5–7], posing great threat to the world's sustainable development [8,9]. Thus, it raises continuous competition for land between urban development and natural conservation [10], which is especially relevant to China due to its limited land resources and huge population living in urban areas [11]. Currently, Chinese local governments have put forward a series of administrative regulations and spatial planning policies, including major function-oriented zoning, land-use planning, urban spatial regulations, ecological redlines and so on. These policies and regulations have played a vital role in restraining the decrease in ecological and agricultural land and guiding urban compact development [12–14], although illegal urban encroachment on protected areas frequently occurs [15,16]. Therefore, simulation of future urban expansion and encroachment incorporating different planning policies is of great significance for authorities and planners to make effective decisions in optimizing urban spatial patterns and achieving urban sustainable development.

Since the first application of Tobler in 1979 [17], cellular automata (CA) and its derived models have been widely used in urban growth modeling and land-use simulation due to their simple and flexible structure, among which the transition rule is a core part [7,18,19]. The process of urban growth and urban land change is influenced by multiple drives across different scales, including socio-economic factors, geophysical factors, proximity factors, environmental factors and planning factors [18,20–22]. In order to link these multiple driving factors with the transition rules of the CA model, scholars have explored various methods including multivariate statistics methods, intelligent algorithms, agent-based models and multi-criterion decision analysis [3,23–25]. Although the combination of these models and cellular automata has made many important achievements in urban land-use simulation, the applicability of these models in reproducing urban patterns resembling reality and solving actual planning problems in complex urban systems is still challenging [18,20].

In recent decades, many researchers have increasingly recognized the important influence of urban planning and other policies on urban growth and land-use change [15,26,27]. Thus, scenario simulation of urban land use incorporated with planning information and conventional urban CA models has attracted considerable attention from scholars and planners. Through the constraint layers or exclusion layers in urban CA models, policy information including supportive planning and restrictive planning can be implemented to simulate and predict urban growth by either a Boolean-constraint mechanism or a gradual mechanism [27–29]. With regard to Boolean-constraint, for example, He et al. consider the primary farmland protection area as mandatory constraints to prohibit urban encroachment [12]; Zhou et al. set water areas, historical protection areas, restricted development areas, prohibited development areas and large-scale parks in planning as constraint layers to simulate urban spatial patterns under the planning intervention scenario [3]. While the binary planning constraint design is too absolute and simplistic [30], some studies have attempted to simultaneously consider the gradual constraints into their urban CA models. Onsteda et al. incorporated zoning policy information into the CA-based SLEUTH model by assigning gradual values to the different zoning categories, which obviously improves model performance with respect to a model without zoning [26]. Liang et al. proposed a random seeding mechanism to simulate the potential effect of planning development zones [20]. Wang et al. integrated a series of urban spatial planning to construct synthetical planning constraint layers in urban CA model by a multi-criteria evaluation (MCE)-based weighting method [31]. These studies have achieved remarkable results and provide good references for building a realistic planning constraint mechanism.

However, in previous studies, the planning constraints in a planning zone are often set to a unified value, ignoring the spatial heterogeneity of different cells in the same planning zone [32,33]. In the real complex urban systems, the planning influence on the cells in a specific planning zone is often not homogeneous. The application of planning constraints in urban CA models should not be restricted to considering the zone effects; it should take the natural environment and socio-economic factors based on the land-use cell into account [34–36]. For example, in the planned ecological protection area, cells with high ecological importance tend to be retained, while cells with low ecological importance located in urban fringe areas are more likely to be occupied. Currently, the influence effects of planning policies based on the trade-off between urban growth and conservation has received little attention. In addition, existing studies often set the restriction or impact coefficients of different planning zones based on subjective or historical experience [27,37]. Once these parameters are determined, they will not change with time in the simulation process, which is not conducive to the dynamic urban growth scenario simulation.

This study aims to explore the potential influence of integrated urban spatial planning on urban growth and realize the planning-constrained urban multi-scenario simulation. Thus, we propose an innovative synthetical planning-constrained strategy to quantitatively characterize the influence of integrated urban spatial planning on realistic urban growth process by combining the cell-based trade-off between urban growth and conservation with a zoning-based planning implementation mechanism. By incorporating the strategy into urban CA models, a planning-constrained CA model is developed. Taking Wuhan city, one of the mega cities in China, as a case study, we applied the planning-constrained CA model to assess the impact of urban planning implementation on the urban growth process and simulate the urban spatial patterns under different future scenarios. The results should help planners and authorities evaluate the guiding effects of urban planning implementation.

The rest of the sections of this article are organized as follows. Section 2 describes the study area and the datasets used in this study. Section 3 introduces in detail the methodology of the planning-constrained CA model. Section 4 presents the multi-scenario simulation results under planning constraints. The discussion and conclusion are given in Sections 5 and 6.

2. Study Area and Datasets

2.1. Study Area

Wuhan, the capital city of Hubei Province, is the key city in Central China and one of the core cities in the Yangtze River economic belt. As shown in Figure 1, Wuhan administers seven central districts (Hanyang, Hongshan, Jiangan, Jianghan, Qiaokou, Qingshan, Wuchang) and six suburban districts (Caidian, Dongxihu, Hannan, Huangpi, Jiangxia, Xinzhou), covering a total area of 8569.15 km^2. In past decades, Wuhan has experienced an unprecedented urbanization process. By the end of 2020, the total population of Wuhan was 12.32 million, of whom 10.39 million were urban residents. Rapid urbanization has led to a large amount of urban land expansion and encroachment on other lands, resulting in strong contradictions between urban growth and natural conservation [38,39]. From 2002 to 2019, the area of urban construction land increased by 98.7%, from 1338.75 km^2 to 2646.26 km^2. Therefore, Wuhan has formulated a series of spatial planning policies, which have greatly affected the process of urban growth [11,13,28]. In this study, we will specifically target the Wuhan Urban Development Area (WUDA), which was designated as the centralized area of urban development in the 2010–2020 Wuhan Master Plan. The WUDA spans a total area of 3261 km^2, consists of all central districts and parts of the six suburban districts.

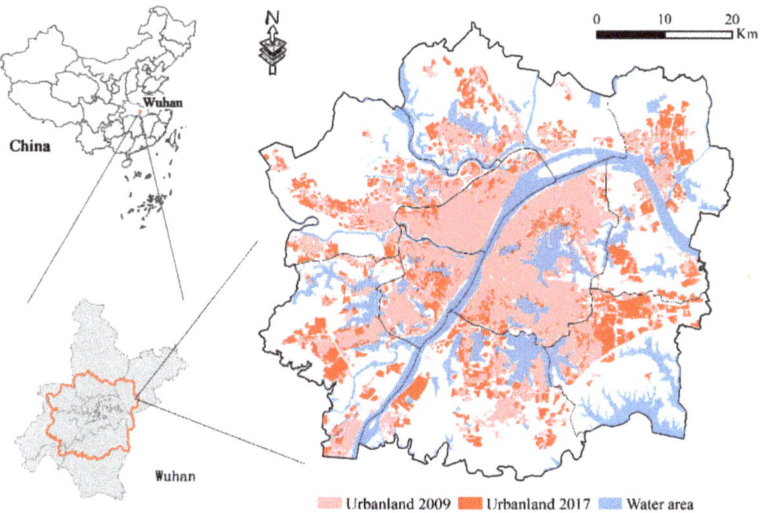

Figure 1. Location of the Wuhan Urban Development Area (WUDA).

2.2. Datasets

The datasets used in this study consist of land-use, socioeconomic, natural eco-environment and planning datasets, including the following:

1. Land-use maps of 2009 and 2017 in vector format were obtained from Wuhan Natural Resources and Planning Information Center. The original land-use maps were categorized into 38 classes and, then, reclassified into nine land-use types: urban land, rural settlements, cropland, orchard land, forests, grassland, waterbodies, traffic land and other land.
2. Socioeconomic datasets were collected from various sources. The demographic and GDP data of 2015 were downloaded from the Resource and Environment Data Cloud Platform website http://www.data.ac.cn (accessed on 20 October 2018). POI data were obtained from Baidu Map website http://map.baidu.com (accessed on 20 December 2017). The vector road network and park data were derived from Wuhan Geographical Conditions Monitoring Datasets, which are produced by Wuhan Geomatics Institute in 2017. Other socioeconomic datasets, including land value, urban centers, commercial centers and subway stations, were downloaded in picture format from the Wuhan Natural Resources and Planning Bureau website http://zrzyhgh.wuhan.gov.cn/ (accessed on 6 January 2019) and, then, digitized in ArcGIS 10.2 software. These socioeconomic datasets were mainly used to process the driving factors of the urban CA model.
3. Natural eco-environment datasets include topographical, meteorological, soil, vegetation and farmland quality data. The topographical information was derived from a digital elevation model (DEM) in raster format with a resolution of 30 m $\times$ 30 m, supplied by Geospatial Data Cloud website http://www.giscloud.cn/ (accessed on 30 January 2017). Meteorological data for 2015 obtained from China Meteorological Data Sharing Service System website http://data.cma.cn/site/index.html (accessed on 9 March 2016) include monthly air temperature, monthly rainfall and monthly radiation. Soil data, such as soil type, soil particle proportion and soil organic matter, were extracted from the Harmonized World Soil Database version 1.2 website http://westdc.westgis.ac.cn (accessed on 30 August 2019). Vegetation data, incorporating normalized difference vegetation index (NDVI) and leaf area index (LEI), were obtained from MODIS products. Farmland quality data were derived from the grade evaluation result of cultivated land quality in 2015, produced by Wuhan Natural Resources and Planning Information Center. These natural eco-environment datasets were mainly utilized to evaluate the cell-based conservation priority.
4. Planning datasets, consisting of the Wuhan City Master Plan (2010–2020), Wuhan Land Use Master Plan (2006–2020) and its adjustment and improvement result (2017), Wuhan Basic Farmland Protection Plan (2006–2020), Wuhan Comprehensive Transportation Plan (2009–2020) and Wuhan Ecological Protection Plan (2012), were acquired from Wuhan Natural Resources and Planning Bureau website http://zrzyhgh.wuhan.gov.cn/ (accessed on 30 January 2019). These planning datasets were integrated to produce urban growth scenarios based on dynamic planning constraints.

All these datasets were preprocessed and rasterized with a spatial resolution of 30 m in ArcGIS 10.2 software as input of the urban CA model.

3. Methodology

A planning-constrained CA model is developed in this study to simulate urban growth and encroachment under different scenarios. Figure 2 shows the model procedure and flowchart. Firstly, cell-based conservation priority and urban growth potential can be calculated by multi criteria analysis and a logistic regression model, respectively. In terms of conservation priority, it consists of ecological land conservation priority and cultivated land conservation priority. The ecological priority is mainly based on relevant indicators such as ecosystem functional importance and ecological vulnerability, while the cultivated land conservation priority mainly depends on the agricultural production suitability of cultivated land. The urban development potential is determined by natural, socioeconomic and location-based factors. Secondly, we defined the synthetical planning-constrained mechanism by integrating the cell-based tradeoff between urban growth potential and conservation priority with zoning-based planning preference. By adjusting the preference parameters of different planning zones, multiple planning-constrained urban growth scenarios can be generated. Thirdly, we combine the synthetical planning-constrained mechanism with urban development potential, neighborhood effect and random disturbance to construct the transition rule of urban CA model. Markov chain is used to predict the amount of urban land-use change. Kappa and FoM coefficients are used to calibrate and verify the model. Finally, future urban growth pattern under different scenarios are simulated. These phases are presented separately as following.

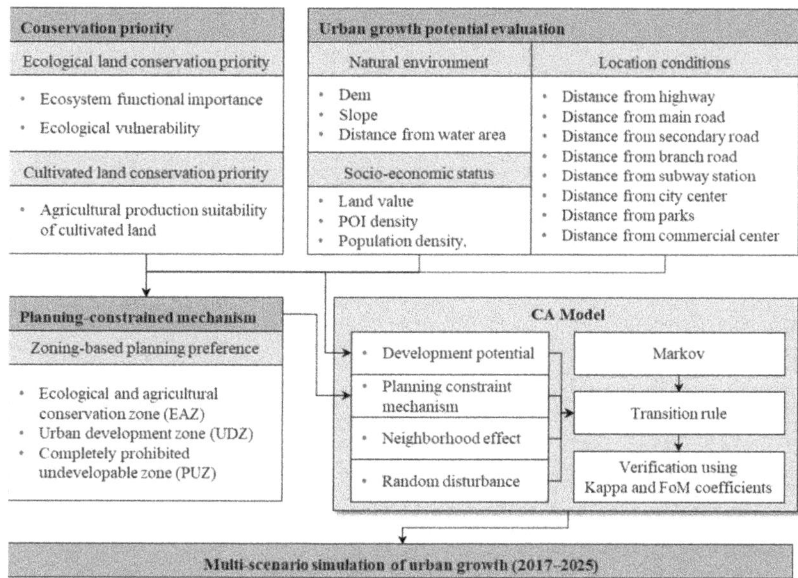

Figure 2. Flowchart for planning-constrained CA model.

3.1. Cell-Based Conservation Priority and Urban Growth Potential

3.1.1. Conservation Priority Evaluation

Urban growth is often at the expense of the loss of ecological land and cultivated land [12,40,41]. In this study, conservation priority includes ecological land conservation priority (ELCP) and cultivated land conservation priority (CLCP). For an ecological land unit such as forest land, grassland, water area and garden land, its conservation priority depends on relevant factors of ecosystem functional importance and ecological vulnerability [42,43], while for a cultivated land unit, its conservation priority should also consider agricultural production suitability [44].

The ecosystem functional importance refers to the ecological conditions and utility formed in the ecosystem and its ecological process, which are conducive to human survival and development [45,46]. According to the ecological status of the study area, biodiversity conservation (BC), carbon storage (CS), water conservation (WR), soil conservation (SC) and flood regulation (FR) are selected as the factors for the evaluation of ecosystem functional importance. Ecological vulnerability refers to the sensitive response and self-recovery ability of ecosystem relative to external interference at a specific time and space scale, wherein a higher vulnerability indicates worse self-recovery ability of the ecosystem [47]. Here, ecological vulnerability includes soil erosion sensitivity (SES), land desertification sensitivity (LDS) and stony desertification sensitivity (SDS). The acquisition and calculation of ecological related factors are mainly based on the Technical Guide for Delimitation of Ecological Protection Red Line issued by the Ministry of Environment Protection of China in 2015 and partly adjusted according to the study area. Table 1 presents the methods for ecosystem functional importance evaluation. Cultivated land quality grade data were used to describe the agricultural production suitability. Then, through ArcGIS 10.2 software, each factor was reclassified to values ranging from 0 to 1, by using the standardized function. The higher the value, the higher the conservation priority. The weight of each factor is obtained by the analytic hierarchy process (AHP) model. Finally, the conservation priority equation can be expressed as:

$$C_i = w_e \times (ELCP) + w_c \times CLCP \tag{1}$$

$$ELCP = \sum_{k=1}^{M} b_k x_k \tag{2}$$

where C_i is the conservation priority for cell i, ranging from 0 to 1; w_e and w_c are the weights of $ELCP$ and $CLCP$, respectively; w_e = 1.0 and w_c = 0.0 for an ecological land unit, while w_e = 0.5 and w_c = 0.5 for a cultivated land unit; x_k represents the value of k_{th} factor, and b_k is the corresponding weight. M is the total number of variables. Figure 3 illustrates the results for conservation priority evaluation.

Table 1. Methods for ecosystem functional importance evaluation.

Indicators	Models	Parameter Explanation
Biodiversity conservation (BC)	The habitat quality model of InVEST: $Q_{xi} = H_i(1 - (D_{xi}^2 \ (D_{xi}^2 + k^2))\ )$	Q_{xi} is the habitat quality index of patch group x in $LULC_i$(dimensionless). H_i and D_{xi} are the habitat suitability score and the total stress level of grid x in $LULC_i$, respectively. k is the scale factor (constant).
Carbon storage (CS)	Carnegie–Ames–Stanford approach model (CASA): $NPP_{xt} = \mathrm{APAR}_{xt} \times \varepsilon_{xt}$; $APAR_{xt} = SOL_{xt} \times FRAP_{xt} \times 0.5$; $\varepsilon_{xt} = T1_{xt} \times T2_{xt} \times W_{xt} \times \varepsilon_{max}$	NPP_{xt} is the ecosystem net primary productivity of pixel x in tst month ($\mathrm{gC{\cdot}m^{-2}{\cdot}month^{-1}}$), $APAR_{xt}$ and ε_{xt} are its corresponding absorbed photosynthetic effective radiation and actual light energy utilization, respectively. SOL_{xt}, $FRAP_{xt}$,and W_{xt} are the total solar radiation, the absorption ratio of vegetation layer to incident photosynthetic effective radiation, the influence coefficient of water stress of pixel x in tst month, respectively. $T1$ and $T2$ indicate the stress effect of low and high temperature on light energy utilization, ε_{max} is the maximum light energy utilization under ideal conditions.
Water conservation (WR)	Water balance equation: $\mathrm{WR} = \sum_{i=1}^{n} WR_i; WR_i = (P_i - R_i - ET_i) \times A_i \times 10^3$	WR_i is the water conservation capacity of type i (m^3), and n is the number of ecosystem types. P_i, R_i, ET_i and A_i are the rainfall, surface runoff, evapotranspiration and ecosystem area of type i, respectively.
Soil conservation (SC)	Revised universal soil loss equation (RUSLE): $SC = S_p - S_a; S_p = R \times K \times LS$; $S_a = R \times K \times LS \times C \times P$	SC is the quantity of soil conservation ($\mathrm{t/hm^2{\cdot}a}$), S_p and S_a are the quantity of potential and actual soil erosion, respectively. R, K, LS, C and P are the rainfall erosivity, soil erosion factor, slope factor, vegetation cover factor and the soil conservation practices factor, respectively.
Flood regulation (FR)	The assessment is based on the water level and area of the water body.	The main rivers of the Yangtze River and Han River, large lakes (>1 km^2), large- and medium-sized reservoirs are extremely important flood control ecological function areas; others belong to important areas.
Soil erosion sensitivity (SES)	$SES = \sqrt[4]{R \times K \times LS \times C}$	SES is the sensitivity index of soil erosion (dimensionless). R, K, LS and C are the sensitivity grades of rainfall erosivity, soil erosion, slope factor and vegetation cover, respectively.
Land desertification sensitivity (LDS)	$LDS = \sqrt[4]{I \times W \times K \times C}$	LDS is the sensitivity index of land desertification (dimensionless). I, W, K and C are the sensitivity grades of dryness index, sand-driving wind days, soil texture and vegetation cover, respectively.
Stony desertification sensitivity (SDS).	$SDS = \sqrt[3]{D \times S \times C}$	SDS is the sensitivity index of stony desertification (dimensionless). D, S and C are the sensitivity grades of exposed area percentage of carbonate stony, terrain slope and vegetation coverage, respectively.

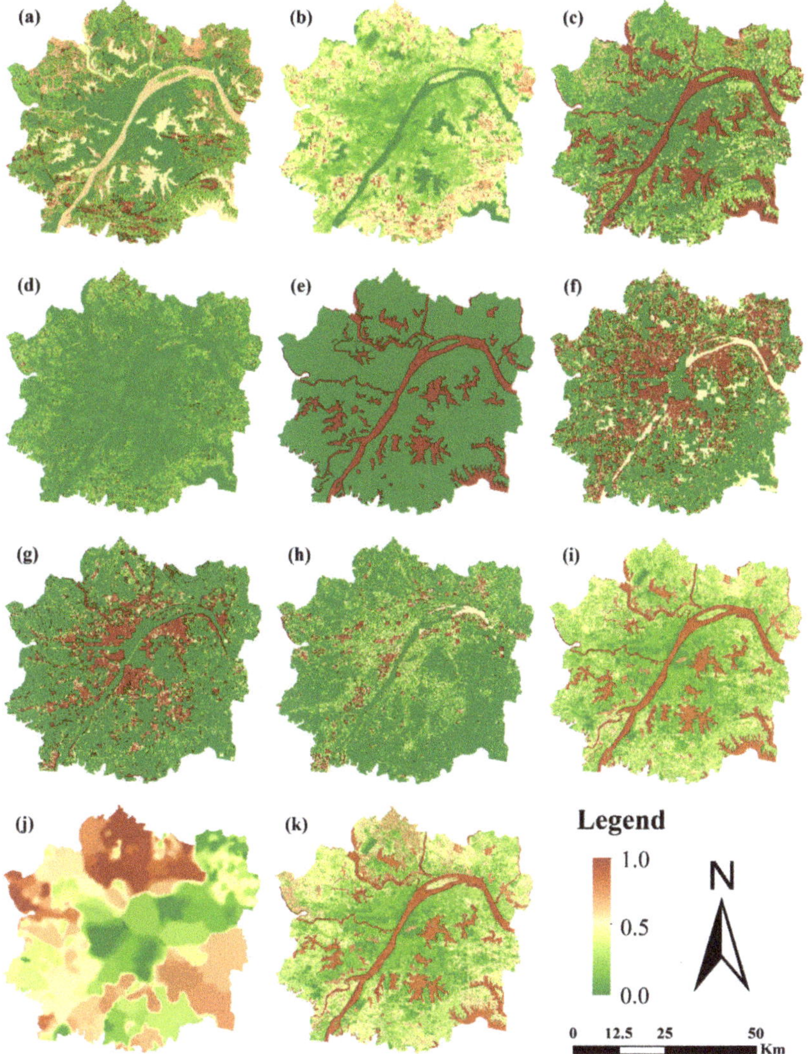

Figure 3. Conservation priority evaluation indicators and results. (**a**) biodiversity conservation (BC); (**b**) carbon storage (CS); (**c**) water conservation (WR); (**d**) soil conservation (SC); (**e**) flood regulation (FR); (**f**) soil erosion sensitivity (SES); (**g**) land desertification sensitivity (LDS); (**h**) stony desertification sensitivity (SDS); (**i**) ecological land conservation priority (ELCP); (**j**) cultivated land conservation priority (CLCP); (**k**) conservation priority.

3.1.2. Urban Growth Potential Evaluation

While urban growth is affected by various factors, the literature shows that the following factors are crucial: natural environment, socio-economic status and location conditions [22,34,48]. A variety of methods, such as multivariate statistics, intelligent algorithms and multi-criterion analysis, were used by researchers to define the contribution of diverse driving factors to historical urban growth [18,49–51]. In this study, a widely used binary logistic regression (BLR) method [6] was used to produce the urban growth potential map. The BLR equation can be expressed as:

$$U_i = \frac{exp\left(b_0 + \sum_{k=1}^{N} b_k x_k\right)}{1 + exp\left(b_0 + \sum_{k=1}^{N} b_k x_k\right)} \tag{3}$$

where U_i is the urban growth potential for cell i, ranging from 0 to 1. b_0 is the constant, x_k represents the k_{th} independent variable, and b_k is the corresponding weight. N is the total number of independent variables.

Based on previous researches [15,51], we selected 14 driving factors, such as dem, slope, distance from water area, distance from highway, distance from main road, distance from secondary road, distance from branch road, distance from subway station, distance from city center, distance from parks, distance from commercial center, land value, POI density and population density, to evaluate their impacts on Wuhan urban growth and produce the urban growth potential map. In addition to the population density data from 2015, other factor layers were acquired in 2017. We mapped all factors using ArcGIS 10.2 software and processed them by standardized methods. Thirty thousand sample points were extracted from the actual urban land-use change map (2009–2017) and the variable maps by a stratified random sampling method to construct the logistic regression equation. Table 2 presents the variables and their coefficients of BLR model. Figure 4 illustrates their spatial distributions and the final urban growth potential map.

Table 2. Independent variables and coefficients of logistic regression model.

Independent Variables	Coefficients	Sig.
Constant	1.788	0.000
Dem	−2.507	0.000
Slope	−0.801	0.000
Distance from water area	1.763	0.000
Distance from highway	−1.557	0.000
Distance from main road	−6.135	0.000
Distance from secondary road	−2.627	0.000
Distance from branch road	−1.481	0.000
Distance from subway station	−0.299	0.000
Distance from city center	−2.424	0.000
Land value	6.358	0.000
Population density	8.528	0.000

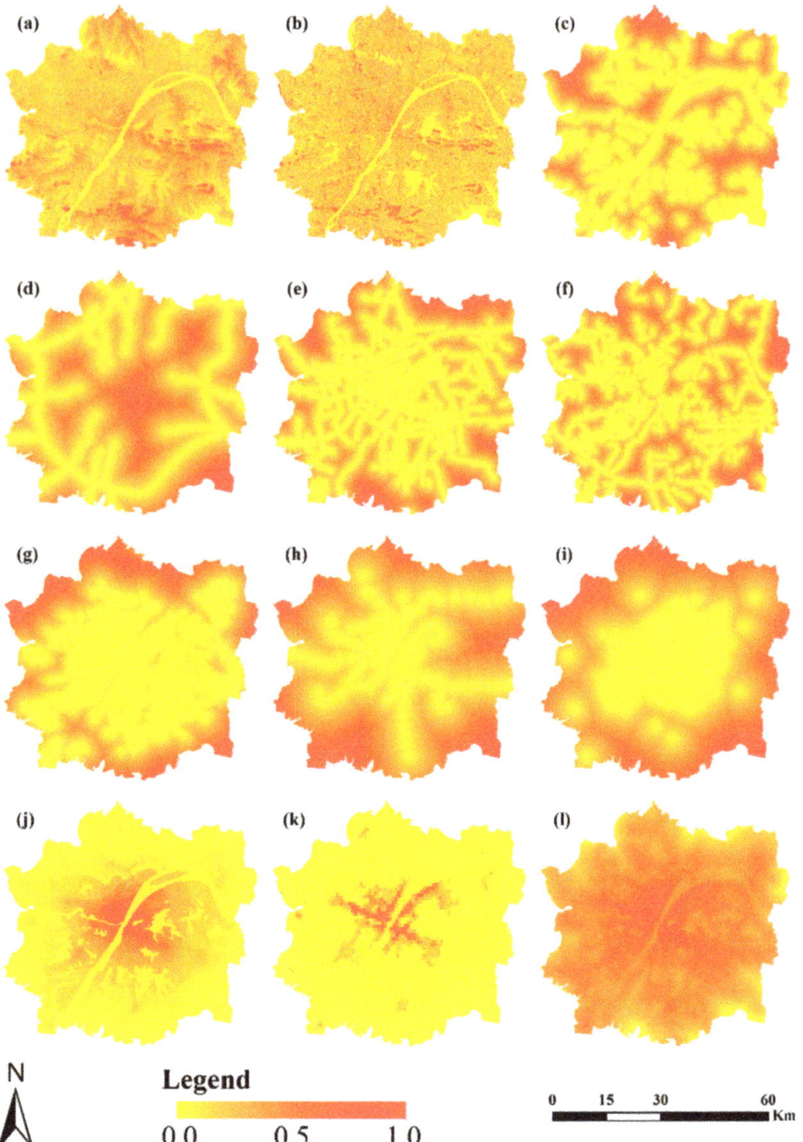

Figure 4. Urban growth potential evaluation indicators and results. (**a**) dem; (**b**) slope; (**c**) distance from water area; (**d**) distance from highway; (**e**) distance from main road; (**f**) distance from secondary road; (**g**) distance from branch road; (**h**) distance from subway station; (**i**) distance from city center; (**j**) land value; (**k**) population density; (**l**) urban growth potential.

3.2. Planning-Constrained Mechanism

Urban spatial planning, widely used in urban growth management, has an obvious impact on guiding and controlling the urban development [3,20]. Planning constraints have become an indispensable part of urban simulation. Different planning zones represent the different preferences of the government and planners for urban development and natural conservation. Previous studies established the planning constraint mechanism through binary planning constraints or preset planning zoning preferences [32,33], ignoring the cell-based trade-off between urban growth and conservation in the same planning zone [35]. In this study, by integrating the cell-based tradeoff between urban growth potential and conservation priority with zoning-based planning preference, a synthetical planning-constrained mechanism was defined as follows.

$$PC_i = f(U_i, C_i, Z_j) = \frac{U_i \times Z_j}{U_i \times Z_j + C_i \times (1 - Z_j)} \tag{4}$$

where PC_i is the planning constraints for cell i, ranging from 0 to 1. U_i and C_i are the value of urban growth potential and conservation priority, respectively. Z_j represents the planning preference on urban growth in planning zone j, ranging from 0 to 1. Z_j = 0 means planning zone j is completely unbuildable, Z_j = 1 means planning zone j has no planning restrictions on urban development.

In past decades, Chinese local governments have issued a series of spatial planning policies to restrain the decrease in ecological and agricultural land and guide urban compact development, among which Overall Land Use Planning, Urban Master Planning, Primary Farmland Protection Planning and Ecological Protection Planning played extremely important roles [13,28,52]. By integrating these plans of the study area, we classified the study area into three categories of planning zones: the urban development zone (UDZ), ecological and agricultural conservation zone (EAZ) and completely prohibited undevelopable zone (PUZ). The UDZ refers to areas where cities and towns are intensively developed and constructed and can meet the needs of urban production and life. The value of the ZUDZ is set to 1, considering a higher preference on urban growth in the UDZ. The PUZ includes primary farmland protection area, mountains, water bodies and natural reserves wherein urban development is strictly prohibited. Thus, the value of ZPUZ is set to 0 in this study. The EAZ refers to agricultural and ecological areas besides the PUZ that need to retain their original appearance, strengthen natural conservation and ecological construction and restrict urban development. The value range of the ZEAZ is [0,1]. By adjusting the planning preference parameter in the EAZ, various planning constraints can be generated. Furthermore, considering the timeliness of planning, planning in different time stages should be applied to build the corresponding planning zones. In this study, the planning zones of 2009–2017 and 2017–2025 were established, respectively, and applied to simulate the urban growth process of corresponding phases.

3.3. Planning-Constrained CA Model

CA is a dynamic system with a finite set of state elements, which evolves in discrete time intervals based on transition rules [53,54]. CA-based models have been widely used in urban growth modeling and land-use simulation due to its simple and flexible structure. In general, the conceptual formula of the urban CA model can be described as follows [48]:

$$S_i^{t+1} = f(S_i^t, U_i, \Omega_i^t, Con, R) \tag{5}$$

where S_i^{t+1} and S_i^t denote the state of the cell i at the moment of $t+1$ and t, respectively; U_i is the urban growth potential of cell i, Ω_i^t is the neighborhood effect, Con is a constraint function, R is the random perturbation, and f represents the function of transition rules.

Specifically, the urban growth potential has been described in Section 3.1.2. The neighborhood effect is defined and calculated based on the ratio of adjacent developed cells by using a regular $n \times n$ Moore neighborhood [25], which is expressed in Equation (6). Con can be replaced by PC_i, which has been introduced in Section 3.2. R represents the uncertainties and random perturbations in the urban land evolution process, which is calculated by Equation (7).

$$\Omega_i^t = \frac{c}{n \times n - 1} \tag{6}$$

$$R = 1 + (-ln(\gamma))^{\alpha} \tag{7}$$

where c is the number of adjacent developed cells, γ is a random number ranging from 0 to 1, and α is a parameter for adjusting the stochastic magnitude, which was set to 1 to represent a small probability of randomness in this study.

By combining the aforementioned planning-constrained mechanism with urban development potential, neighborhood effect and random disturbance, the final transition probability TP_i^t of urban CA model can be expressed as follows:

$$TP_i^t = U_i \times \Omega_i^t \times PC_i \times R \tag{8}$$

Furthermore, the amount of urban land development was calculated by the Markov chain model, which has been widely and successfully used in the quantitative prediction of land-use change [3,55]. The Kappa index [56] and *FoM* coefficient [57] were used to calibrate the urban CA model and assess the simulation outcomes. By a pixel-by-pixel comparison between the simulation result and reference map, the Kappa index and *FoM* coefficient can be calculated as follows:

$$Kappa = \frac{P_o - P_c}{1 - P_o} \tag{9}$$

$$P_c = \frac{a_0 \times b_0 + a_1 \times b_1}{N^2} \tag{10}$$

$$FoM = \frac{B}{A + B + C} \tag{11}$$

where P_O is the observation consistency, which refers to the ratio of the number of correctly simulated cells to the total number of cells; P_C is the expected consistency; N is the total number of cells; a_0 and a_1 are the number of non-urban and urban cells, respectively; b_0 and b_1 are the number of cells simulated as non-urban land and urban land, respectively; A is the number of error cells observed as urban but simulated as persisted non-urban; B is the number of correctness cells observed and predicted as change; C is the number of error cells observed persistence but simulated to be urban land.

3.4. Multi-Scenario Design of Planning Constraints on Urban Growth

Multi-scenario analysis of planning constraints on the urban growth process has been conducted in this study. By adjusting the planning preference parameters of different planning zones, multiple planning-constrained urban growth scenarios can be derived. Considering the actual situation, the planning preference parameters of the UDZ and PUZ were preset to 1 and 0, respectively. To model the actual planning preference in the EAZ, 11 values, $Z_{EAZ} \in \{0,0.1,0.2,0.3,0.4,0.5,0.6,0.7,0.8,0.9,1\}$, were selected to generate different planning constraints. Among these values, 0 means corresponding planning zone is completely unbuildable, and 1 means corresponding planning zone has no planning restrictions on urban development. Then, multiple scenarios of urban land patterns were simulated for Wuhan in 2025 through integrating the multiple planning constraints into the urban CA model.

4. Model Application and Results

4.1. Urban Growth and Encroachment in 2009–2017

Based on the land-use maps in 2009 and 2017, we obtained the spatial pattern of urban growth in the WUDA from 2009 to 2017. Then, by superimposing it with the integrated urban spatial planning map, the spatial distribution of urban growth in different planning zones was produced, as shown in Figure 5. In addition, through further statistical analysis, the area and proportion of urban growth and encroachment on other lands in different planning zones are calculated, as shown in Table 3. During 2009–2017, the urban growth of the WUDA presented obvious infilling and edge sprawling characteristics, with a total area of 26,351.21 ha. At the same time, the expansion of urban land has led to the occupation of cultivated land, forest land, garden land, grassland and water areas. Among them, the occupied area of cultivated land accounts for 57.96% of the increased area of urban land, the occupied area of water area accounts for 23.69%, and the occupied areas of forest land, garden land and grassland account for 4.25%, 1.96% and 1.96%, respectively. These show that the urban growth in the WUDA during 2009–2017 is mainly at the cost of the reduction in cultivated land and water area. In addition, we found that the integrated urban spatial planning of Wuhan has an obvious guiding and restricting effect on the expansion of urban land at this stage, with 73.26% of the new urban land occurring in the UDZ. However, the urban land expansion beyond the preset UDZ should not be ignored, with 23.53% of the new urban land appearing in the EAZ, and even 3.21% in the PUZ.

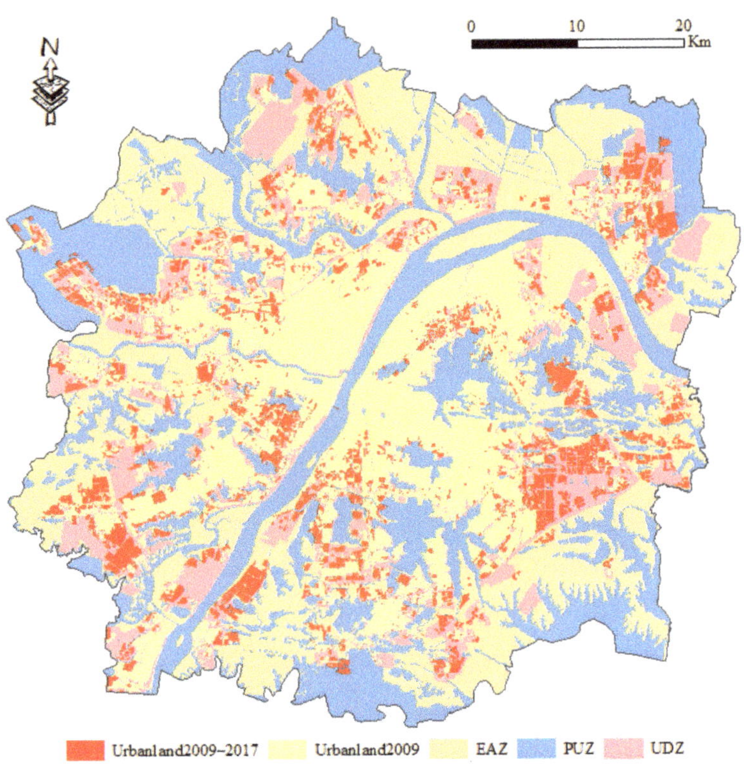

Figure 5. Spatial distribution of urban growth in different planning zones.

Table 3. Area and percent of different land occupied in different planning zones.

	EAZ		UDZ		PUZ		Total	
	Area (km^2)	**Percent (%)**	**Area (km^2)**	**Percent (%)**	**Area (km^2)**	**Percent (%)**	**Area (km^2)**	**Percent (%)**
Cultivated land	33.41	12.68	116.46	44.20	2.87	1.09	152.73	57.96
Garden land	1.17	0.44	3.95	1.50	0.05	0.02	5.17	1.96
Forest land	2.74	1.04	7.69	2.92	0.77	0.29	11.20	4.25
Grassland	0.59	0.22	4.54	1.72	0.04	0.01	5.17	1.96
Water areas	14.31	5.43	44.56	16.91	3.56	1.35	62.43	23.69
Rural settlement	9.41	3.57	14.03	5.33	1.14	0.43	24.58	9.33
Unused land	0.39	0.15	1.81	0.69	0.03	0.01	2.22	0.84
Total	62.02	23.53	193.05	73.26	8.45	3.21	263.51	100.00

4.2. Calibration and Validation of the Simulation Model

This paper analyzes the impact of urban spatial planning policies on urban growth during 2009–2017. By adjusting the planning preference coefficients of different planning zones in this time stage, we obtain the corresponding planning-constrained mechanism. Considering the practical significance of planning zones, the planning preference parameters of the UDZ and PUZ were preset to 1 and 0, respectively, while 11 values ranging from 0.0 to 1.0 on 0.1 intervals are used to generate different planning-constrained scenarios in the EAZ. By using the planning-constrained CA model described in Section 3.3, the urban growth process and spatial pattern from 2009 to 2017 under different planning constraints were simulated. Finally, the actual land-use map of 2017 is used to verify the simulation results.

Figure 6 shows the changes in the Kappa index and FoM coefficient with the adjustment of planning constraints during the verification process. The results show that with the increase in planning preference coefficient of AEZ planning zone, Kappa index and FoM coefficient increase first and, then, decrease, which means that the simulation accuracy has the same change trend. The planning preference coefficient indicates the preference degree of different planning zones for urban growth relative to ecological or agricultural protection. The results show that the actual urban growth violates the urban spatial planning to a certain extent and pays more attention to urban development than ecological and agricultural protection. When the Z_{EAZ} value is 0.7, the accuracy of the model is the highest. This shows that from 2009 to 2017, the urban growth of the WUDA is based on the development strategy that urban growth takes precedence over ecological and agriculture protection. Although this development strategy considers the trade-off relationship between urban development and ecological and agricultural protection to a certain extent, it obviously pays more attention to the economic benefits brought by urban development and largely ignores the ecosystem functional importance, ecological sensitivity and agricultural production suitability in the EAZ. This will cause higher risk of ecological and agricultural loss and further lead to the deterioration of ecological environment and the reduction in grain production.

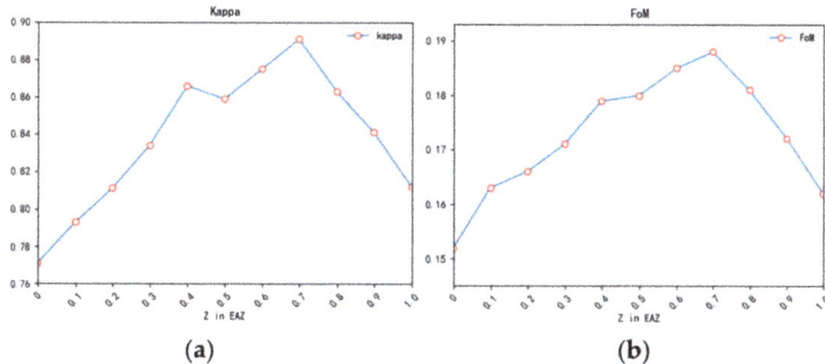

Figure 6. Simulation accuracy assessment. (**a**) Kappa index; (**b**) FoM coefficient.

4.3. Multi-Scenario Simulation of Urban Growth Based on Planning Constraints

Figure 7 shows the spatial distribution of urban growth probability under different planning constraints, which is calculated by the method described in Section 3.2. It can be found that with the adjustment of planning zoning preference parameters, the transition probability map of urban growth shows obvious differences. Overall, with the increase in Z_{AEZ}, the area of land with high transition probability increases significantly, showing obvious influence of planning constraints. From the perspective of the planning zone, the land located in the UDZ shows significantly higher transition probability and remains unchanged with the increase in planning zoning preference parameters, mainly because the planned UDZ is often located around the existing urban land, and it is not affected by the spatial containment function of the planning zone. The land in the EAZ shows obvious spatial heterogeneity, which is mainly located at the urban fringe and suburban area. The farther away from the urban center, the lower the transition probability. At the same time, with the reduction in the planning zone preference parameters, the effect of planning constraints enhanced, resulting in lower transition probability. When the $Z_{EAZ} = 0$, it means that the EAZ is subject to the strictest planning constraint, and the urban growth probability of all land located in this area is 0. When $Z_{EZA} = 1$, it means that the EAZ is not affected by the planning constraint; thus, its transition probability only depends on its urban growth potential. When the $Z_{EAZ} < 1$, it means that the transition probability of the EAZ area depends on the joint action of conservation priority and urban growth potential, and the land with higher urban growth potential and lower conservation priority is easier to convert to urban land. The transition probability of the land located in the PUZ is always 0, because it is subject to the strictest planning constraint.

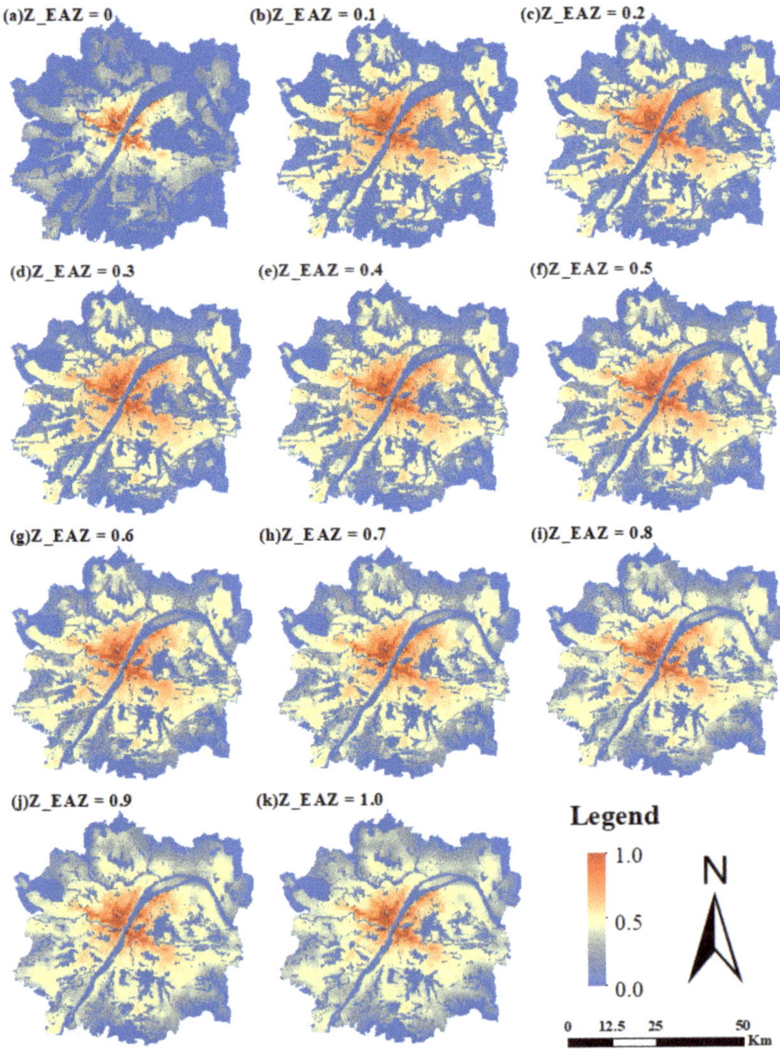

Figure 7. Urban growth probability under different planning constraints.

Based on the urban growth potential map and the planning-constrained CA model introduced in Section 3.3, we simulate the spatial distribution of the WUDA urban growth in 2025 under different planning-constrained scenarios. We used the Markov chain model to predict the urban land demand in 2025, showing that the total area of urban land would be 1342.21 km^2. Figure 8 illustrates the simulation results under different planning constraints. During 2017–2025, urban growth in the WUDA continued to occur around the existing urban land, showing obvious characteristics of edge and infilling expansion. At the same time, the urban growth in this stage is more distributed in the new town groups far from the urban center. The central urban area shows weak urban growth due to strict urban spatial planning policies and few land areas for urban growth. With continuous changes in the Z_{EAZ} value, the urban spatial pattern evolved in a steady and gradient mode; however, a sharp transformation was observed with significant changes to the parameter. In comparison to the weak planning-constrained scenarios (when $Z > 0.5$), the strong planning-constrained scenarios (when $Z < 0.5$) showed an obvious reduction in urban encroachment on farmland with high agricultural production suitability and ecological land with high ecological importance. This shows that the strong planning constraint scheme proposed in this paper can effectively avoid the occupation of cultivated land and ecological land with high conservation priority, which is conducive to the urban sustainable development.

Table 4 presents the landscape indexes of urban growth simulation results under different planning constraints, including number of patch (NP), mean patch size (MPS), mean Euclidean nearest neighbor distance (ENN_MN), mean perimeter area ratio (PARA_MN) and aggregation index (AI). Overall, NP and PARA_MN decrease first and, then, increase, MPS, AI and ENN_MN show the opposite trend. When the $Z_{EAZ} \leq 0.5$, NP and PARA_MN values decrease gradually, MPS, AI and ENN_MN values increased gradually with the increase in the Z_{EAZ}. When the $Z_{EAZ} > 0.5$, with the increase in the Z_{EAZ}, NP and PARA_MN values increase slowly, while MPS, AI and ENN_MN values decrease gradually. This indicates that when there is no obvious preference on urban growth or ecological land and cultivated land protection in the EAZ, the patch number of urban lands is the least, the patch shape is more regular, and the spatial distribution is the most concentrated. With the increase in preference on urban growth or ecological land and cultivated land protection in the EAZ, the patch number of urban lands increases, the patch shape is more irregular, and its spatial distribution is more fragmented. Further, the urban spatial distribution under weak planning constraints is more concentrated than that under strong planning constraints.

Table 4. Landscape index of different planning scenarios.

Z_{EAZ}	NP	MPS	PARA_MN	ENN_MN	AI
0	130	1031.6834	310.3234	202.342	98.3563
0.1	124	1081.3902	309.9919	203.5717	98.3664
0.2	112	1197.3471	281.6773	242.9079	98.3771
0.3	101	1326.1127	260.7926	267.2124	98.377
0.4	99	1354.608	266.4246	270.452	98.3699
0.5	94	1425.4162	241.3943	274.8021	98.3691
0.6	94	1424.8101	258.2396	265.5334	98.3561
0.7	96	1391.0121	269.231	275.8715	98.358
0.8	99	1354.4294	258.9014	276.8843	98.3501
0.9	104	1287.8022	286.3029	234.4195	98.3556
1	107	1251.6569	307.4349	224.5641	98.3379

Based on the spatial structure of three main urban areas (Wuchang, Hankou and Hanyang) and six new city clusters (North, East, Southeast, South, Southwest and West) in the WUDA determined by the master plan of Wuhan, we have made zoning statistics on the simulation results of urban growth in the study area in 2025. Table 5 shows the area of urban growth and its proportion in the UDZ across different regions under multi-scenarios.

With the increase in Z value in the EAZ, the proportion of urban growth in the UDZ will decrease in 2025, from 100% to 68.24%, which clearly shows the impact of planning constraints proposed in this paper on urban growth. In addition, we note that the South and Southwest new city clusters show more obvious changes, while the Hanyang main urban area has the least change. This shows that with the weakening of the impact of planning constraints, the urban growth of the South and Southwest new city cluster is more likely to exceed the UDZ preset in the urban spatial plans, while the urban growth of Hanyang tends to occur in the UDZ.

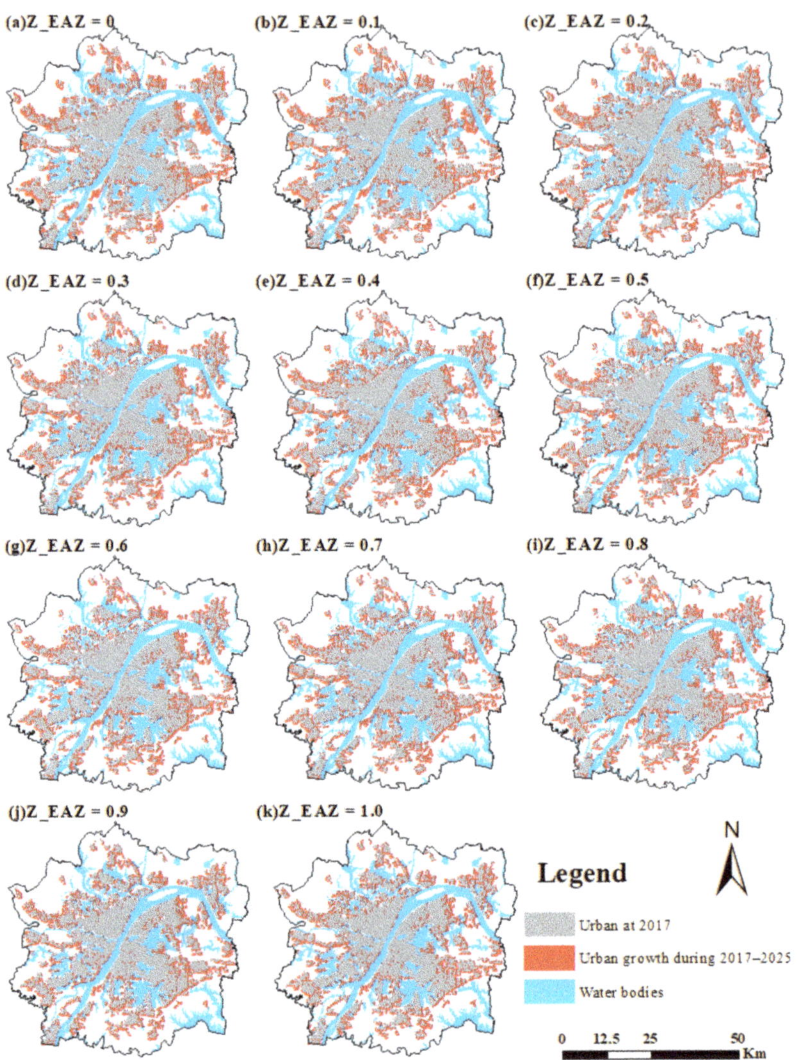

Figure 8. Urban growth pattern under different planning-constrained scenarios.

Table 5. Urban growth area and percent in the UDZ across different regions.

	Z_{EAZ} = 0		Z_{EAZ} = 0.2		Z_{EAZ} = 0.5		Z_{EAZ} = 0.8		Z_{EAZ} = 1.0	
	Area (km^2)	Percent (%)	Area (km^2)	Percent (%)	Area (km^2)	Percent (%)	Area (km^2)	Percent (%)	Area (km^2)	Percent (%)
Wuchang	16.68	100.00	22.66	68.32	22.28	66.01	22.24	65.37	21.71	66.09
Hankou	2.80	100.00	3.53	76.28	3.76	71.14	3.79	69.20	3.79	70.06
Hanyang	8.48	100.00	9.07	89.06	8.73	88.91	8.69	87.98	8.47	88.19
North	40.69	100.00	35.71	91.79	36.78	79.56	36.88	74.82	38.39	71.78
East	45.32	100.00	44.26	79.18	43.06	70.30	42.24	67.63	41.10	67.14
Southeast	26.40	100.00	25.84	84.05	25.22	74.37	25.64	71.86	25.07	70.68
South	36.93	100.00	38.35	77.96	40.91	64.18	41.51	60.08	42.77	58.85
Southwest	38.20	100.00	37.01	78.34	36.39	66.88	37.49	61.94	37.22	61.09
West	44.80	100.00	43.88	85.16	43.18	79.18	41.82	78.21	41.79	77.51
Total	260.31	100.00	260.31	81.46	260.31	72.30	260.31	69.25	260.31	68.24

5. Discussion

CA-based models have been widely used to simulate urban growth and land-use change in past decades. Although there are many practices of using technologically sophisticated methods to construct the urban CA model, the incorporation of planning constraints into modeling has not been adequately addressed so far. By combining the cell-based trade-off between urban growth and conservation with a zoning-based planning implementation mechanism, we proposed a new planning-constrained CA model to simulate urban growth and encroachment under different planning scenarios and explore the potential influence of integrated urban spatial planning on urban growth process. In comparison to the earlier publications [15,26–28], our research incorporated the synthetical urban planning constraints into the CA model, so as to make the predicted future urban scenario closer to reality and provide a useful evaluation tool for the urban planning implementation.

Taking the Wuhan Urban Development Area as a case study, the planning-constrained CA model was applied to simulate the spatial pattern of urban growth from 2009 to 2017 and predict the urban scenarios under different planning constraints in 2025. We find that a higher simulation accuracy can be achieved by considering planning constraints in the simulation. When the Z_{EAZ} value is 0.7, the accuracy of the model is the highest, which indicates that the actual urban growth of the WUDA from 2009 to 2017 violates the urban spatial planning to a certain extent and pays more attention to urban development than ecological and agricultural protection. In future scenarios, with the weakening of planning constraints, urban growth tends to occupy more ecological and agricultural land with high conservation priority. With the increase in preference on urban growth or ecological land and cultivated land protection in the EAZ, the spatial distribution of future urban land becomes more fragmented. This shows that the planning constraint scheme proposed in this paper can effectively guide urban compact development and protect the ecological environment and food security. Furthermore, in the simulation process, the location and quantity of new urban land beyond the planned urban development area can be captured, which will provide certain early warning.

At present, the Chinese government and local governments are formulating a series of new spatial planning policies, one of which is to optimize the regional spatial pattern. It mainly includes the delineation of the three lines (ecological protection red line, permanent basic farmland protection red line and urban expansion boundary line), as well as the determination of planning zoning such as urban development area, rural development area, ecological protection area and agricultural protection area. Policymakers are faced with the problem of how to evaluate spatial planning policies and how to integrate ecological priority into urban spatial planning [32,33]. The methods and models proposed in this study can provide effective support.

However, there are several limitations in our work. First, the Markov model was used to predict the quantity of future urban land in this study, neglecting the impact of different policies and planning schemes, which should be addressed in further research. Second, this study mainly considers the natural and socio-economic driving factors to construct the transformation rules of the urban CA model, ignoring the impact of people's behaviors and decisions. Future research needs to focus on multi-objective human decision-making behaviors and their impact on the urban growth process. Finally, this study predicts the spatial pattern of urban growth based on the 30 m grid, ignoring that the actual land development process is based on the plot. Future research can consider building the conversion rules of urban CA model based on the land-use plot, so as to produce more accurate simulation results.

6. Conclusions

Spatial planning policies are important factors affecting urban development. How urban growth is impacted by urban spatial planning policies needs to be understood in order to better simulate the future urban patterns [28]. Many other publications have made significant contributions to the urban CA model and application [18]. However, so far, the scenario simulation considering the actual urban spatial planning policies has not been adequately addressed, which weakens the ability of CA to provide insight into the future urban spatial patterns and inform the sustainable development strategy [20]. In this paper, we developed a planning-constrained CA model to simulate urban growth and encroachment under different planning scenarios by integrating cellular automata and actual urban spatial planning. In the planning-constrained CA model, a synthetical planning-constrained mechanism was proposed by combing the cell-based tradeoff between urban growth potential and conservation priority with zoning-based planning preference. By adjusting the preference parameters of different planning zones, multiple planning-constrained scenarios can be generated. Taking the Wuhan Urban Development Area as a case study, the planning-constrained CA model was applied to simulate current (2017) and future (2025) urban scenarios. The main conclusions were:

1. The planning-constrained CA model demonstrated a higher simulation accuracy compared to the model without planning constraints.
2. The simulation result of 2017 shows that a weak planning-constrained urban development was consistent with the actual situation.
3. With the weakening of planning constraints, urban growth tends to occupy more ecological and agricultural land with high conservation priority. With the increase in preference on urban growth or ecological land and cultivated land protection in the EAZ, the future urban land pattern becomes more fragmented.
4. Location and quantity of new urban land beyond the planned urban development area can be captured in future urban scenarios, which will provide certain early warning.

The method proposed in this study can provide effective support for planners and decisionmakers to evaluate the impact of actual urban spatial planning policies on urban growth and generate urban future scenarios in which urban development and natural protection are coordinated, so as to realize the urban sustainable development.

Author Contributions: Conceptualization, H.W. and Y.L.; methodology, H.W.; software, H.W. and Y.W.; formal analysis, H.W.; data curation, H.W. and J.Z.; writing—original draft preparation, H.W.; writing—review and editing, H.W. and Y.L.; visualization, H.W. and G.Z.; All authors have read and agreed to the published version of the manuscript.

Funding: This research was funded by the National Key Research and Development Program (Grant No: 2017YFB0503505).

Institutional Review Board Statement: Not applicable.

Informed Consent Statement: Not applicable.

Data Availability Statement: The data presented in this study are available on request from the corresponding author.

Conflicts of Interest: The authors declare no conflict of interest.

References

1. Bai, X.; Shi, P.; Liu, Y. Society: Realizing China's urban dream. *Nature* **2014**, *509*, 158–160. [CrossRef]
2. Mcphearson, T. Scientists must have a say in the future of cities. *Nature* **2016**, *538*, 165–166. [CrossRef]
3. Zhou, L.; Dang, X.; Sun, Q.; Wang, S. Multi-scenario simulation of urban land change in Shanghai by random forest and CA-Markov model. *Sustain. Cities Soc.* **2020**, *55*, 102045. [CrossRef]
4. Seto, K.C.; Güneralp, B.; Hutyra, L.R. Global forecasts of urban expansion to 2030 and direct impacts on biodiversity and carbon pools. *Proc. Natl. Acad. Sci. USA* **2012**, *109*, 16083–16088. [CrossRef]
5. Grimm, N.B.; Faeth, S.H.; Golubiewski, N.E.; Redman, C.L.; Wu, J.; Bai, X.; Briggs, J.M. Global Change and the Ecology of Cities. *Science* **2008**, *319*, 756–760. [CrossRef]
6. Grêt-Regamey, A.; Altwegg, J.; Sirén, E.A.; van Strien, M.J.; Weibel, B. Integrating ecosystem services into spatial planning—A spatial decision support tool. *Landsc. Urban Plan.* **2017**, *165*, 206–219. [CrossRef]
7. Li, Y.; Ma, Q.; Song, Y.; Han, H. Bringing conservation priorities into urban growth simulation: An integrated model and applied case study of Hangzhou, China. *Resour. Conserv. Recycl.* **2019**, *140*, 324–337. [CrossRef]
8. McCormick, K.; Anderberg, S.; Coenen, L.; Neij, L. Advancing sustainable urban transformation. *J. Clean. Prod.* **2013**, *50*, 1–11. [CrossRef]
9. Godwin, C.; Chen, G.; Singh, K.K. The impact of urban residential development patterns on forest carbon density: An integration of LiDAR, aerial photography and field mensuration. *Landsc. Urban Plan.* **2015**, *136*, 97–109. [CrossRef]
10. Haberl, H. Competition for land: A sociometabolic perspective. *Ecol. Econ.* **2015**, *119*, 424–431. [CrossRef]
11. Ke, X.; van Vliet, J.; Zhou, T.; Verburg, P.H.; Zheng, W.; Liu, X. Direct and indirect loss of natural habitat due to built-up area expansion: A model-based analysis for the city of Wuhan, China. *Land Use Policy* **2018**, *74*, 231–239. [CrossRef]
12. He, J.; Liu, Y.; Yu, Y.; Tang, W.; Xiang, W.; Liu, D. A counterfactual scenario simulation approach for assessing the impact of farmland preservation policies on urban sprawl and food security in a major grain-producing area of China. *Appl. Geogr.* **2013**, *37*, 127–138. [CrossRef]
13. Zhang, Y.; Liu, Y.; Zhang, Y.; Kong, X.; Jing, Y.; Cai, E.; Zhang, L.; Liu, Y.; Wang, Z.; Liu, Y. Spatial Patterns and Driving Forces of Conflicts among the Three Land Management Red Lines in China: A Case Study of the Wuhan Urban Development Area. *Sustainability* **2019**, *11*, 2025. [CrossRef]
14. Long, Y.; Han, H.; Tu, Y.; Shu, X. Evaluating the effectiveness of urban growth boundaries using human mobility and activity records. *Cities* **2015**, *46*, 76–84. [CrossRef]
15. Tong, X.; Feng, Y. How current and future urban patterns respond to urban planning? An integrated cellular automata modeling approach. *Cities* **2019**, *92*, 247–260. [CrossRef]
16. Li, X.; Lao, C.; Liu, Y.; Liu, X.; Chen, Y.; Li, S.; Ai, B.; He, Z. Early warning of illegal development for protected areas by integrating cellular automata with neural networks. *J. Environ. Manag.* **2013**, *130*, 106–116. [CrossRef]
17. Tobler, W.R. Cellular Geography. In *Philosophy in Geography*; Gale, S., Olsson, G., Eds.; Springer: Dordrecht, The Netherlands, 1979; pp. 379–386. [CrossRef]
18. Liu, Y.; Batty, M.; Wang, S.; Corcoran, J. Modelling urban change with cellular automata: Contemporary issues and future research directions. *Prog. Hum. Geogr.* **2021**, *45*, 3–24. [CrossRef]
19. Berberoğlu, S.; Akın, A.; Clarke, K.C. Cellular automata modeling approaches to forecast urban growth for Adana, Turkey: A comparative approach. *Landsc. Urban Plan.* **2016**, *153*, 11–27. [CrossRef]
20. Liang, X.; Liu, X.; Li, D.; Zhao, H.; Chen, G. Urban growth simulation by incorporating planning policies into a CA-based future land-use simulation model. *Int. J. Geogr. Inf. Sci.* **2018**, *32*, 2294–2316. [CrossRef]
21. Zhang, Q.; Su, S. Determinants of urban expansion and their relative importance: A comparative analysis of 30 major metropolitans in China. *Habitat Int.* **2016**, *58*, 89–107. [CrossRef]
22. Li, J.; Deng, X.; Seto, K.C. Multi-level modeling of urban expansion and cultivated land conversion for urban hotspot counties in China. *Landsc Urban Plan.* **2012**, *108*, 131–139. [CrossRef]
23. Tian, G.; Ma, B.; Xu, X.; Liu, X.; Xu, L.; Liu, X.; Xiao, L.; Kong, L. Simulation of urban expansion and encroachment using cellular automata and multi-agent system model—A case study of Tianjin metropolitan region, China. *Ecol. Indic.* **2016**, *70*, 439–450. [CrossRef]
24. Lagarias, A. Urban sprawl simulation linking macro-scale processes to micro-dynamics through cellular automata, an application in Thessaloniki, Greece. *Appl. Geogr.* **2012**, *34*, 146–160. [CrossRef]
25. Liao, J.; Tang, L.; Shao, G.; Su, X.; Chen, D.; Xu, T. Incorporation of extended neighborhood mechanisms and its impact on urban land-use cellular automata simulations. *Environ. Model. Softw.* **2016**, *75*, 163–175. [CrossRef]
26. Onsted, J.A.; Chowdhury, R.R. Does zoning matter? A comparative analysis of landscape change in Redland, Florida using cellular automata. *Landsc Urban Plan.* **2014**, *121*, 1–18. [CrossRef]
27. Yin, H.; Kong, F.; Yang, X.; James, P.; Dronova, I. Exploring zoning scenario impacts upon urban growth simulations using a dynamic spatial model. *Cities* **2018**, *81*, 214–229. [CrossRef]
28. Li, L.; Song, Y.; Wei, X.; Dong, J. Exploring the impacts of urban growth on carbon storage under integrated spatial regulation: A case study of Wuhan, China. *Ecol. Indic.* **2020**, *111*, 106064. [CrossRef]
29. Xi, F.; He, H.S.; Hu, Y.; Bu, R.; Chang, Y.; Wu, X.; Liu, M.; Shi, T. Simulating the impacts of ecological protection policies on urban land use sustainability in Shenyang-Fushun, China. *Int. J Urban Sustain. Dev.* **2010**, *1*, 111–127. [CrossRef]

30. Wang, W.; Jiao, L.; Jia, Q.; Liu, J.; Mao, W.; Xu, Z.; Li, W. Land use optimization modelling with ecological priority perspective for large-scale spatial planning. *Sustain. Cities Soc.* **2021**, *65*, 102575. [CrossRef]
31. Wang, H.; Xia, C.; Zhang, A.; Deng, Y. Scenario simulation and control of metropolitan outskirts urban growth based on constrained CA: A case study of Jiangxia District of Wuhan City. *Prog. Geogr.* **2016**, *35*, 793–805. [CrossRef]
32. Yeh, A.G.O.; Li, X.; Xia, C. Cellular Automata Modeling for Urban and Regional Planning. In *Urban Informatics*; Shi, W., Goodchild, M.F., Batty, M., Kwan, M.-P., Zhang, A., Eds.; Springer: Singapore, 2021; pp. 865–883. [CrossRef]
33. Hersperger, A.M.; Grădinaru, S.; Oliveira, E.; Pagliarin, S.; Palka, G. Understanding strategic spatial planning to effectively guide development of urban regions. *Cities* **2019**, *94*, 96–105. [CrossRef]
34. Dadashpoor, H.; Azizi, P.; Moghadasi, M. Analyzing spatial patterns, driving forces and predicting future growth scenarios for supporting sustainable urban growth: Evidence from Tabriz metropolitan area, Iran. *Sustain. Cities Soc.* **2019**, *47*, 101502. [CrossRef]
35. Shoemaker, D.A.; BenDor, T.K.; Meentemeyer, R.K. Anticipating trade-offs between urban patterns and ecosystem service production: Scenario analyses of sprawl alternatives for a rapidly urbanizing region. *Comput. Environ. Urban Syst.* **2019**, *74*, 114–125. [CrossRef]
36. Olagunju, A.; Gunn, J.A.E. Integration of environmental assessment with planning and policy-making on a regional scale: A literature review. *Environ. Impact Assess. Rev* **2016**, *61*, 68–77. [CrossRef]
37. York, A.M.; Munroe, D.K. Urban encroachment, forest regrowth and land-use institutions: Does zoning matter? *Land Use Policy* **2010**, *27*, 471–479. [CrossRef]
38. Du, N.; Ottens, H.; Sliuzas, R. Spatial impact of urban expansion on surface water bodies—A case study of Wuhan, China. *Landsc. Urban Plan.* **2010**, *94*, 175–185. [CrossRef]
39. Jin, G.; Shi, X.; He, D.W.; Guo, B.S.; Li, Z.H.; Shi, X.B. Designing a spatial pattern to rebalance the orientation of development and protection in Wuhan. *J. Geog. Sci.* **2020**, *30*, 569–582. [CrossRef]
40. Long, H.; Liu, Y.; Hou, X.; Li, T.; Li, Y. Effects of land use transitions due to rapid urbanization on ecosystem services: Implications for urban planning in the new developing area of China. *Habitat Int.* **2014**, *44*, 536–544. [CrossRef]
41. Kraxner, F.; Nordström, E.-M.; Havlík, P.; Gusti, M.; Mosnier, A.; Frank, S.; Valin, H.; Fritz, S.; Fuss, S.; Kindermann, G.; et al. Global bioenergy scenarios—Future forest development, land-use implications, and trade-offs. *Biomass Bioenergy* **2013**, *57*, 86–96. [CrossRef]
42. Haase, D.; Larondelle, N.; Andersson, E.; Artmann, M.; Borgstrom, S.; Breuste, J.; Gomez-Baggethun, E.; Gren, A.; Hamstead, Z.; Hansen, R.; et al. A quantitative review of urban ecosystem service assessments: Concepts, models, and implementation. *Ambio* **2014**, *43*, 413–433. [CrossRef]
43. Nguyen, A.K.; Liou, Y.-A.; Li, M.-H.; Tran, T.A. Zoning eco-environmental vulnerability for environmental management and protection. *Ecol. Indic.* **2016**, *69*, 100–117. [CrossRef]
44. Wang, W.; Jiao, L.; Zhang, W.; Jia, Q.; Su, F.; Xu, G.; Ma, S. Delineating urban growth boundaries under multi-objective and constraints. *Sustain. Cities Soc.* **2020**, *61*, 102279. [CrossRef]
45. Peng, J.; Liu, Y.; Wu, J.; Lv, H.; Hu, X. Linking ecosystem services and landscape patterns to assess urban ecosystem health: A case study in Shenzhen City, China. *Landsc. Urban Plan.* **2015**, *143*, 56–68. [CrossRef]
46. Luederitz, C.; Brink, E.; Gralla, F.; Hermelingmeier, V.; Meyer, M.; Niven, L.; Panzer, L.; Partelow, S.; Rau, A.-L.; Sasaki, R.; et al. A review of urban ecosystem services: Six key challenges for future research. *Ecosyst. Serv* **2015**, *14*, 98–112. [CrossRef]
47. Hong, W.; Jiang, R.; Yang, C.; Zhang, F.; Su, M.; Liao, Q. Establishing an ecological vulnerability assessment indicator system for spatial recognition and management of ecologically vulnerable areas in highly urbanized regions: A case study of Shenzhen, China. *Ecol. Indic.* **2016**, *69*, 540–547. [CrossRef]
48. Xia, C.; Zhang, B. Exploring the effects of partitioned transition rules upon urban growth simulation in a megacity region: A comparative study of cellular automata-based models in the Greater Wuhan Area. *GIScience Remote Sens.* **2021**, *58*, 693–716. [CrossRef]
49. Feng, Y.; Tong, X. Dynamic land use change simulation using cellular automata with spatially nonstationary transition rules. *Giscience Remote Sens.* **2018**, *55*, 678–698. [CrossRef]
50. Li, X.; Chen, Y.; Liu, X.; Xu, X.; Chen, G. Experiences and issues of using cellular automata for assisting urban and regional planning in China. *Int. J. Geogr. Inf. Sci.* **2017**, *31*, 1606–1629. [CrossRef]
51. He, Q.; Tan, S.; Yin, C.; Zhou, M. Collaborative Optimization of Rural Residential Land Consolidation and Urban Con-struction Land Expansion: A Case Study of Huangpi in Wuhan, China. *Comput. Environ. Urban Syst.* **2019**, *74*, 218–228. [CrossRef]
52. Bai, Y.; Wong, C.P.; Jiang, B.; Hughes, A.C.; Wang, M.; Wang, Q. Developing China's Ecological Redline Policy using ecosystem services assessments for land use planning. *Nat. Commun.* **2018**, *9*, 3034. [CrossRef]
53. Batty, M.; Xie, Y.; Sun, Z. Modeling urban dynamics through GIS-based cellular automata. *Comput. Environ. Urban Syst.* **1999**, *23*, 205–233. [CrossRef]
54. Li, X.; Yeh, A.G.-O. Modelling sustainable urban development by the integration of constrained cellular automata and GIS. *Int. J. Geogr. Inf. Sci.* **2000**, *14*, 131–152. [CrossRef]
55. Aburas, M.M.; Ho, Y.M.; Ramli, M.F.; Ash'Aari, Z.H. The simulation and prediction of spatio-temporal urban growth trends using cellular automata models: A review. *Int. J. Appl. Earth Obs. Geoinf.* **2016**, *52*, 380–389. [CrossRef]

56. Pontius, R.G. Quantification Error versus Location Error in Comparison of Categorical Maps. *Photogramm. Eng. Remote Sens.* **2000**, *66*, 1011–1016.
57. Pontius, R.G.; Boersma, W.; Castella, J.-C.; Clarke, K.; de Nijs, T.; Dietzel, C.; Duan, Z.; Fotsing, E.; Goldstein, N.; Kok, K.; et al. Comparing the input, output, and validation maps for several models of land change. *Ann. Reg. Sci.* **2008**, *42*, 11–37. [CrossRef]

Article

Hybrid Economic-Environment-Ecology Land Planning Model under Uncertainty—A Case Study in Mekong Delta

Yuxiang Ma , Min Zhou *, Chaonan Ma, Mengcheng Wang and Jiating Tu

College of Public Administration, Huazhong University of Science and Technology, Number 1037, Luoyu Road, Hongshan District, Wuhan 430074, China; myx0996@163.com (Y.M.); 17863935943@163.com (C.M.); wmc@hust.edu.cn (M.W.); tujiating1999@163.com (J.T.)

* Correspondence: shijieshandian00@163.com; Tel./Fax: +86-27-87543047

Abstract: The research on land natural resources as the leading factor in the Mekong Delta (MD) is insufficient. Facing the fragile and sensitive ecological environment of MD, how to allocate limited land resources to different land use types to obtain more economic benefits is a challenge that local land managers need to face. Three uncertainties in land use system, interval uncertainty, fuzzy uncertainty, and random uncertainty, are fully considered and an interval probabilistic fuzzy land use allocation (IPF-LUA) model is proposed and applied to multiple planning periods for MD. IPF-LUA considers not only the crucial socio-economic factors (food security, output of wood products, etc.) but also the ecological/environmental constraints in agricultural production (COD discharge, BOD5 discharge, antibiotic consumption, etc.). Therefore, it can effectively reflect the interaction among different aspects of MD land use system. The degree of environmental subordination is between 0.51 and 0.73, the net benefit of land system is between USD 23.31 $\times 10^9$ and USD 24.24 $\times 10^9$ in period 1, and USD 25.44 $\times 10^9$ to 25.68 $\times 10^9$ in period 2. The results show that the IPF-LUA model can help the decision-makers weigh the economic and ecological benefits under different objectives and work out an optimized land use allocation scheme.

Keywords: land planning; uncertainty; interval probabilistic fuzzy; land use allocation; Mekong Delta

Citation: Ma, Y.; Zhou, M.; Ma, C.; Wang, M.; Tu, J. Hybrid Economic-Environment-Ecology Land Planning Model under Uncertainty—A Case Study in Mekong Delta. *Sustainability* **2021**, *13*, 10978. https://doi.org/10.3390/su131910978

Academic Editors: Qingsong He, Jiayu Wu, Chen Zeng and Linzi Zheng

Received: 26 August 2021
Accepted: 30 September 2021
Published: 3 October 2021

1. Introduction

Land use planning is a comprehensive economic and technical measure for the development, utilization, governance and protection of land resources in time and space according to the requirements of economic development in the future and regional development conditions. Its purpose is to promote sustainable land use. Among them, land use allocation (LUA) is the core content and key component of land use planning [1,2]. LUA has great significance under the background of a growing economy and population [3]. On the one hand, LUA is an effective means of ensuring the sustainable development of regional land resources [4–7]. LUA can not only maintain the coordinated development of the regional economy and ecological protection [2,8,9] but also provides a healthy and comfortable living environment for residents [10–12]. On the other hand, there must be conflicts of interest among farmers, developers, government, and other stakeholders. Thus, LUA faces the challenge of balancing the interests of all parties as much as possible [13–17]. A reasonable LUA can effectively alleviate the conflicts between cultivated land production and urban construction land, and coordinate the conflicts between economic development and ecological protection [18–20]. Therefore, in order to maximize the benefits of the land system, managers should take various factors into consideration and obtain as many economic and ecological benefits of the system as possible when they conduct LUA [21–23].

In the past decades, many models, which are mainly divided into four types, have been used to solve LUA problems. (1) mathematical programming models [7,24–29], like the following: Kumar et al. [24] established a mixed-integer linear programming (MILP) model, which can analyze various factors affecting land use and sprawl measures; Turk and Zwick [27]. developed an inquired land assignment model (ILAM) which combines binary integer planning with the geographic information system (GIS); Medeiros et al. [7] used a multi-objective linear programming to analyze the environmental and economic benefits of forest land production systems, and optimized the LUA of forest land; (2) spatial optimization models [30–35], for example, Sante et al. [31] proposed a parallel simulated annealing algorithm for land use spatial allocation based on irregular spatial structure, which optimizes the allocation of land use classification on the cadastral parcel map; Rall et al. [33] proposed the consumption of public participation geographic information system (PPGIS) to urban green infrastructure (UGI) planning, which can support multifunctional assessment; Ramya and Devadas [34] built a geographic information system-multi-criteria decision making (GIS-MCDM) model, which considered natural resource conditions, labor, urban location and other cost factors into consideration to solve the problems of industrial land planning; (3) simulative prediction models [36–42], for example, Liu et al. [39] put forward a future land use simulation (FLUS) model with full consideration of various climatic and social-economic factors, which can be applied to simulate land use change in different scenarios in the future; Wang et al. [41] proposed a land use prediction model that integrates the geographic information system (GIS) and the artificial neural network (ANNs), which is conducive to improving the accuracy and timeliness of decision-makers in formulating LUA policies; Liang et al. [38] incorporated planning policies into the future land use simulation model (FLUS) based on cellular automata (CA) to accurately predict and explain the development trend of cities; Xu et al. [42] proposed an artificial neural network–cellular automata–Markov chain (ANN–CA–MC) model to analyze the limitation of the traditional CA-MC model, and verified that the result of this model of accurately simulating and forecasting urban expansion was better than that of other CA-MC coupled models; (4) intelligent algorithm models [3,43–47], for example, Masoomi et al. [3] used multi-objective particle swarm optimization algorithm considering multiple objectives and constraints and tried to optimize multiple objectives at the same time; Mi et al. [43] combined the advantages of the genetic algorithm (GA) with the ant colony algorithm (ACA) and put forward the genetic ant colony algorithm (GACA), which was applied to determine the optimal space use allocation of the limited development ecological zone (LDEZ); Li and Parrott [45] proposed an improved genetic algorithm (GA) for multi-site land use allocation that could meet the different management objectives of decision-makers; Huang and Song [47] developed the LUA model based on a multi-agent system—"Multi-agent shuffled frog leaping algorithm" (MASFLA), which can effectively solve the optimized allocation problems of spatial structure and quantity of regional land use under the conditions of multi-agent, multi-objective and multi-constraints.

We can see from existing models that researchers are constantly improving and developing the LUA model from single-objective programming to multi-objective programming, from developing a single structural or spatial optimization model to developing multiple model couplings of structural and spatial models. These models can effectively deal with the LUA problems, such as location, urban expansion in the process of urban and agricultural land use.

However, the above models still have some shortcomings: (1) The uncertainties in the land system should be more fully considered. There are many uncertain factors in the actual land use system, such as policy uncertainty: the government's development, construction, and investment in land are random; economic uncertainty: the demand for land with different uses is uncertain in the regional economic development; social uncertainty: the change of population, the demand for food increase and the uncertainty of land use allocation; environmental uncertainty: there is uncertainty in the change of climatic and hydrological conditions, which will have an uncertain impact on agricultural production and industrial construction, as well as the uncertain external environmental impacts of excessive consumption of water resources on the land use allocation structure; ecological uncertainty: wetlands, soil erosion and the consumption of chemical fertilizers will bring uncertain impacts to LUA [22,48–50]. These uncertain factors may greatly affect the accuracy of the calculation results of the above-mentioned land use allocation models [22]; (2) in studying land use systems, the above models mainly consider policies and economic factors, and there is still a lack of consideration for some important ecological and environmental factors, including the consumption of chemical fertilizers in agricultural land, the discharge of waste-water and solid waste in construction land [46], etc. How to integrate the uncertain factors in many land systems, such as policy, economy, society, ecology, and environment into a more complete model system for comprehensive analysis will be the focus of this study.

In addition, most of the world's deltas have become major economic development zones of various regions due to their unique natural resource conditions. At the same time, in the process of land use, these regions have caused acute environmental problems, such as urban sewage and solid waste discharge, water quality decline, and climate change [51–54].

However, the current research on LUA in the Mekong Delta region is relatively few. Therefore, this study proposes an interval probabilistic fuzzy programming model to simulate the LUA of the MD. Compared with previous LUA models, this model has advantages as follows: (1) A systematic and scientific quantitative analysis is conducted by comprehensively considering various uncertain factors in the land use system. (2) The proposed model supplements and improves existing LUA models in terms of constraints and modeling methods. (3) This study will analyze land use issues of the MD, such as constraints on water product output from aquaculture land, COD and BOD5 emissions, and wood product output from forest land. MD's LUA is quantitatively analyzed by the IPF-LUA model, which provides a reliable basis for managers to make planning decisions. At the same time, it can also provide theoretical and technical support for managers in other countries of the world to solve LUA problems in the delta regions.

2. Study Area

The Mekong Delta (MD) is located in southwestern Vietnam from 8°33′ N~11°01′ N and 103°50′ E~106°50′ E (Figure 1). There are 13 provinces and cities in the region, with an area of over 40,500 km^2, covering part of southwestern Vietnam. With a population of about 21 million (2019) and a population density of 530 people/km^2, the MD is one of the most densely populated delta regions in the world. Based on the geographical and climatic characteristics of the MD and considering MD's development strategy and regional development planning, the study area is divided into three sub-zones (Figure 1): The deep-water flood disaster sub-zone, the central sub-zone, the coastal and island sub-zone. Deep-water flood disaster sub-zone: the main uses of the region are flood management and freshwater storage, freshwater aquaculture, and watershed establishment; central sub-zone: this is a shallow water area with favorable soil conditions, so it is mainly used for developing diversified and specialized agricultural production areas; coastal and island sub-zone: the coastal areas of the region are vulnerable to seawater erosion and used to shift conventional agricultural production and aquaculture towards sustainable woodland (mangrove) ecosystems and to build and develop a sustainable tourism and service center.

The MD benefits from natural geographical conditions: the terrain is flat, with rich water resources, fertile land, and a dense river network, turning the region into veritable a "land of fish and rice" from an almost uninhabited jungle. Moreover, the MD is the largest agricultural production center in Vietnam's agricultural land, accounting for 25% of the total agricultural land. Rice, aquaculture, and fruit accounts for 54%, 65%, and 70% of the country's total, respectively, with about 95% of rice and 60% of aquaculture exported. Furthermore, the MD is called a "biological treasure house", since thousands of creatures live there.

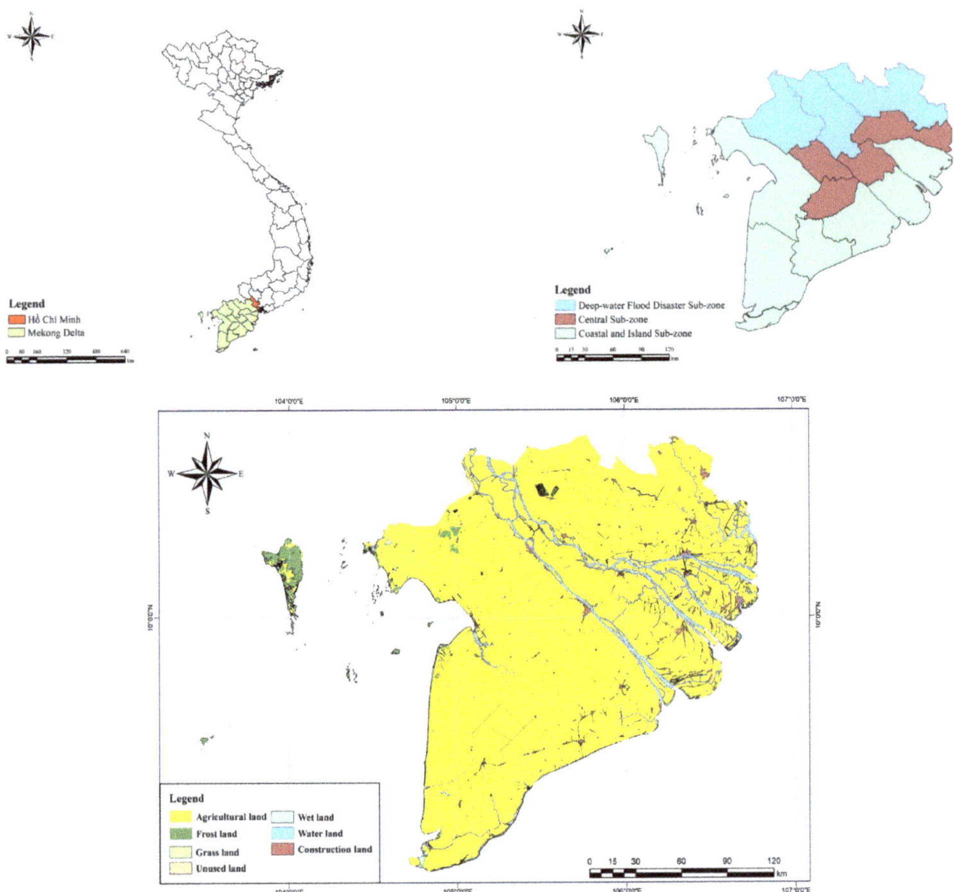

Figure 1. The study area.

However, the rapid pace of industrialization and urbanization has made MD's already sensitive natural conditions more fragile. According to the environmental management report of the MD provinces, enterprises in industrial zones and industrial clusters directly discharge about 220,000 tons of solid waste into the environment every year. Meanwhile, the amount of solid waste brought directly into rivers and canals day and night by livestock is about 22,500 tons. In recent years, while the agricultural land of the MD is estimated at 22,500 tons. In recent years, while the agricultural land of the MD has been reduced, the agricultural productivity has been steadily increasing, which is related to the use of a large quantity of pesticides, fertilizers, and aquaculture antibiotics. Relevant studies have shown that in the two largest rice-producing provinces of the MD, An Giang Province and Kien Giang Province, rice farmers used 20–30% more fertilizers than recommended. These behaviors are also one of the important causes for the deterioration of surface water quality in this region. These problems have led to the low level of ecological environment in the MD, threatening the life quality of the local people and the sustainable development of the region.

In summary, the regional characteristics of the MD can be summarized as follows: obvious advantages in agricultural economic development conditions, serious environmental pollutions, and ecological deterioration, which need to be solved urgently. The region needs effective land-use allocation plans to solve the problem. In view of the above challenges, the corresponding land use allocation model will be proposed in the next part of this study.

3. IPF-LUA for MD

The land use system of the MD in the study area of this paper is defined as an uncertain system. Based on the factors analyzed by typical land use allocation system and taking the land use characteristics of the MD region into full consideration, this paper analyzes the following uncertain factors in the MD land use system: (1) Economic factors: MD's development requirements of agriculture, forestry, and fishery industries, such as grain and aquatic product demand, investment and other factors; (2) social factors: The development of the MD industry is affected by water supply and labor force, for example, agricultural water supply and labor force supply; (3) environmental factors: MD mainly focuses on agriculture, forestry, and fishery industries, accompanied by the production of sewage, solid waste, and other pollutants, such as wastewater, COD and BOD5 factors; (4) ecological factors: The biochemical substances used by farmers in the production process of MD paddy and aquaculture, such as antibiotics, pesticides and fertilizers. Based on the MD land use characteristics and the above factors, we applied this IPF-LUA model to land-use allocation in the MD which leads to an IPF-LUA (interval probabilistic fuzzy land-use allocation model).

3.1. Objective Function

The objective function is the net benefit from land-use system of the MD, and it equals all benefits from cultivated land, forest land, aquaculture land, and construction land minus cost from all types of land uses. Specifically, the benefits come from industries in the land use system. For example, the benefit of industries such as grain comes from cultivated land; the benefit of industries such as timber and furniture comes from forest land; the benefit of industries such as fish, shrimp, and aquatic products comes from aquaculture land; the benefit of industries such as coal, petroleum, and rubber come from construction land. Cost is from maintenance fees of all types of land uses. For example, we must provide cultivated land and aquaculture land with water; furthermore, these two land uses will generate wastewater and solid waste which are needed to handle. Based on the above settings, our objective function can be expressed as:

$$
\begin{aligned}
\text{MaxNBL}^{\pm} \cong & \sum_{i=1}^{3}\sum_{t=1}^{2}\left(\text{CLB}^{\pm}_{i,j=1,t} \times x^{\pm}_{i,j=1,t}\right) + \sum_{i=1}^{3}\sum_{t=1}^{2}\left(\text{FLB}^{\pm}_{i,j=2,t} \times x^{\pm}_{i,j=2,t}\right) \\
& + \sum_{i=1}^{3}\sum_{t=1}^{2}\left(\text{ALB}^{\pm}_{i,j=3,t} \times x^{\pm}_{i,j=3,t}\right) + \sum_{i=1}^{3}\sum_{t=1}^{2}\left(\text{CLP}^{\pm}_{i,j=4,t} \times x^{\pm}_{i,j=4,t}\right) \\
& - \sum_{i=1}^{3}\sum_{t=1}^{2}\left[\left(\text{UWTC}^{\pm}_{i,j=1,t} + \text{USTC}^{\pm}_{i,j=1,t}\right) \times x^{\pm}_{i,j=1,t}\right] \\
& - \sum_{i=1}^{3}\sum_{t=1}^{2}\left[\left(\text{UWTC}^{\pm}_{i,j=3,t}\right) \times x^{\pm}_{i,j=3,t}\right] \\
& - \sum_{i=1}^{3}\sum_{t=1}^{2}\left[\left(\text{UWTC}^{\pm}_{i,j=4,t} + \text{USTC}^{\pm}_{i,j=4,t}\right) \times x^{\pm}_{i,j=4,t}\right] \\
& - \sum_{i=1}^{3}\sum_{t=1}^{2}\left(\text{UWMC}^{\pm}_{i,j=5,t} \times x^{\pm}_{i,j=5,t}\right) \\
& - \sum_{i=1}^{3}\sum_{t=1}^{2}\left(\text{UUDC}^{\pm}_{i,j=6,t} \times x^{\pm}_{i,j=6,t}\right)
\end{aligned}
$$

where NBL is objective function which means the net benefit from land-use system (USD); "$\pm$" = interval values; "@" = fuzzy equal; x is the decision variable; i represents the name of district, where $i = 1$ for Deep-water flood disaster sub-zone, $i = 2$ for central sub-zone, and $i = 3$ for coastal and island sub-zone; j represents type of land use, where $j = 1$ for cultivated land, $j = 2$ for forest land, $j = 3$ for aquaculture land, $j = 4$ for construction land, $j = 5$ for water land, $j = 6$ for unused land; t represents time of planning, where $t = 1$ for period 1 (2021), $t = 2$ for period 2 (2022); $\text{CLB}_{i,j=1,t}$ = unit benefit of cultivated land (USD/ha),

$FLB_{i,j=2,t}$ = unit benefit of forest land (USD/ha); $ALB_{i,j=3,t}$ = unit benefit of aquaculture land (USD/ha); $CLP_{i,j=4,t}$ = unit benefit of construction land (USD/ha); $UWTC_{i,t}$ = unit waste-water-tackling cost; USTC*i,t* = unit solid waste-tackling cost; $UWMC_{i,j=5,t}$ = unit maintenance costs of water land; $UUDC_{i,j=6,t}$ = unit developing costs of unused land.

3.2. Social-Economic Constraints

(i) Government investment constraints:

In the MD, all costs are paid by government investment, so the government investment constraints can be expressed as:

$$\begin{aligned}
&\sum_{i=1}^{3}\sum_{t=1}^{2}\left[\left(UWTC^{\pm}_{i,j=1,t}+USTC^{\pm}_{i,j=1,t}\right)\times x^{\pm}_{i,j=1,t}\right]\\
&-\sum_{i=1}^{3}\sum_{t=1}^{2}\left[\left(UWTC^{\pm}_{i,j=3,t}\right)\times x^{\pm}_{i,j=3,t}\right]\\
&-\sum_{i=1}^{3}\sum_{t=1}^{2}\left[\left(UWTC^{\pm}_{i,j=4,t}+USTC^{\pm}_{i,j=4,t}\right)\times x^{\pm}_{i,j=4,t}\right]\\
&-\sum_{i=1}^{3}\sum_{t=1}^{2}\left(UWMC^{\pm}_{i,j=5,t}\times x^{\pm}_{i,j=5,t}\right)-\sum_{i=1}^{3}\sum_{t=1}^{2}\left(UUDC^{\pm}_{i,j=6,t}\times x^{\pm}_{i,j=6,t}\right)\lesssim MGI^{\pm}
\end{aligned}$$

where MGI = maximum government investment (USD).

(ii) Grain input-output constraints:

In the MD, the core industry in food production is the main pillar of local economic development. In our model, the food supply is aimed at meeting the needs of both local residents and economic development (export of food products), food products are mainly produced by paddy cultivated land:

$$\sum_{i=1}^{3}\sum_{t=1}^{2}\left(UGP^{\pm}_{i,t}\times x^{\pm}_{i,j=1,t}\right)\gtrsim GD^{\pm}$$

where $UGP_{i,t}$ = unit grain production from cultivated land (ton/ha); GD = demand grain production (ton).

(iii) Water production input-output constraints:

In the MD, aquaculture is an important pillar of local economic development. Therefore, in this model, the supply of aquatic products should not only meet the living needs of local residents but meet the needs of economic development (export of aquatic products). Aquatic products are mainly produced by aquaculture land:

$$\sum_{i=1}^{3}\sum_{t=1}^{2}\left(UWP^{\pm}_{i,t}\times x^{\pm}_{i,j=3,t}\right)\gtrsim WD^{\pm}$$

where $UWP_{i,t}$ = unit water production from aquaculture land (ton/ha); WD = demand water production (ton).

(iv) Wooden production input-output constraints:

Vietnam is the largest exporter of wood and wood products in Southeast Asia. The wooden production industry not only brings economic benefits but provides a large number of jobs. Therefore, wooden production should meet the economic development needs of the MD:

$$\sum_{i=1}^{3}\sum_{t=1}^{2}\left(UWOP^{\pm}_{i,t}\times x^{\pm}_{i,j=3,t}\right)\gtrsim WOD^{\pm}$$

where $UWOP_{i,t}$ = unit wooden production from forest land (m^3/ha); WOD = demand wooden production (m^3).

(v) Agricultural water consumption constraints:

MD's main water sources are from the Mekong River Basin and rainfall; MD's 70% of regional water supply is used for agricultural production. Effective water supply has a decisive impact on MD's most important industry, food production, and aquaculture. Therefore, the total agricultural water consumption should not exceed the total regional agricultural water supply capacity:

$$\sum_{i=1}^{3}\sum_{t=1}^{2}\left(\mathrm{WCC}_{i,t}^{\pm} \times x_{i,j=1,t}^{\pm}\right) + \sum_{i=1}^{3}\sum_{t=1}^{2}\left(\mathrm{WCA}_{i,t}^{\pm} \times x_{i,j=3,t}^{\pm}\right) \lesssim \mathrm{RWSA}_{t}^{\pm}$$

where $\mathrm{WCC}_{i,t}$ = Water consumption per unit of cultivated land (m^3/ha); $\mathrm{WCA}_{i,t}$ = Water consumption per unit of aquaculture land (m^3/ha); RWSA_t = regional water supply planning for agriculture in the MD (ton).

(vi) Available labor constraints:

Similarly, all land industries require labor, however, available labor is limited in the MD:

$$\sum_{i=1}^{3}\sum_{j=1}^{6}\sum_{t=1}^{2}\left(\mathrm{LC}_{i,t}^{\pm} \times x_{i,j,t}^{\pm}\right) \lesssim \mathrm{AL}^{\pm}$$

where $\mathrm{LC}_{i,t}$ = labor in a unit land area (person/ha); AL = available labor (person).

3.3. Environmental Constraints

(i) BOD5 emissions constraints:

The intensive and industrialized aquaculture of the MD has brought rich aquatic benefits, but at the same time, it has also brought a large amount of aquatic wastewater, which contains BOD5, COD, and other substances. In order to ensure the continuous growth of the quality and quantity of MD fish, shrimp, and other aquatic products, the emissions of BOD5 in MD aquaculture wastewater should not exceed the emissions allowed by the MD environment:

$$\sum_{i=1}^{3}\sum_{t=1}^{2}\left(\mathrm{DB}_{i,t}^{\pm} \times x_{i,j=3,t}^{\pm}\right) \leq \mathrm{MTDB}^{p}$$

where $\mathrm{DB}_{i,t}$ = discharge amount of BOD5 for unit aquaculture land (ton/ha); MTDB = maximum total BOD5 discharge (ton); p = probability of violating the constraints of environmental capacities, and $p \in [0,1]$.

(ii) COD emissions constraints

Similarly, the discharge of COD in the MD aquaculture wastewater should not exceed the allowable discharge in the MD environment:

$$\sum_{i=1}^{3}\sum_{t=1}^{2}\left(\mathrm{DC}_{i,t}^{\pm} \times x_{i,j=3,t}^{\pm}\right) \leq \mathrm{MTDC}^{p}$$

where $\mathrm{DC}_{i,t}$ = discharge amount of COD for unit aquaculture land (ton/ha); MTDC = maximum total COD discharge (ton).

(iii) Wastewater treatment capacity constraints

In the model, the wastewater generated from aquaculture land and construction land should not exceed the wastewater treatment capacity of the MD:

$$\sum_{i=1}^{3}\sum_{t=1}^{2}\left(\mathrm{WDA}_{i,t}^{\pm} \times x_{i,j=3,t}^{\pm}\right) + \sum_{i=1}^{3}\sum_{t=1}^{2}\left(\mathrm{WDB}_{i,t}^{\pm} \times x_{i,j=4,t}^{\pm}\right) \leq \mathrm{WTPC}^{p}$$

where $WDA_{i,t}$ = wastewater discharging factor of aquaculture land (ton/ha); $WDB_{i,t}$ = wastewater discharging factor of construction land (ton/ha); WTPC = wastewater treatment capacity in the MD (ton).

(iv) Solid-waste treatment capacity constraints

The discharge of solid waste should not exceed the solid waste treatment capacity of the MD:

$$\sum_{i=1}^{3}\sum_{t=1}^{2}\left(SDC_{i,t}^{\pm} \times x_{i,j=1,t}^{\pm}\right) + \sum_{i=1}^{3}\sum_{t=1}^{2}\left(SDB_{i,t}^{\pm} \times x_{i,j=4,t}^{\pm}\right) \leq STC^{p}$$

where $SDC_{i,t}$ = solid-waste discharging factor of cultivated land (ton/ha); $SDB_{i,t}$ = solid-waste discharging factor of construction land (ton/ha); STC = solid-waste treatment capacity in the MD (ton).

3.4. Ecological Constraints

(i) Antibiotic consumption constraints

In the past ten years, the MD has vigorously developed industrialized aquaculture and constructed a large amount of industrial aquaculture land. In the production process, the fishermen did not use antibiotics scientifically and reasonably, which led to the destruction of the ecosystem. The extensive use of antibiotics pollutes the water land of the MD and has a toxic effect on most aquatic organisms. Therefore, the consumption of antibiotics should not exceed the maximum consumption amount:

$$\sum_{i=1}^{3}\sum_{t=1}^{2}\left(AC_{i,t}^{\pm} \times x_{i,j=3,t}^{\pm}\right) \leq MAC^{p}$$

where $AC_{i,t}$ = antibiotics consumption for unit aquaculture land (ton/ha); MAC = maximum antibiotics consumption (ton).

(ii) Fertilizer consumption constraints

Due to the favorable natural conditions of the MD, its crops such as rice are planted twice or three times a year, which makes farmers need to continuously apply fertilizers to maintain and improve the soil quality as much as possible. However, the use of large amounts of fertilizers will seriously damage MD's fragile environment. Therefore, the consumption of fertilizers should not exceed the maximum consumption amount:

$$\sum_{i=1}^{3}\sum_{t=1}^{2}\left(FC_{i,t}^{\pm} \times x_{i,j=1,t}^{\pm}\right) \leq MFC^{p}$$

where $FC_{i,t}$ = fertilizer consumption for unit cultivated land (ton/ha); MFC = maximum fertilizer consumption (ton).

(iii) Pesticide consumption constraints

Similarly, the pesticide consumption in cultivated land should not exceed the maximum consumption amount:

$$\sum_{i=1}^{3}\sum_{t=1}^{2}\left(PC_{i,t}^{\pm} \times x_{i,j=1,t}^{\pm}\right) \leq MPC^{p}$$

where $PC_{i,t}$ = pesticide consumption for unit cultivated land (ton/ha); MPC = maximum pesticide consumption (ton).

3.5. Technical Constrains

(i) Total land areas constraints:

$$\sum_{i=1}^{3}\sum_{j=1}^{6}\sum_{t=1}^{2} x_{i,j,t}^{\pm} = \text{TLA}_{i,j,t}^{\pm}$$

where TLA is total land areas constrain in period t.

(ii) Non-negative constraints:

$$x_{i,j,t}^{\pm} \geq 0$$

3.6. Data Collection

The parameters of the IPF-LUA model include four types: beneficial and cost parameters (Table 1); economic and social parameters (Table 2); environmental and ecological parameters (Table 3). Beneficial and cost parameters can be obtained from land evaluation. Economic and social parameters can be obtained through index forecasting models. Environmental and ecological parameters can be obtained through stochastic models.

Table 1. Benefits and cost for different land-use types (USD/ha).

Land-Use Type	Symbol	Period			
		$t = 1$		$t = 2$	
		Lower	Upper	Lower	Upper
Benefits of land use	$CLB_{i=1,j=1}$ (10^3)	4.74	6.42	4.89	6.61
	$CLB_{i=2,j=1}$ (10^3)	3.69	4.99	3.80	5.14
	$CLB_{i=3,j=1}$ (10^3)	2.87	3.88	2.95	3.99
	$FLB_{i=1,j=2}$ (10^3)	3.02	4.09	3.12	4.21
	$FLB_{i=2,j=2}$ (10^3)	6.76	9.15	6.96	9.42
	$FLB_{i=3,j=2}$ (10^3)	2.16	2.93	2.23	3.01
	$ALB_{i=1,j=3}$ (10^3)	9.83	13.30	10.13	13.70
	$ALB_{i=2,j=3}$ (10^3)	10.73	14.51	11.05	14.95
	$ALB_{i=3,j=3}$ (10^3)	8.94	12.10	9.21	12.46
	$CLP_{i=1,j=4}$ (10^3)	14.85	20.09	15.89	21.50
	$CLP_{i=2,j=4}$ (10^3)	25.68	34.75	27.48	37.18
	$CLP_{i=3,j=4}$ (10^3)	12.70	17.19	13.59	18.39
Costs of land use	$UWTC_{i=1,j=1}$	0.76	1.14	0.93	1.29
	$USTC_{i=1,j=1}$	95.71	103.05	104.03	117.25
	$UWTC_{i=1,j=3}$	1.02	1.51	1.24	1.71
	$UWTC_{i=1,j=4}$	77.68	83.99	86.00	93.81
	$USTC_{i=1,j=4}$	707.60	747.47	877.03	956.76
	$UWMC_{i=1,j=5}$	398.65	428.55	508.28	538.18
	$UUDC_{i=1,j=6}$	847.13	896.96	976.69	1245.78

3.7. Model Solving

According to the algorithm of the IPF-LUA model which is introduced in Appendix A, the IPF-LUA model can be transformed into two definite sub-models, which correspond to the upper and lower limits of the expected objective function values at different p levels. Then, we can calculate these two linear models in Lingo 12 software. The framework of the IPF-LUA model is shown in Figure 2.

Table 2. Economic, social, environmental, and technical parameters.

Symbol	Period			
	$t = 1$		$t = 2$	
	Lower	Upper	Lower	Upper
MGI_t (10^6USD)	23.9	32.3	28.0	37.8
$UGP_{i=1}$ (ton/ha)	9.84	13.32	9.76	13.20
GD (10^6ton)	1.50	2.03	1.51	2.05
$UWP_{i=1}$ (ton/ha)	57.54	77.84	59.14	80.02
WD (10^3ton)	494.52	669.05	499.46	675.74
$UWOP_{i=1}$ (m^3/ha)	4.70	6.36	4.68	6.34
WOD (10^3m^3)	872.78	1180.82	898.96	1216.25
$WCC_{i=1}$ (10^3/ha)	1.79	2.42	1.89	2.56
$WCA_{i=1}$ (10^3m^3/ha)	8.50	11.50	9.01	12.19
RWSA (10^9m^3)	0.45	0.61	0.48	0.65
$LC_{i=1}$ (people/ha)	2.29	3.09	2.29	3.09
AL (10^6people)	8.65	11.71	8.64	11.69
$DB_{i=1,j=3}$ (ton/ha)	0.23	0.31	0.24	0.33
$DC_{i=1,j=3}$ (ton/ha)	0.41	0.56	0.43	0.59
$WDA_{i=1,j=3}$ (10^3ton/ha)	5.15	6.97	5.41	7.32
$WDB_{i=1,j=4}$ (10^3ton/ha)	0.14	0.19	0.15	0.20
$SDC_{i=1,j=1}$ (ton/ha)	9.86	13.34	9.73	13.16
$SDB_{i=1,j=4}$ (ton/ha)	8.14	11.01	8.95	12.11
$AC_{i=1}$ (kg/ha)	13.02	17.62	13.28	17.97
$FC_{i=1}$ (kg/ha)	663.00	897.00	595.00	805.00
$PC_{i=1}$ (kg/ha)	5.60	7.60	5.10	6.90

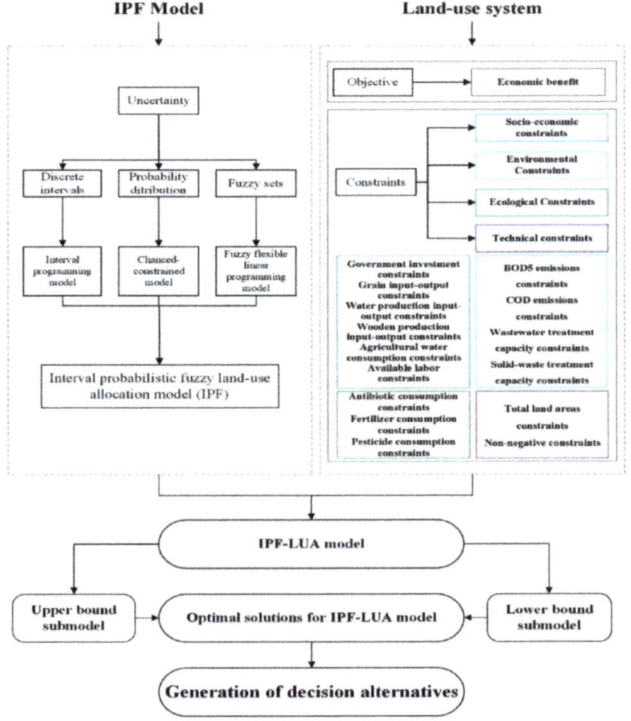

Figure 2. Framework of the IPF-LUA model.

Table 3. Eco-environmental capacity under different *p* levels.

Symbol	Period							
	$t = 1$				$t = 2$			
	$p = 0.01$	$p = 0.05$	$p = 0.10$	$p = 0.15$	$p = 0.01$	$p = 0.05$	$p = 0.10$	$p = 0.15$
MTDB (10^3 ton)	296.92	304.72	314.48	324.24	306.76	314.96	325.21	335.45
MTDC (10^3 ton)	365.03	379.50	397.59	415.67	383.28	398.47	417.47	436.46
WTPC (10^9 ton)	4.90	5.09	5.33	5.58	5.14	5.35	5.60	5.86
STPC (10^6 ton)	4.52	4.70	4.92	5.15	4.97	5.17	5.42	5.66
MAC (10^3 ton)	8.04	8.36	8.76	9.16	8.20	8.53	8.93	9.34
MFC (10^6 ton)	2.06	2.14	2.24	2.34	2.06	2.14	2.24	2.34
MPC (10^3 ton)	174.42	181.33	189.93	198.52	174.56	181.47	190.06	198.78

4. Result and Discussion

4.1. Optimized Land-Use Patterns under Different p Levels during Two Periods

Figures 3–5 present the optimization solutions to land allocation for decision variables obtained through the IPF-LUA model under different *p* levels. The results show that any change of *p* will result in different environmental capacity which leads to different LUA modes. At the same time, in the land use system, economic development, environmental and ecological changes, and pollutant emissions will also lead to different LUA modes. In the case of excess waste, allotments to forest land for environmental protection should be assigned firstly and then to construction land, while allotments to cultivated land, aquaculture land, and unused land are mainly due to policy constraints. Analysis of the modeling solutions is provided below.

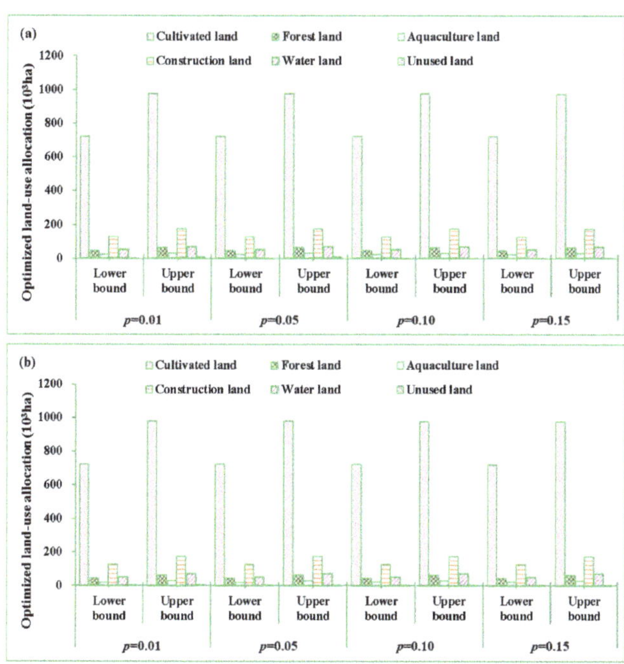

Figure 3. (**a**) $t = 2021$ and (**b**) $t = 2022$ Optimized land-use allocation in the Deep-water flood disaster sub-zone under *p*-levels during two periods.

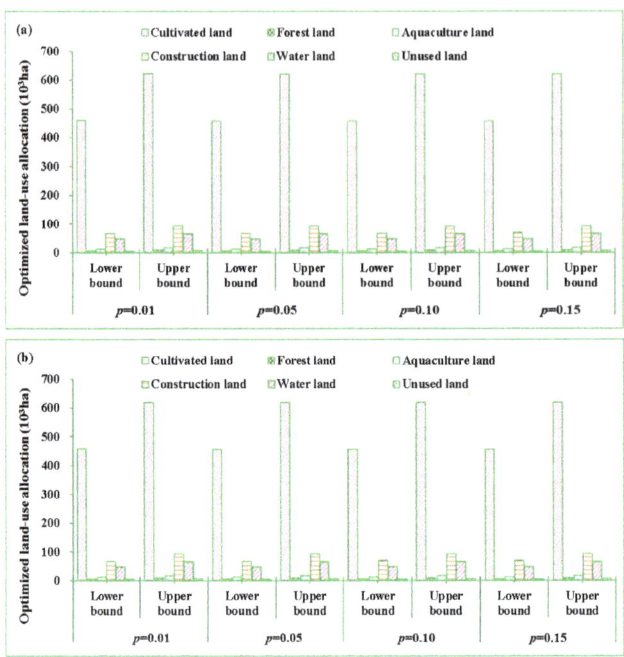

Figure 4. (**a**) t = 2021 and (**b**) t = 2022 optimized land-use allocation in the central sub-zone under different p-levels during two periods.

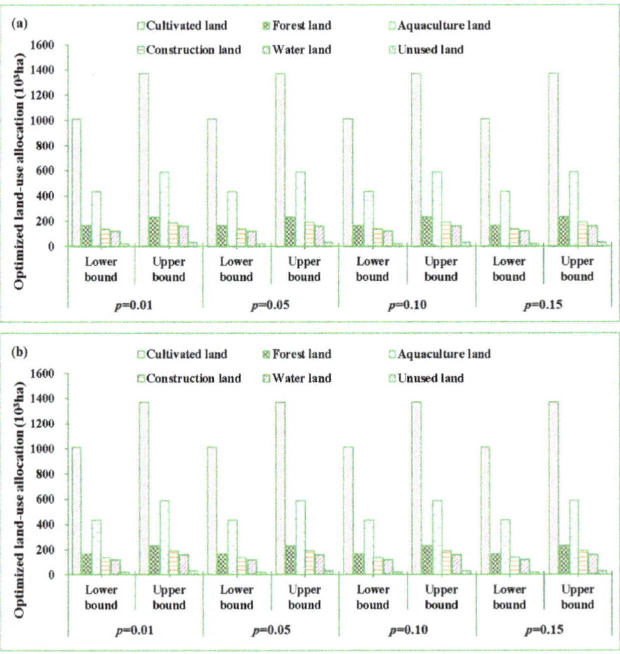

Figure 5. (**a**) t = 2021 and (**b**) t = 2022 optimized land-use allocation in the coastal and island sub-zone under different p-levels during two periods.

The optimized allocation in the deep-water flood disaster sub-zone under varied risk levels of violating environmental capacity constraints during two periods is presented in Figure 3 and Table 4. We can clearly see that in period 1, with the increase of *p*-value, the area of forest land, aquaculture land and construction land in the deep-water flood disaster sub-zone also increased. The *p*-level indicates the probability of environmental/ecological constraints being violated. When the *p*-value increases, the probability of violating environmental/ecological constraints will increase, but at the same time, the environmental/ecological capacity will expand. Therefore, the IPF-LUA model tends to allocate more land types with economic benefits, such as forest land, aquaculture land, and construction land. Based on this, in general, the cultivated land should also increase, but in the overall development plan of the MD, the cultivated land will turn to fruit planting and aquaculture. This feature of MD's regional development has been considered in the constraints of the IPF-LUA model. Therefore, in the land optimization results of the MD, the cultivated land area will be slightly decreased. This also shows that the IPF-LUA model has the characteristics of adjusting measures to local conditions. Compared with period 2, under the situation that MD's development planning policy remains unchanged, there is no obvious change in the way of land allocation in the deep-water flood disaster sub-zone. The agricultural industry represented by cultivated land is the regional leading industry of the MD. The model is based on MD's macro-policy planning. First, it is in order to meet the local people's food demand and food export plan. Second, it is constrained by MD's economic development requirements and natural ecological/environmental conditions. Cultivated land should be transferred to land types with more economic benefits, such as fruit, forest, and aquaculture land, to develop diversified agricultural industries. As a result, the cultivated land area continued to decrease, while the forest, aquaculture, and construction land area continued to increase.

Table 4. Optimized land-use allocation in the deep-water flood disaster sub-zone under different *p*-levels during two periods.

Variable		x (*j* = 1)	x (*j* = 2)	x (*j* = 3)	x (*j* = 4)	x (*j* = 5)	x (*j* = 6)
t = 1	***p* = 0.01**	[721,709, 976,430]	[44,985, 60,862]	[20,594, 27,862]	[127,872, 173,003]	[49,930, 67,552]	[899, 1216]
	***p* = 0.05**	[721,114, 975,625]	[45,155, 61,092]	[20,849, 28,207]	[128,042, 173,233]	[50,015, 67,667]	[814, 1101]
	***p* = 0.10**	[720,604, 974,935]	[45,325, 61,322]	[21,104, 28,552]	[128,212, 173,463]	[50,100, 67,782]	[729, 986]
	***p* = 0.15**	[720,179, 974,360]	[45,495, 61,552]	[21,274, 28,782]	[128,297, 173578]	[50,185, 67,897]	[644, 871]
t = 2	***p* = 0.01**	[721,284, 975,855]	[45,070, 60,977]	[20,764, 28,092]	[128,042, 173,233]	[50,015, 67,667]	[814, 1101]
	***p* = 0.05**	[720,774, 975,165]	[45,240, 61,207]	[21,019, 28,437]	[128,127, 173,348]	[50,100, 67,782]	[729, 986]
	***p* = 0.10**	[720,349, 974,590]	[45,410, 61,437]	[21,189, 28,667]	[128,297, 173,578]	[50,185, 67,897]	[644, 871]
	***p* = 0.15**	[719,924, 974,015]	[45,580, 61,667]	[21,359, 28,897]	[128,382, 173,693]	[50,270, 68,012]	[559, 756]

The optimized allocation in the other two sub-zone under various *p* levels during two periods is presented in Figures 4 and 5. Figure 4 indicates that in the central sub-zone, the area of cultivated land and unused land decreases with the *p*-level and the area of other land types will decrease. Unlike the other two sub-zones affected by floods and coastal environment, the central sub-zone is less affected by adverse natural conditions. Therefore, the area of cultivated land in the region has diverted more towards other agricultural lands, aquaculture land and construction land, with a slight increase in forest land. Generally speaking, the area less restricted by adverse natural environmental conditions should allocate more land to the land types with higher economic benefits, and the central sub-zone should be allocated to construction land first, followed by aquaculture land. However, according to the development advantages and natural characteristics of the region, the model shows some constraints. For example, according to the construction land and labor population density, the labor force constraint is put forward; according to the characteristics of pollutant discharge during aquaculture, COD and BOD5 discharge constraints are established, along with other constraints. These constraints cause the model to fail when simply increasing the area of high-benefit land types in the process of land allocation.

For example, the benefits of construction land are higher than that of aquaculture land and other agricultural land. However, based on the above conditions, the area of other agricultural land and aquaculture land increases more than that of construction land in the central sub-zone. In Figure 5, the land allocation pattern of the coastal and island sub-zone is similar to that of the deep-water flood disaster sub-zone.

However, we can see that the aquaculture area in the three districts has increased more. The coastal and island sub-zone is close to the ocean, which is vulnerable to natural disasters such as coastal landslides. However, at the same time, the sub-zone is rich in water resources and has the largest woodland area in the entire MD. Considering these factors, the model allocates more land to aquaculture land and less land to forest land and construction land.

The optimization results in Figures 3–5 also show the interval solution generated by the IPF-LUA model. The interval solution can provide an effective scheme for regional land use allocation. For example, in the central sub-zone, combining the lower bound for cultivated land, construction land and aquaculture land with the upper bound for forest land, unused land and water land corresponds to lower land system economic benefits. However, this combined allocation pattern can meet the needs of people's lives and ensure a high-quality ecological environment. It is a trade-off between economic benefits and the quality of the ecological environment and is a relatively conservative land use management strategy. When the upper bound of cultivated land, construction land, and aquaculture land are combined with the lower bound of forest land, unused land, and water land, a higher economic benefit of the land system can be achieved. This kind of combined allocation pattern can obtain as many economic benefits as possible. At the same time, it also means that the probability of ecological and environmental damage increased. It represents a more radical economic strategy. The optimization results of the model provide a system benefit interval solution in the planning scheme. For example, when $p = 0.01$, the land system benefit of the whole MD is USD [20.8, 28.1] $\times 10^9$ in period 1. The net benefits of the land system in this region range from USD 20.8 $\times 10^9$ to USD 28.1 $\times 10^9$, which means that the actual net profits brought by different land use planning patterns change in the upper and lower bound. Generally speaking, the allocation with lower system benefits has a lower risk of violating system constraints, and this region has higher ecological environment quality.

4.2. Optimized Ecological/Environmental Pollutant Discharge and Eco-Environmental Policy Analysis under Different p Value Levels

Figures 6 and 7 show the BOD5 discharge and aquaculture antibiotic consumption after IPF-LUA optimization. The lower bound of BOD5 discharge and antibiotic consumption correspond to lower system benefits, but maintain a higher level of eco-environmental quality, implying a more conservative environmental protection strategy. The values of BOD5 and antibiotics increase correspondingly when the manager wants to obtain more net benefits from the system. For the whole system, the risks of water pollution and unhealthy aquatic products increase.

4.3. Trade-Off between Economic Objective and Eco-Environmental Constraints

The p-value level represents the probability of violating ecological/environmental capacity. Figures 6 and 7 also reflect that different levels of pollutants (such as BOD5 and antibiotics) will be produced under different p-values, thus the land system will produce different levels of economic benefits. For example, in period 1, when $p = 0.05$, the net profit of the land system is USD [20.78, 28.11] $\times 10^9$. In contrast, when $p = 0.15$, the net profit of the land system is USD [20.82, 28.16] $\times 10^9$. Obviously, the area of the land type that brings higher economic benefits will increase with an increase in the p-value. An increase in the value of the violation probability p means that the model tends to have a more relaxed environmental capacity. Based on this, the model will allocate more land area to the land with higher net profits. Accordingly, the area of land with low profits or costs will be reduced. Figure 8 shows the relationship between violation probability p and economic profits of the land system.

4.4. Trade-Off between System Benefit and Membership λ and Constraints

Through calculation, the value of optimized membership λ is [0.51,0.73]. The value of membership λ represents the membership satisfying all objective functions and constraints. As shown in Figure 9, when λ value increases, the system profit increases corresponding to more aggressive land use policies, and each constraint condition of the model is more relaxed. When the value of λ decreases, the profit of the system decreases, and each constraint condition of the model is stricter corresponding to the more conservative land use policies. The value of λ represents the compromise between the objective function of the land system and all constraints and reflects the relationship between economic benefits and social benefits, ecological benefits, and environmental benefits. Through calculation, the value of optimized membership λ is [0.51,0.77]. The value of membership λ represents the membership satisfying all objective functions and constraints. As shown in Figure 9, when λ value increases, the system profit increases corresponding to more aggressive land use policies, and each constraint condition of the model is more relaxed. When the value of λ decreases, the profit of the system decreases, and each constraint condition of the model is stricter corresponding to the more conservative land use policies. The value of λ represents the compromise between the objective function of the land system and all constraints and reflects the relationship between economic benefits and social benefits, ecological benefits, and environmental benefits. For example, more water resources, more fertilizers, and antibiotics correspond to a larger λ, thus generating more systematic economic benefits, as shown in Figure 10.

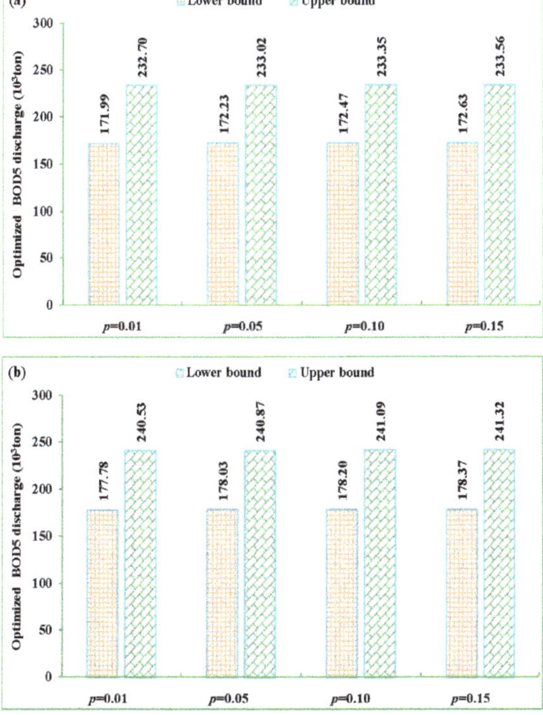

Figure 6. (**a**) t = 2021 and (**b**) t = 2022 relationship between p and optimized BOD5 discharge in MD during two periods.

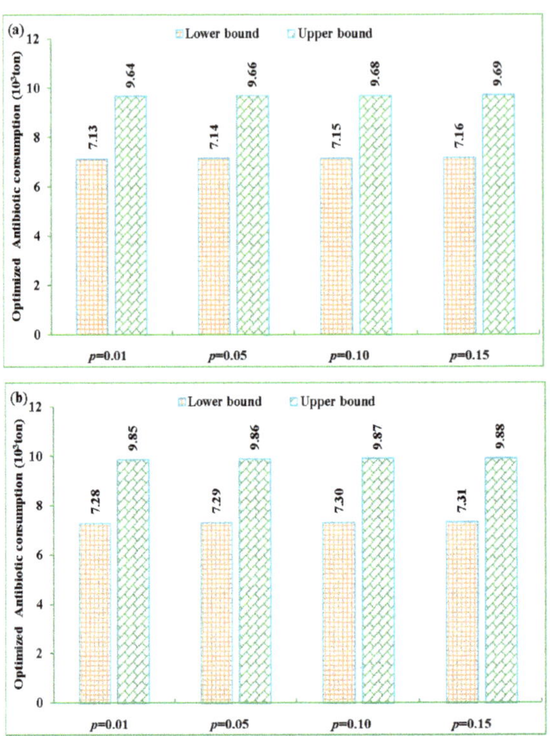

Figure 7. (**a**) t = 2021 and (**b**) t = 2022 relationship between p and optimized Antibiotic consumption in MD during two periods.

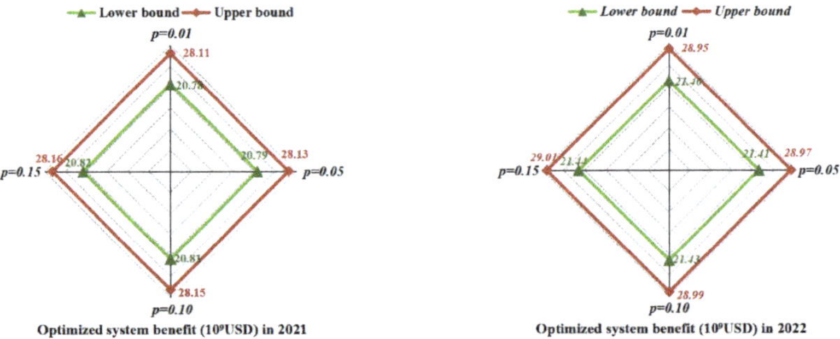

Figure 8. Relationship between p and optimized system benefit in MD during two periods.

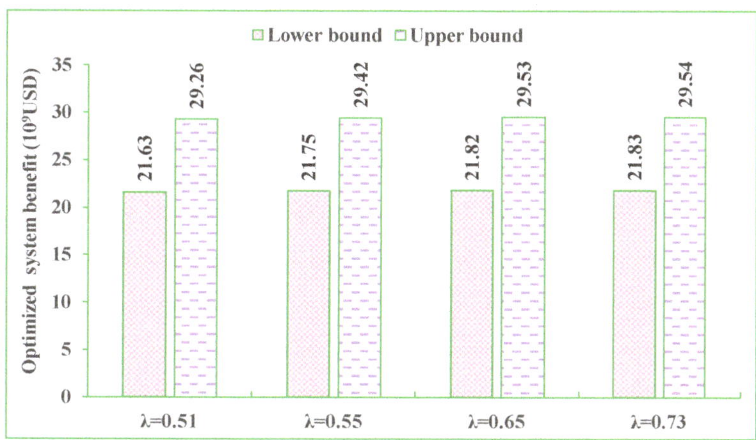

Figure 9. Relationship between λ and optimized system benefit.

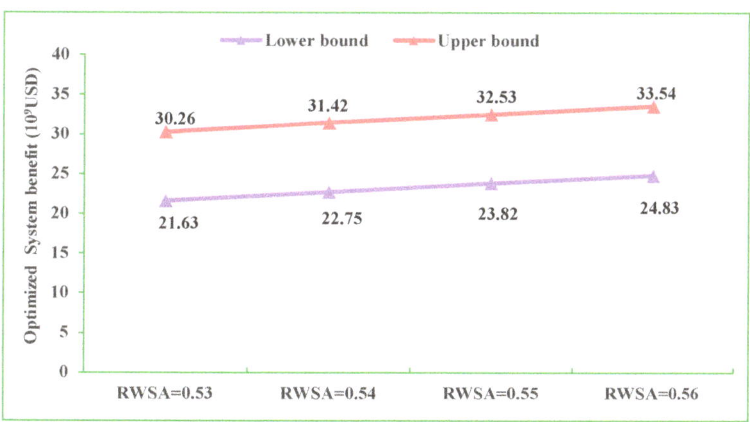

Figure 10. Relationship between regional water supply planning and optimized system benefit.

5. Conclusions

In this paper, the interval probabilistic fuzzy land-use allocation model (IPF-LUA) based on uncertainty is proposed and applied to the Mekong River Delta. The existing studies in this region mainly focus on water resources, and few studies focus on land. At the same time, the ecological environment of the MD is extremely fragile and sensitive. The IPF-LUA model proposed in this paper establishes a reasonable land use optimization model by combining interval parametric programming, probabilistic programming, and fuzzy linear programming, and fully considering the ecological and environmental factors of the MD. This method can help decision-makers to analyze the relationship among economic, ecological, and environmental benefits. By dismantling the model, calculating the upper and lower expectations, and choosing different land allocation methods, the decision-makers can keep the balance between economic and ecological benefits. The IPF-LUA model can solve the problem of allocation of land use quantity structure in an LUA: by analyzing the current land use situation, development conditions and objectives of the research area, and fully considering the uncertain factors in the regional land use system, the objective function and constraint conditions are constructed to make the optimization results more flexible. Moreover, combining the quantitative structure optimization results of IPF-LUA with the LUCC simulation model or spatial layout optimization model can simultaneously provide support and help for land use planning from both quantitative structure and spatial layout.

This method has been applied to the LUA case study in the MD of Vietnam. The results of four different *p*-value scenarios in two planning periods show the quantitative relationship between economic benefits and the ecological environment. It provides decision-makers with a variety of planning schemes that give priority to economic benefits or conservation of the ecological environment. Collectively, the results show an optimization method that compromises the economic benefits of land system and ecological environment management.

The IPF-LUA model in this paper can effectively tackle the optimization problem in LUA. The results of model optimization would be more rational if other key ecological and environmental data, as well as stakeholder-related data, were available. In addition, the model has limitations: IPF-LUA is a method to optimize the quantitative structure of land use, which focuses on the optimization of the quantitative structure, but fails to provide help for land use planning in the aspect of spatial layout optimization. At the same time, the optimization of land use quantity structure in LUA is not only a linear programming problem, but also a nonlinear programming problem which is not considered in this study. Therefore, our next work is to couple IPF-LUA with the land use spatial layout optimization model and build a coupling model that can solve both quantitative and spatial optimization, so as to solve the more complex LUA optimization problem in reality. In addition, there are a large number of uncertain factors in the land use system. We will consider and design more constraints, such as climate change and policy constraints, to further improve the IPF-LUA model and make it more perfect.

Author Contributions: M.Z. and Y.M. conceived and designed the research; C.M. and M.W. collected, managed, and verified the data; M.Z., Y.M., J.T. calculated and analyzed the data and the results; M.Z. and Y.M. wrote the manuscript. All authors have read and agreed to the published version of the manuscript.

Funding: This research was funded by "the Fundamental Research Funds for the Central Universities" (No. 2021WKYXZD006).

Institutional Review Board Statement: Not applicable.

Informed Consent Statement: Not applicable.

Data Availability Statement: Not applicable.

Conflicts of Interest: The authors declare that they have no conflict of interest.

Ethics Approval: Not applicable.

Appendix A

Introduction of the ICCF (interval chance-constrained fuzzy) model [48,55]:

$$Maxf^{\pm} \cong C^{\pm}x^{\pm} \tag{A1}$$

Subject to:

$$C^{\pm}x^{\pm} \gtrsim b^{\pm}_{opt} \tag{A2}$$

$$A^{\pm}_{i}x^{\pm} \lesssim b^{\pm}_{i} \;\; i = 1,2,\ldots,m, i \neq s \tag{A3}$$

$$A^{\pm}_{s}x^{\pm} \lesssim b^{(p_s)}_{s} \;\; s = 1,2,\ldots,m, s \neq i \tag{A4}$$

where x is an $n \times 1$ alternative set; C is a $1 \times n$ coefficient of an objective function; A_i is an $m \times n$ matrix of coefficients of constraints and b_i is an $m \times 1$ matrix (right-hand sides, RHS). "$\pm$" express intervals; "$\cong$" represents fuzzy equality; "$\gtrsim$" and "$\lesssim$" represent fuzzy inequality. p_s denotes the probability that the constraints s are violated. $b_s^{(p_s)}$ represent corresponding values given the cumulative distribution function of bs and the probability of violating constraint s (p_s).

On the basis of the principle of fuzzy flexible programming, let $l^{\pm}$ value corresponds to the membership grade of satisfaction for a fuzzy decision. Specifically, the flexibility in the constraints and fuzziness in the system objective, which are represented by fuzzy sets and denoted as "fuzzy constraints" and "fuzzy goal", are expressed as membership grades $l^{\pm}$ corresponding to the degrees of overall satisfaction for the constraints/objective. Thus, above model can be converted to:

$$Max\ l^{\pm} \tag{A5}$$

Subject to:

$$C^{\pm}x^{\pm} \lesssim f_{opt}^{-} + (1 - l^{\pm})(f_{opt}^{+} - f_{opt}^{-}) \tag{A6}$$

$$A_i^{\pm}x^{\pm} \leq b_i^{-} + (1 - l^{\pm})(b_i^{+} - b_i^{-}) \quad i = 1, 2, \ldots, m, i \neq s \tag{A7}$$

$$A_s^{\pm}x^{\pm} \lesssim b_s^{(p_s)} \quad s = 1, 2, \ldots, m, s \neq i \tag{A8}$$

$$x^{\pm} \geq 0 \tag{A9}$$

$$0 \leq l^{\pm} \leq 1 \tag{A10}$$

where f_{opt}^{+} and f_{opt}^{-} denote the upper and lower bounds of the objective's aspiration level as designated by decision-makers; $l^{\pm}$ denotes the control decision variable corresponding to the degree (membership grade) to which $x^{\pm}$ solution fulfills the fuzzy objective or constraints. Model (A5) can be solved through a two-step method where a sub-model corresponding to l^{-} is first formulated and solved. In the second step, the other sub-model corresponding to l^{+} can then be formulated supported by the solution of the first submodel. If $b_i^{+} \geq 0$ and $f_i^{+} \geq 0$, the sub-model corresponding to l^{-} can be formulated as follows:

$$Max \qquad l^{-} \tag{A11}$$

Subject to:

$$\sum_{j=1}^{k_1} C_j^{+}x_j^{+} + \sum_{j=k_1+1}^{n} C_j^{+}x_j^{-} \leq f_{opt}^{-} + (1 - l^{\pm})(f_{opt}^{+} - f_{opt}^{-}) \tag{A12}$$

$$\sum_{j=1}^{k_1} \left|a_{ij}\right|^{-} sign(a_{ij}^{-})x_j^{+} + \sum_{j=k_1+1}^{n} \left|a_{ij}\right|^{+} sign(a_{ij}^{+})x_j^{-} \leq b_i^{-} + (1 - l^{-})(b_i^{+} - b_i^{-}),\ \forall i \tag{A13}$$

$$\sum_{j=1}^{k_1} \left|a_{sj}\right|^{-} sign(a_{sj}^{-})x_j^{+} + \sum_{j=k_1+1}^{n} \left|a_{sj}\right|_{+} sign(a_{sj}^{+})x_j^{-} \leq b_s^{p_s},\ \forall s,\ s \neq i \tag{A14}$$

$$x_j^{-} \geq 0, j = 1, 2, \ldots, k_1 \tag{A15}$$

$$x_j^{+} \geq 0, j = k_1 + 1, k_2 + 2, \ldots, n \tag{A16}$$

$$0 \leq l^{\pm} \leq 1 \tag{A17}$$

where Sign is a signal function, which is defined as:

$$sign(x^{\pm}) = \left\{ \begin{array}{c} 1,\ \text{if } x^{\pm} \geq 0 \\ -1,\ \text{if } x^{\pm} \leq 0 \end{array} \right\}$$

Let x_{jopt}^{+} ($j = 1, 2, \ldots, k_1$) and x_{jopt}^{-} ($j = k_1 + 1, k_2 + 2, \ldots, n$) be solutions of sub-model (3). Then, the second sub-model corresponding to l^{+} can be formulated supported by the solution of sub-model (A18):

$$Max \qquad l^{+} \tag{A18}$$

Subject to:

$$\sum_{j=1}^{k_1} C_j^- x_j^- + \sum_{j=k_1+1}^{n} C_j^- x_j^+ \leq f_{opt}^- + (1 - l^{\pm})(f_{opt}^+ - f_{opt}^-) \tag{A19}$$

$$\sum_{j=1}^{k_1} \left|a_{ij}\right|^+ sign(a_{ij}^+) x_j^- + \sum_{j=k_1+1}^{n} \left|a_{ij}\right|^- sign(a_{ij}^-) x_j^+ \leq b_i^- + (1 - l^+)(b_i^+ - b_i^-),\ \forall i \tag{A20}$$

$$\sum_{j=1}^{k_1} \left|a_{sj}\right|^+ sign(a_{sj}^+) x_j^- + \sum_{j=k_1+1}^{n} \left|a_{sj}\right|^- sign(a_{sj}^-) x_j^+ \leq b_s^{p_s},\ \forall s,\ s \neq i \tag{A21}$$

$$x_{jopt}^+ \geq x_j^- \geq 0,\ j = 1, 2, \ldots, k_1 \tag{A22}$$

$$x_j^+ \geq x_{jopt}^-,\ j = k_1 + 1, k_2 + 2, \ldots, n \tag{A23}$$

$$0 \leq l^{\pm} \leq 1 \tag{A24}$$

Let x_{jopt}^- ($j = 1, 2, \ldots, k_1$) and x_{jopt}^+ ($j = k_1 + 1, k_2 + 2, \ldots, n$) be solutions of sub-model (4). Thus, we can obtain the interval solutions as follows:

$$l_{opt}^{\pm} = \left[l_{opt}^-, l_{opt}^+\right] \tag{A25}$$

$$x_{jopt}^{\pm} = \left[x_{jopt}^-, x_{jopt}^+\right], \forall j \tag{A26}$$

Then, the optimized objectivef_{opt}^- and f_{opt}^+ can be calculated as follows:

$$f_{opt}^-, \sum_{j=1}^{k_1} C_j^- x_j^- + \sum_{j=k_1+1}^{n} C_j^- x_j^+ \tag{A27}$$

$$f_{opt}^-, \sum_{j=1}^{k_1} C_j^+ x_j^+ + \sum_{j=k_1+1}^{n} C_j^+ x_j^- \tag{A28}$$

Thus, we have:

$$f_{jopt}^{\pm} = \left[f_{jopt}^-, f_{jopt}^+\right], \forall j \tag{A29}$$

References

1. Arata, L.; Diluiso, F.; Guastella, G.; Pareglio, S.; Sckokai, P. Willingness to pay for alternative features of land-use policies: The case of the lake Garda region. *Land Use Policy* **2021**, *100*, 10. [CrossRef]
2. Zheng, W.W.; Ke, X.L.; Xiao, B.Y.; Zhou, T. Optimising land use allocation to balance ecosystem services and economic benefits—A case study in Wuhan, China. *J. Environ. Manag.* **2019**, *248*, 10. [CrossRef]
3. Masoomi, Z.; Mesgari, M.S.; Hamrah, M. Allocation of urban land uses by Multi-Objective Particle Swarm Optimization algorithm. *Int. J. Geogr. Inf. Sci.* **2013**, *27*, 542–566. [CrossRef]
4. Hersperger, A.M.; Oliveira, E.; Pagliarin, S.; Palka, G.; Verburg, P.; Bolliger, J.; Gradinaru, S. Urban land-use change: The role of strategic spatial planning. *Glob. Environ. Chang. Hum. Policy Dimens.* **2018**, *51*, 32–42. [CrossRef]
5. Liu, Y.S.; Zhang, Z.W.; Zhou, Y. Efficiency of construction land allocation in China: An econometric analysis of panel data. *Land Use Policy* **2018**, *74*, 261–272. [CrossRef]
6. Jha, M.K.; Singh, L.K.; Nayak, G.K.; Chowdary, V.M. Optimization modeling for conjunctive use planning in Upper Damodar River basin, India. *J. Clean Prod.* **2020**, *273*, 16. [CrossRef]
7. Medeiros, G.; Florindo, T.; Talamini, E.; Fett, A.; Ruviaro, C. Optimising Tree Plantation Land Use in Brazil by Analysing Trade-Offs between Economic and Environmental Factors Using Multi-Objective Programming. *Forests* **2020**, *11*, 22. [CrossRef]
8. Bai, Y.; Wong, C.P.; Jiang, B.; Hughes, A.C.; Wang, M.; Wang, Q. Developing China's Ecological Redline Policy using ecosystem services assessments for land use planning. *Nat. Commun.* **2018**, *9*, 13. [CrossRef]
9. Hu, M.M.; Wang, Y.F.; Xia, B.C.; Jiao, M.Y.; Huang, G.H. How to balance ecosystem services and economic benefits?—A case study in the Pearl River Delta, China. *J. Environ. Manag.* **2020**, *271*, 13. [CrossRef]
10. Barton, H. Land use planning and health and well-being. *Land Use Policy* **2009**, *26*, S115–S123. [CrossRef]

11. Navarrete-Hernandez, P.; Laffan, K. A greener urban environment: Designing green infrastructure interventions to promote citizens' subjective wellbeing. *Landsc. Urban Plan.* **2019**, *191*, 14. [CrossRef]
12. Nieuwenhuijsen, M.J. Urban and transport planning pathways to carbon neutral, liveable and healthy cities; A review of the current evidence. *Environ. Int.* **2020**, *140*, 10. [CrossRef]
13. Law, E.A.; Bryan, B.A.; Meijaard, E.; Mallawaarachchi, T.; Struebig, M.J.; Watts, M.E.; Wilson, K.A. Mixed policies give more options in multifunctional tropical forest landscapes. *J. Appl. Ecol.* **2017**, *54*, 51–60. [CrossRef]
14. Karner, K.; Cord, A.F.; Hagemann, N.; Hernandez-Mora, N.; Holzkamper, A.; Jeangros, B.; Lienhoop, N.; Nitsch, H.; Rivas, D.; Schmid, E.; et al. Developing stakeholder-driven scenarios on land sharing and land sparing—Insights from five European case studies. *J. Environ. Manag.* **2019**, *241*, 488–500. [CrossRef]
15. Maruna, M.; Crncevic, T.; Milojevic, M.P. The Institutional Structure of Land Use Planning for Urban Forest Protection in the Post-Socialist Transition Environment: Serbian Experiences. *Forests* **2019**, *10*, 28. [CrossRef]
16. De Bisthoven, L.J.; Vanhove, M.P.M.; Rochette, A.J.; Huge, J.; Verbesselt, S.; Machunda, R.; Munishi, L.; Wynants, M.; Steensels, A.; Malan-Meerkotter, M.; et al. Social-ecological assessment of Lake Manyara basin, Tanzania: A mixed method approach. *J. Environ. Manag.* **2020**, *267*, 15. [CrossRef]
17. Namatama, N. An assessment of stakeholders' participation in land use planning process of Luapula Province Planning Authority. *Land Use Policy* **2020**, *97*, 21. [CrossRef]
18. Tao, T.; Tan, Z.; He, X. Integrating environment into land-use planning through strategic environmental assessment in China: Towards legal frameworks and operational procedures. *Environ. Impact Assess. Rev.* **2007**, *27*, 243–265. [CrossRef]
19. Lichtenberg, E.; Ding, C.R. *Assessing Farmland Protection Policy in China*; Lincoln Inst Land Policy: Cambridge, MA, USA, 2007; p. 101.
20. Abrantes, P.; Fontes, I.; Gomes, E.; Rocha, J. Compliance of land cover changes with municipal land use planning: Evidence from the Lisbon metropolitan region (1990–2007). *Land Use Policy* **2016**, *51*, 120–134. [CrossRef]
21. Gaaff, A.; Reinhard, S. Incorporating the value of ecological networks into cost-benefit analysis to improve spatially explicit land-use planning. *Ecol. Econ.* **2012**, *73*, 66–74. [CrossRef]
22. Ou, G.L.; Tan, S.K.; Zhou, M.; Lu, S.S.; Tao, Y.H.; Zhang, Z.; Zhang, L.; Yan, D.P.; Guan, X.L.; Wu, G. An interval chance-constrained fuzzy modeling approach for supporting land-use planning and eco-environment planning at a watershed level. *J. Environ. Manag.* **2017**, *204*, 651–666. [CrossRef]
23. Pennington, D.N.; Dalzell, B.; Nelson, E.; Mulla, D.; Taff, S.; Hawthorne, P.; Polasky, S. Cost-effective Land Use Planning: Optimizing Land Use and Land Management Patterns to Maximize Social Benefits. *Ecol. Econ.* **2017**, *139*, 75–90. [CrossRef]
24. Kumar, P.; Rosenberger, J.M.; Iqbal, G.M.D. Mixed integer linear programming approaches for land use planning that limit urban sprawl. *Comput. Ind. Eng.* **2016**, *102*, 33–43. [CrossRef]
25. Mic, P.; Koyuncu, M.; Hallak, J. Primary Health Care Center (PHCC) Location-Allocation with Multi-Objective Modelling: A Case Study in Idleb, Syria. *Int. J. Environ. Res. Public Health* **2019**, *16*, 23. [CrossRef] [PubMed]
26. Najafabadi, M.M.; Ziaee, S.; Nikouei, A.; Borazjani, M.A. Mathematical programming model (MMP) for optimization of regional cropping patterns decisions: A case study. *Agric. Syst.* **2019**, *173*, 218–232. [CrossRef]
27. Turk, E.; Zwick, P.D. Optimization of land use decisions using binary integer programming: The case of Hillsborough County, Florida, USA. *J. Environ. Manag.* **2019**, *235*, 240–249. [CrossRef] [PubMed]
28. Bacca, E.J.M.; Knight, A.; Trifkovic, M. Optimal land use and distributed generation technology selection via geographic-based multicriteria decision analysis and mixed-integer programming. *Sust. Cities Soc.* **2020**, *55*, 12.
29. Gosling, E.; Reith, E.; Knoke, T.; Paul, C. A goal programming approach to evaluate agroforestry systems in Eastern Panama. *J. Environ. Manag.* **2020**, *261*, 13. [CrossRef]
30. Mohammadi, M.; Nastaran, M.; Sahebgharani, A. Development, application, and comparison of hybrid meta-heuristics for urban land-use allocation optimization: Tabu search, genetic, GRASP, and simulated annealing algorithms. *Comput. Environ. Urban Syst.* **2016**, *60*, 23–36. [CrossRef]
31. Sante, I.; Rivera, F.F.; Crecente, R.; Boullon, M.; Suarez, M.; Porta, J.; Parapar, J.; Doallo, R. A simulated annealing algorithm for zoning in planning using parallel computing. *Comput. Environ. Urban Syst.* **2016**, *59*, 95–106. [CrossRef]
32. Li, X.; Ma, X.D. An improved simulated annealing algorithm for interactive multi-objective land resource spatial allocation. *Ecol. Complex.* **2018**, *36*, 184–195. [CrossRef]
33. Rall, E.; Hansen, R.; Pauleit, S. The added value of public participation GIS (PPGIS) for urban green infrastructure planning. *Urban For. Urban Green.* **2019**, *40*, 264–274. [CrossRef]
34. Ramya, S.; Devadas, V. Integration of GIS, AHP and TOPSIS in evaluating suitable locations for industrial development: A case of Tehri Garhwal district, Uttarakhand, India. *J. Clean Prod.* **2019**, *238*, 14. [CrossRef]
35. Marzouk, M.; Othman, A. Planning utility infrastructure requirements for smart cities using the integration between BIM and GIS. *Sust. Cities Soc.* **2020**, *57*, 14. [CrossRef]
36. ul Hussnain, M.Q.; Waheed, A.; Anjum, G.A.; Naeem, M.A.; Hussain, E.; Wakil, K.; Pettit, C.J. A framework to bridge digital planning tools' utilization gap in peri-urban spatial planning; lessons from Pakistan. *Comput. Environ. Urban Syst.* **2020**, *80*, 12. [CrossRef]
37. Fu, X.; Wang, X.H.; Yang, Y.J. Deriving suitability factors for CA-Markov land use simulation model based on local historical data. *J. Environ. Manag.* **2018**, *206*, 10–19. [CrossRef]

38. Liang, X.; Liu, X.P.; Li, D.; Zhao, H.; Chen, G.Z. Urban growth simulation by incorporating planning policies into a CA-based future land-use simulation model. *Int. J. Geogr. Inf. Sci.* **2018**, *32*, 2294–2316. [CrossRef]
39. Liu, X.P.; Liang, X.; Li, X.; Xu, X.C.; Ou, J.P.; Chen, Y.M.; Li, S.Y.; Wang, S.J.; Pei, F.S. A future land use simulation model (FLUS) for simulating multiple land use scenarios by coupling human and natural effects. *Landsc. Urban Plan.* **2017**, *168*, 94–116. [CrossRef]
40. Mondal, B.; Das, D.N.; Bhatta, B. Integrating cellular automata and Markov techniques to generate urban development potential surface: A study on Kolkata agglomeration. *Geocarto Int.* **2017**, *32*, 401–419. [CrossRef]
41. Wang, L.Z.; Pijanowski, B.; Yang, W.S.; Zhai, R.X.; Omrani, H.; Li, K. Predicting multiple land use transitions under rapid urbanization and implications for land management and urban planning: The case of Zhanggong District in central China. *Habitat Int.* **2018**, *82*, 48–61. [CrossRef]
42. Xu, T.T.; Gao, J.; Coco, G. Simulation of urban expansion via integrating artificial neural network with Markov chain—cellular automata. *Int. J. Geogr. Inf. Sci.* **2019**, *33*, 1960–1983. [CrossRef]
43. Mi, N.; Hou, J.W.; Mi, W.B.; Song, N.P. Optimal spatial land-use allocation for limited development ecological zones based on the geographic information system and a genetic ant colony algorithm. *Int. J. Geogr. Inf. Sci.* **2015**, *29*, 2174–2193. [CrossRef]
44. Yang, L.N.; Sun, X.; Peng, L.; Shao, J.; Chi, T.H. An improved artificial bee colony algorithm for optimal land-use allocation. *Int. J. Geogr. Inf. Sci.* **2015**, *29*, 1470–1489. [CrossRef]
45. Li, X.; Parrott, L. An improved Genetic Algorithm for spatial optimization of multi-objective and multi-site land use allocation. *Comput. Environ. Urban Syst.* **2016**, *59*, 184–194. [CrossRef]
46. Hakli, H.; Uguz, H. A novel approach for automated land partitioning using genetic algorithm. *Expert Syst. Appl.* **2017**, *82*, 10–18. [CrossRef]
47. Huang, Q.; Song, W. A land-use spatial optimum allocation model coupling a multi-agent system with the shuffled frog leaping algorithm. *Comput. Environ. Urban Syst.* **2019**, *77*, 19. [CrossRef]
48. Zhou, M. An interval fuzzy chance-constrained programming model for sustainable urban land-use planning and land use policy analysis. *Land Use Policy* **2015**, *42*, 479–491. [CrossRef]
49. Ma, S.H.; Wen, Z.Z. Optimization of land use structure to balance economic benefits and ecosystem services under uncertainties: A case study in Wuhan, China. *J. Clean Prod.* **2021**, *311*, 16. [CrossRef]
50. Rao, Y.X.; Zhou, M.; Ou, G.L.; Dai, D.Y.; Zhang, L.; Zhang, Z.; Nie, X.; Yang, C. Integrating ecosystem services value for sustainable land-use management in semi-arid region. *J. Clean Prod.* **2018**, *186*, 662–672. [CrossRef]
51. Yuen, K.W.; Hanh, T.T.; Quynh, V.D.; Switzer, A.D.; Teng, P.; Lee, J.S.H. Interacting effects of land-use change and natural hazards on rice agriculture in the Mekong and Red River deltas in Vietnam. *Nat. Hazards Earth Syst. Sci.* **2021**, *21*, 1473–1493. [CrossRef]
52. Luo, D.; Liang, L.W.; Wang, Z.B.; Chen, L.K.; Zhang, F.M. Exploration of coupling effects in the Economy-Society-Environment system in urban areas: Case study of the Yangtze River Delta Urban Agglomeration. *Ecol. Indic.* **2021**, *128*, 14.
53. Dou, P.F.; Zuo, S.D.; Ren, Y.; Rodriguez, M.J.; Dai, S.Q. Refined water security assessment for sustainable water management: A case study of 15 key cities in the Yangtze River Delta, China. *J. Environ. Manag.* **2021**, *290*, 13. [CrossRef]
54. Zhou, M.; Wang, X.; Lin, X.T.; Yang, S.; Zhang, J.; Chen, J. Automobile exhaust particles retention capacity assessment of two common garden plants in different seasons in the Yangtze River Delta using open-top chambers. *Environ. Pollut.* **2020**, *263*, 9.
55. Zhou, M.; Chen, Q.; Cai, Y.L. Optimizing the industrial structure of a watershed in association with economic-environmental consideration: An inexact fuzzy multi-objective programming model. *J. Clean Prod.* **2013**, *42*, 116–131. [CrossRef]

Article

An Innovative Framework on Spatial Boundary Optimization of Multiple International Designated Land Use

Hei Gao [1,2], Yubing Weng [3], Yutian Lu [4] and Yan Du [5,*]

[1] The Architectural Design & Research Institute of Zhejiang University Co., Ltd., Zhejiang University, Hangzhou 310028, China; gh1@zuadr.com
[2] Center for Balanced Architecture, Zhejiang University, Hangzhou 310058, China
[3] College of Agriculture and Biotechnology, Zhejiang University, Hangzhou 310058, China; 3190103473@zju.edu.cn
[4] College of Computer Science and Technology, Zhejiang University, Hangzhou 310058, China; yutianlu@zju.edu.cn
[5] Department of Landscape Architecture, Huazhong Agricultural University, Wuhan 430070, China
* Correspondence: yuanscape@mail.hzau.edu.cn; Tel.: +86-027-8728-2135

Citation: Gao, H.; Weng, Y.; Lu, Y.; Du, Y. An Innovative Framework on Spatial Boundary Optimization of Multiple International Designated Land Use. *Sustainability* **2022**, *14*, 587. https://doi.org/10.3390/su14020587

Academic Editor: Tim Gray

Received: 3 December 2021
Accepted: 1 January 2022
Published: 6 January 2022

Abstract: The continuous improvement of international protection awareness has dramatically increased the number of protection organizations and promoted various reserve-naming methods. However, the existing global natural reserves have either fully or partially overlapped, thereby allowing the same region to hold various international titles, resulting in serious issues, which are especially manifested in the boundary delimitation process of natural reserves. Therefore, delimiting the titles of reserve borders will become an enormous challenge in protected-area governance worldwide. This study conducted an in-depth investigation of the technical methods for delineating the spatial boundaries of natural reserves. Taking Jiangshan Nature Reserve in China as the case object, the Candidate Area–Natural background–Heritage Resource–Construction (C-NHC) framework was constructed, and the boundaries of the new reserves were delineated. This study has changed the status quo of the spatial overlap of the reserve through the quantitative evaluation of the conflict patches and the triple optimization of the boundary of the reserve. The area of the new reserve is 150.524 km^2, which is 6.682 km^2 larger than the original one. The original reserves are all included within the scope of the new one. This study provides guidance and new insights into the boundary delineation of integrated nature reserves worldwide.

Keywords: protected areas; boundary optimization; heritage resource; China

1. Introduction

Industrialization and urbanization continue to expand globally, along with the continuous development of the human society [1,2]. Excessive resource utilization has exacerbated the disappearance and fragmentation of habitats [3,4], thereby inducing soil erosion, environmental pollution [5,6], and biodiversity loss [7]. The International Union for Conservation of Nature and its state parties have established numerous natural reserves as powerful tools for protecting natural resources, maintaining biodiversity [8,9], improving ecosystem services [10], and introducing economic benefits to the surrounding areas to cope with the increasingly severe challenges of the ecological environment. The resources of the reserves are still damaged despite the constantly expanding scale of natural reserves and may further degrade the ecosystem protection function due to people who are driven by economic benefits, thus contradicting the original intention of the establishment of natural reserves [11].

However, the continuous improvement of international protection awareness has dramatically increased the number of protection organizations and promoted various reserve naming methods [12] under different objectives, purposes, and management requirements, thereby causing the same region to hold two, three, or even four international titles [13]. Natural reserves with multiple titles reflect the high value of environmental protection and sustainable development. However, the existence of multiple titles also causes several problems, as follows [14]:

First, different titles lead to differences in protection objects and management modes, and the various monitoring and reporting requirements may instigate some conflicts and consequently increase the workload of the reserves. Second, constructing a unified management model in the absence of national regulations and controls is difficult due to a lack of communication and coordination among the institutions and departments involved in the management of multi-title reserves. Third, large spatial differences among the different titles of the same nature reserve may exist; that is, the boundaries of the reserve may be non-overlapping. Therefore, the difficulty in managing the reserves may increase. Fourth, multiple titles elicit confusion in the reserve identification system and weaken the effectiveness of international protection titles, thereby hindering the satisfactory fulfilment of the corresponding roles. Fifth, the international influence created by the multiple titles of natural reserves promotes the development of local tourists but also induces substantial pressure on the management.

The complex correspondence between the various types of protection and resource causes the widespread phenomenon of overlapping protection objects in Jiangshan, and this wastes the resources and aggravates conflicts in protection provisions and departments [15]. Therefore, solving the problem of overlapping boundaries of protected areas caused by the overlapping of multiple international designated areas and delimiting scientific protected area boundaries are of considerable importance for realizing efficient management of natural protected areas in Jiangshan. We conducted an in-depth study of this difficult problem to provide technical support for the boundary demarcation in the integration of protected areas in Jiangshan.

In this paper, a technical framework for the delimitation of nature reserves in Jiangshan is proposed. Different suggestions for the development of multi-title natural reserves from three levels, namely local managers, national institutions, and international organizations, have been proposed by the existing studies. However, most of these suggestions remain in the policy and theoretical guidance levels, fail to clearly define the spatial scope of multi-title nature reserve, and lack the methods for delineating the boundaries of nature reserves. This study proposes an evaluation index system for the integration and optimization of reserve boundaries on the basis of the Candidate Area–Natural background–Heritage Resource–Construction (C-NHC) framework; optimizes the boundaries of natural reserves on the basis of the characteristics of the resource background, heritage resource characteristics, and construction management conditions thrice; and establishes a set of identifiable and popularized technical framework for the demarcation of the boundaries of natural reserves that will exert an extensive influence on the in-depth analysis of the spatial pattern of natural reserves in the future, improvement of the planning and management, and overall promotion of the construction of natural reserves. This technical framework can provide a useful reference for the boundary delimitation of nature reserves in China and even the world.

2. Socioecological Framework

The goal of optimizing the boundary of the reserves must be addressed to solve the problems caused by the fragmentation of the administrative areas [16]. Therefore, the principle of integration and merging of adjacent reserves was established in this study. Adjacent reserves within the same geographical unit that possess strong ecosystem integrity, similar protected objects, and satisfactory management conditions were integrated preferentially [17]. Therefore, the overlapping areas after the integration of the adjacent reserves are processed in accordance with the principle of "no decrease in strength, no decrease in area, and no change in properties".

The delimitation of the boundary of natural reserves should involve the establishment of an effective connection through land space planning and concurrently focus on the coordination of natural reserves with the urban development boundary, permanent basic farmland, and ecological protection red line to conduct element-based, micro, and precise assessments of the suitability of land space utilization and preliminarily delimit the waiting area of natural reserves. In addition, the development and utilization of territorial spaces should match the carrying capacity of the resources and environment to guide the positioning of regional main functions, clarify the order of the territorial space development, and improve the utilization efficiency of the territorial spaces.

The characteristics of the natural resources, such as topographical features, hydrological basins, soils, and flora and fauna in the natural protected areas, are the cornerstone of the natural ecological protection framework. This framework provides a basic analysis unit for the construction and boundary demarcation of the reserves. Most of the existing studies use natural geographical units as the basis for such demarcation [18,19]. The initial boundary is regarded as the reference for the superposition of the vegetation, climate, soil, animal zoning, and other elements to form the spatial boundary of natural reserve lands [20,21]. This study performed the preliminary aggregation and optimization for the boundaries of nature reserves based on their background characteristics.

Integrity is an internationally recognized principle of world heritage protection [22]. In Recommendation concerning the Preservation of Cultural Property endangered by Public or Private Works, which was formulated in 1968, UNESCO mentioned that the preservation of monuments should be an absolute requirement of any well-designed plan for urban redevelopment, especially in historic cities or districts. Similar regulations should cover the area surrounding a scheduled monument or site and its setting to preserve its association and character [23]. While integrity was introduced only recently (2005), it was an implicit quality for many cultural properties even before it was formally named [24]. Recent applications of the integrity principle in the context of heritage conservation place an emphasis on assessing and maintaining the outstanding universal value and complete representation of both natural and cultural heritage features and their attributes [25]. On the basis of protecting ecosystem integrity and biodiversity and maintaining landscape characteristics, this study extracted the three aspects of the spatial elements, namely ecological corridor, patch and matrix, and natural and human landscape elements, as well as other heritage resource characteristics, to delimit the boundary of the reserves according to guidelines and specifications.

From the perspective of practical constraints, the management status and construction conditions of the reserves are important factors that influence the cost of implementing a new space control system and, thus, directly determine whether the delimitation of the boundaries of nature reserves can be strictly and effectively implemented [26]. This study combines and adjusts the specific situation of space control and human development and construction activities [27,28] based on the comprehensive evaluation of the resource background and characteristics of the heritage resources to execute the tertiary optimization of natural reserve boundaries.

A C-NHC framework was constructed in this study to perform the boundary delineation and tertiary optimization of a new reserve. First, the preliminary candidate areas of the reserve range are selected by constructing the evaluation index system. Second, the core elements of the reserve boundary optimization are selected to evaluate the resource background elements for the initial aggregation optimization. Finally, the secondary optimization range is obtained by combining the characteristics of the heritage resources, and the third optimization is completed by connecting the existing construction regulation conditions to generate a new nature-reserve boundary.

3. Methods and Data

3.1. Preliminary Selection of the Candidate Areas within the Reserve

As an important part of the territorial space planning system, the construction of a natural reserve system must strengthen the connection between the natural reserve plan and the national spatial planning and comprehensively evaluate the conflicts among the spatial layout of the natural reserve, urban development boundary, red line for the permanent basic farmland protection, and ecological red line. On this basis, the spatial grid evaluation of natural reserves is performed to evaluate and compare the ecological land use, cultivated land, and urban development needs of the conflicting spots [29]. Grid evaluation can refer to the evaluation model of Foundation-Process Management. To integrate protected areas, the natural factors of the land resources; transportation conditions; spatial location; and the other factors necessary to construct an evaluation index system, including ecological, cultivated land, and construction suitability, must be comprehensively considered. The evaluation unit of this part is the map used in the area. The index is standardized by the extreme value linear standardization method for dimensionless values between 0 and 100. It assigns equivalent weight to the suitability factor and the neighborhood influence factor. The suitability comprehensive index of each land type can be calculated as follows:

$$S_i = C_i \times T_i \times N_i = \prod_{j=1}^{\alpha} c_{i,j} \times \left(\sum_{k=1}^{\beta} t_{i,k}\varpi_{1k}\right)\left(\sum_{l=1}^{\lambda} n_{i,l}\varpi_{2l}\right) \quad (1)$$

where S_i is the suitability of the conflict map for the i-th land type; i represents a certain land type (ecological, cultivated, or construction land); C_i, T_i, and N_i respectively represent the scores of the restrictive, suitability, and neighborhood influence factors; $c_{i,j}$ denotes the restriction type of the j-th restriction factor for the i-th land type, where a value of 0 and 1 represents the restriction of the existence of the i and i-th land types, respectively; $t_{i,k}$ represents the suitability degree of the j-th suitability factor for the i-th land type and for the positive correlation indicators; $t_{i,k}$ is the normalized value of the k-th index, which is the difference between 100 and the normalized value of the k-th index for reverse correlation indicators; and ϖ_{1k} represents the weight of the k-th suitability factor, which is obtained through the analytic hierarchy process (AHP) [30]. AHP is a structured technique for organizing and analyzing complex decisions based on mathematics and psychology (Figure 1). Moreover, $n_{i,l}$ represents the degree of influence of the l-th neighborhood factor on the i-type land; ϖ_{2l} represents the weight of the first neighborhood influence factor for positive correlation indicators, which is obtained through AHP; and j, k, and n represent the number of the restriction, suitability, and neighborhood influence factors, respectively.

On the basis of the calculated land-use competitiveness of conflict maps, the discriminant matrix is used to qualitatively evaluate the suitable land type for the conflict spot. The qualitative evaluation results of competitiveness are divided into three categories, namely high, medium, and low, through the natural breakpoint method (Appendix A, Table A1). The quantitative structure of various types of land-suitability levels is represented by statistical charts, and the distribution of various land-use levels is illustrated by the spatial-distribution-map characteristics. The appropriate land-classification matrix is used to calculate the appropriate land types of each conflict map. Subsequently, the classified conflict map and the original map are merged to generate the distribution maps of the different land types [31].

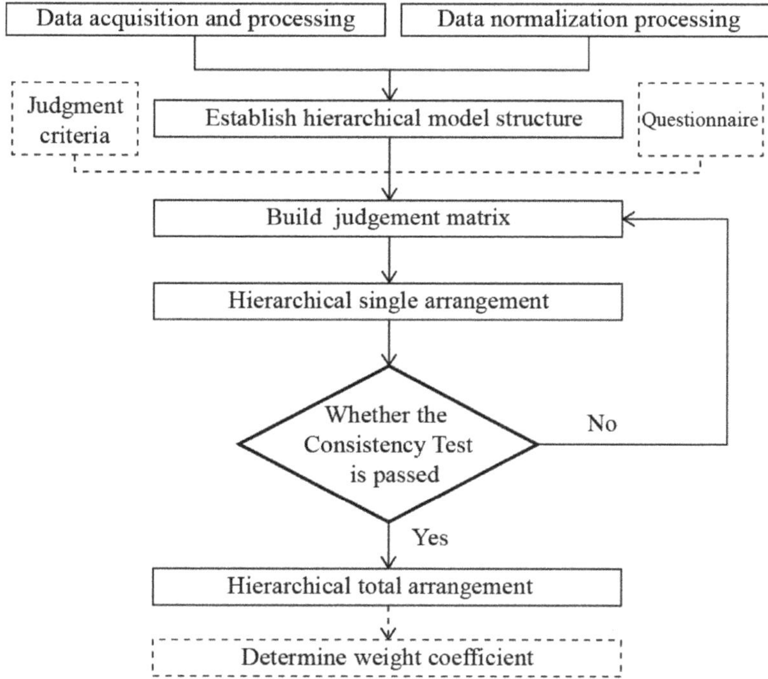

Figure 1. AHP workflow for determining the weight of suitability factors.

3.2. First Optimization: Resource Background Assessment and Initial Aggregation Optimization

After the features of the geomorphology, soil, and vegetation characteristics were comprehensively analyzed, the corresponding analysis units with similar characteristics were clustered and classified, and the identified candidate areas for reserve designation were divided into two areas. These divided areas are PIN and POUT, which refer to the areas within and outside the original protected area, respectively. The average values of each resource of PIN and POUT were calculated according to the graded background characteristics of the resource.

All spots outside the original reserve are reclassified into two categories according to the similarity of the resource background characteristics, namely potential protected and non-protected spots. Furthermore, the preliminary candidate range of the reserves is determined. The similarity is calculated based on Euclidean distance, as follows:

$$Sim(m,T) = 1/1+\sqrt{\sum_{p=1}^{q}(V_{mp}-V_{Tp})^2} \tag{2}$$

where V_{mp} represents the value of the p-th feature of the m-th spot, and V_{Tp} is the average value of the p-th feature of PIN or POUT [32].

3.3. Second Optimization: Intersection of the Characteristic Elements of Heritage Resources and Clustering of Resource Bases

This study optimized the intersection between the extracted characteristic elements of the heritage resources and the clustering results of the resource bases and divided the former into protection objects while maintaining the conditions under control to ensure the authenticity and integrity of the ecosystem and the biodiversity of the natural reserves [22,33]. The elements of the protection objects include all kinds of important natural ecosystems, wild animal and plant habitats, geological relics, natural landscapes, protected values, and geographical distribution in the reserve. The priority of the feature extraction of the heritage resources is determined by combining the main function orientation and core protection objectives of the reserve. This step guarantees the integrity, authenticity, connectivity, and systematic nature of the core protection objectives. Subsequently, the reference elements of the ecosystem and distribution characteristics of important animals and plants and landscape remains were extracted in accordance with the patch–corridor–matrix model. Lastly, the boundary of the reserve was aggregated, smoothed, and re-optimized. Specifically, the important ecological origins of various ecosystems and habitats of important species were first determined. The least-resistance model was then used to compute the internal corridor, which was optimized with adjustments to the computed corridor in combination with the observations on the solid and species migration corridors. The important environmental matrix, which aims to ensure the integrity of the ecosystem and species structure, was extracted in accordance with the network structure comprising patch–corridor after setting a reasonable buffer width. Finally, all the extracted spatial feature elements were integrated on the basis of the following principles: the low-security level obeys the high-security level, and the secondary protection object obeys the primary protection object.

The elements of the regulatory conditions include the spatial distribution and land-use-right information of the development and production activities involving the minerals, forest farms, pastures, orchards, fishponds, and farms within the reserve. These factors were integrated, and the spatial boundaries, which exert a considerable impact on the reserve space control, were extracted or redefined and combined with the adjusted and optimized reserve spatial boundaries. Specifically, the existing construction and zoning control situations in the protected areas should be first coordinated and unified, the implementation of high-level control boundary should be prioritized, and the adjustment of low-level control boundary should be optimized. Then, in combination with the current situation of land use and natural resource development and management, the range dimensions of land, natural resource development and management, and tourism franchise ownerships should be clarified, and the boundary between state-owned property rights and collective or private property rights should be distinguished. Therefore, the development intensity of collective land, the production and development intensity of natural resources from private or collective ownership, and the profitability and the development intensity of existing franchises are evaluated. The appropriate assessment and exit of the protected or core protection areas would be conducted after estimating the space control cost of different intensities of protected areas, and the boundaries of the reserves would be adjusted. Combining the above points, the recommended selection of referable elements for the boundary optimization of protected areas is shown in Table 1:

Table 1. Referable elements for boundary optimization of protected areas.

First Class	Second Class	Third Class	Referable Elements
Resource background	Topographical unit	Geography and geomorphy	Elevation, slope, aspect, ridge line, valley line, river line, forest line, and snow line.
		Geological conditions	Structural lines, faults, and seepage conditions.
	Natural resource zoning	Hydrology conditions	River basin, drainage divide, water conservation area, big lake wetland patch, and groundwater protection zone.
		Soil conditions	Soil zoning, soil thickness, and soil hardness.
		Vegetation conditions	Vegetation zoning, forest coverage, vegetation canopy density, stand structure, and tree age structure.
Heritage resource characteristics	Ecosystem integrity	Ecosystem corridor	Material energy connection channel of ecosystem.
		Ecosystem patch	Ecological source, ecosystem fragile area, and ecosystem sensitive area.
		Ecosystem matrix	Distribution boundary of ecosystem.
	Species diversity	Species conservation patch	Distribution density and habitat of key protected animals and plants (possibly seasonal).
		Species conservation corridor	Migration or retrogressive passage of key protected animals (possibly seasonal).
		Community complexity	Species richness and structural complexity differentiation.
		Biological integrity	Integrity differentiation of various species, such as predator and human species.
	Characteristic landscape relics	Natural heritage and landscape characteristics	Natural relics or natural landscape densely distributed areas, natural landscape connecting corridors, natural landforms, and landscape zoning.
		Humanities landscape characteristics	Cultural landscape densely distributed area, cultural landscape connecting corridor, and cultural ecological zoning.
Construction management conditions	Construction management continuity	Protected area situations	Construction and zoning control of existing protected areas.
		Land ownership	Land (forest land) ownership, collective land-development intensity, and ecological migration cost.
		Natural resource development and management rights	Management right, development intensity, and setting cost of easement of privately or collectively owned natural resources.
		Tourism franchise ownership	Existing privileged management right, intensity of development, and cost of setting up the easement.
		Administrative authority boundary	Border of different administrative regions and spatial boundary of jurisdiction of local government cross-regional cooperative organizations.
	Construction management coordination	Land-use status	Construction of towns and administrative and natural villages. Historic and cultural heritage reserve. Permanent basic farmland. Exploration and mining rights. Ecological red line.
		National territorial space planning	Recent major project planning. Major control line delineation.

3.4. Third Optimization: Connection and Coordination of the Existing Construction Control Conditions

With consideration of the management status and construction conditions of the reserve, the principle of "continuity, stability, conversion, and innovation" is adopted to link and coordinate the construction control elements and the secondary optimization results. The land-use-status information provided in the national land-use survey and the space area of the proposed protected land was superposed and checked to coordinate the residential construction land, historical and cultural site protection areas, permanent basic farmland, ecological protection red line, and exploration and mining rights. Afterward, the conflict area is determined, and the priority rules and compatibility control conditions in conflict processing are established prior to optimizing the boundary again. The superimposed status of important road traffic and linear infrastructure distribution map focus on the analysis of the cutting strength and crossing grade of the linear infrastructure running through the protected land according to the linear infrastructure, with strong cutting boundary function fine-tuning of the protected land boundary.

The preliminary delimitation map of the protected land should be overlapped with the land space and the major project planning maps in equal weight to place the agricultural, mineral, and major project lands from the preliminary boundary of the protected land as far as possible. Then, the boundary should be adjusted according to the important control line defined in the plan to determine the new boundary of the reserve. Figure 2 demonstrates the technical route mentioned above.

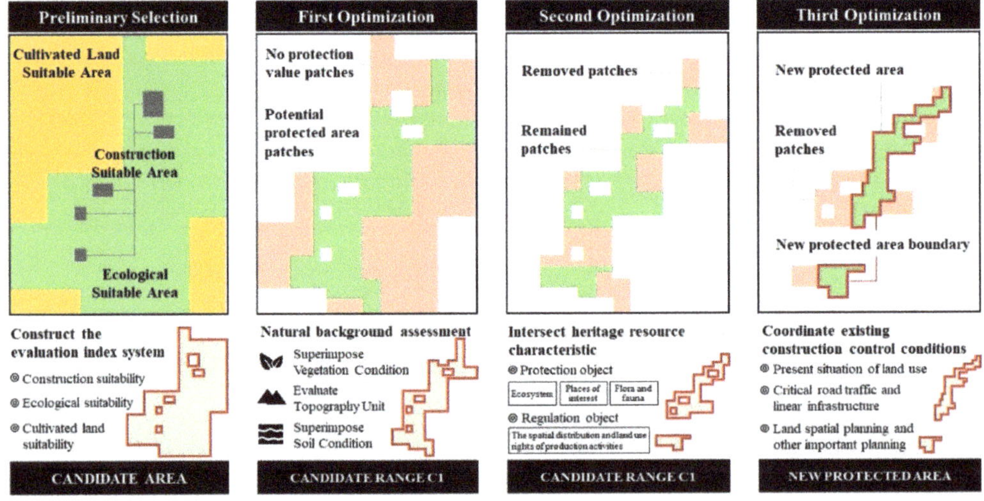

Figure 2. Technical process.

3.5. Study Area

The aforementioned problems are eminent in China [34], which has nine types of natural reserves. Reserves with different names, such as national scenic spots, national nature reserves, and local government-level reserves, exist within the same geographical space, thus resulting in overlapping reserve boundaries. To solve the problem of overlapping boundaries of protected areas caused by the overlapping of multiple international designated areas, the most significant problem is delimiting scientific protected-area boundary, which is of great importance for realizing efficient management of natural protected areas. Jiangshan, which is located in the mountainous area in the southwest part of the Zhejiang province, China, has a total area of 2019.4 km^2 and is the headstream of the Qiantang River. This area possesses abundant natural mountain water resources, with numerous natural reserves. Six natural reserves in Jiangshan, namely Jianglangshan International Scenic Spot, Xianxia National Forest Park, Fugaishan Provincial Geological Park, Jiangshangang Provincial Wetland Park, Xianxialing Provincial Nature Reserves, and the provincial natural reserve of Jindingzi Geological Relics, were selected as research objects to study the boundaries of nature reserves (Table 2). The six natural reserves are spatially distributed in the north–south direction and are mainly concentrated in the southern part of the mountainous and forest area (Figure 3a), which covers a total area of 186.88 km^2. Apart from the provincial natural reserve of Jindingzi, all natural reserves have different overlap degrees with the multi-title situation (Figure 3b). The total area of the six natural reserves without overlap is 145.2 km^2; the overlapped area covers 41.68 km^2, according to statistics.

Table 2. Jiangshan nature reserves.

Name	Area/hm^2	Year Designated
Jianglangshan National Scenic Area	5390	2002
Xianxia National Forest Park	3449.46	2004
Fugaishan Provincial Geological Park	402.96	2014
Jiangshan Port Provincial Wetland Park	2143.75	2015
Xianxialing Provincial Nature Reserve	6992	2015
Provincial natural reserve of Jindingzi	22.84	2015

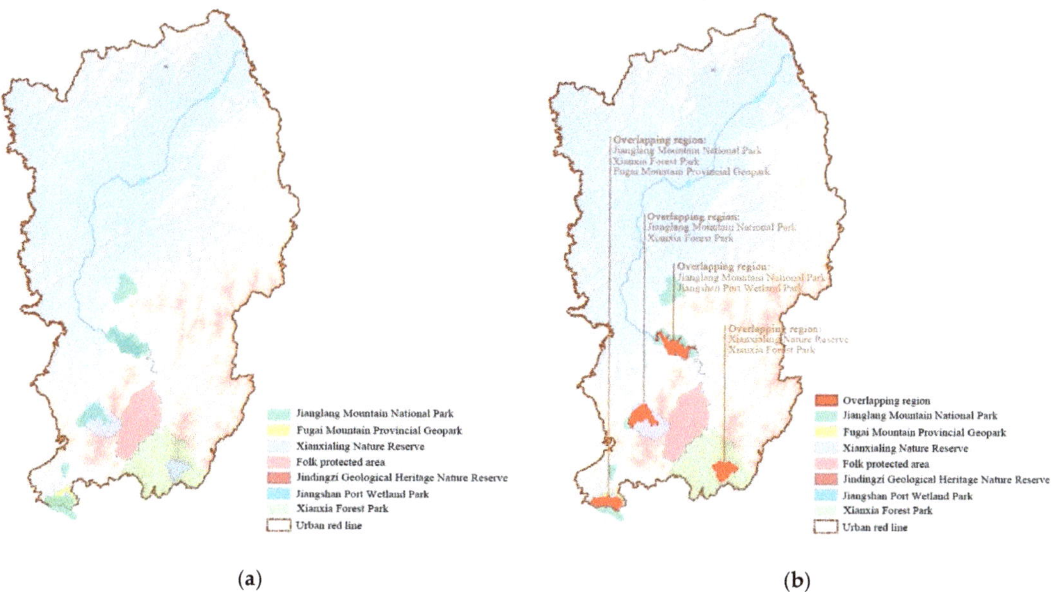

Figure 3. (**a**) Distribution of Jiangshan natural reserves. (**b**) Overlap of Jiangshan natural reserves.

3.6. Data Sources

The data used in this study are shown in Table 3:

Table 3. Data source.

Database	Indicators Used in the Study	Description
The third national land survey	The data include land-use status information, such as land type; location; area and distribution; land ownership and use rights; natural and social conditions of the land; and extractable cultivated land, basic farmland, construction land, and various construction control indicators.	It was conducted from January 2018 to June 2019, and the field survey database construction was completed by the end of 2019.
The second national forest resources survey	The forest land type, location and area, extract vegetation height, diameter at breast height, tree age, tree species structure and other information, and forest landownership and use rights.	The data from the second national forest resources survey from 1994 to 2006. The data on soil and wild animal and plant resources were obtained from the information of the natural geographical environment and ecological factors related to the forest resources in the survey.
Digital elevation model (DEM) data	Elevation.	The DEM data are image data with a resolution of 30 m that were provided by the International Scientific Data Service Platform and Geospatial Data Cloud Platform of the Computer Network Information Center of the Chinese Academy of Sciences (http://www.gscloud.cn/sources/accessdata/310, we last accessed the link on 1 December 2021). Relevant information, such as topography, slope, aspect, and slope position, were extracted from the DEM data.
Various plans	The vector data of the geographical boundaries of the reserves were extracted, and the type, level, main protection object, and construction period were arranged and summarized.	Including the Master Plan of Jianglang Mountain National key scenic spot (2007–2025), Master Plan of Xianxia National Forest Park, Master Plan of Jiangxia Xianxialing Provincial Nature Reserve (2016–2025), Master Plan of Fukaishan Provincial Geological Park (2015–2025), Master Plan of Zhejiang Jiangshangang Provincial Wetland Park (2019–2023), Master Plan of Zhejiang Jiangshangang Provincial Wetland Park (2019–2023), and Master Plan of Jiangxi Xianxialing Provincial Nature Reserve (2016–2025).

4. Results

4.1. Spatial Distribution of Suitable Land Types

Based on the different types of conflicts and combining the data accessibility of Jiangshan, this study utilized the differentiation index (Appendix A, Tables A2–A4) to analyze the reference basis of the different boundary delimitations and the types and characteristics of boundary conflicts. The factors based on the scale of the spots should be selected when spots are used as the object of evaluation, and the factors that can reflect the neighborhood in the conflict spots into the evaluation index system should be considered. Through comprehensively analyzing the restrictive, suitability, and neighboring influencing factors, the comprehensive index of the land suitability for various land types in Jiangshan natural reserves was calculated. In addition, the suitable land types for the conflict spots were qualitatively evaluated, using the corresponding discrimination matrix of the conflict spot (Table 4) to generate the land distribution maps (Figure 4). "Strong", "medium", and "weak" in Figure 4 were determined by the natural breakpoint method.

Table 4. Average value of the background characteristic indices of PIN and POUT.

Index		PIN	POUT
Landform	Topography	3.027929	3.110242
	Slope grade	2.73307	3.135822
	Aspect	3.460949	3.584256
	Slope position	3.158181	3.006121
Soil	Soil thickness	2.439446	2.764656
	Soil name	1.940435	2.023209
	Soil type	1.008651	0.993066
Vegetation	Vegetation coverage	2.592437	2.982063
	Average height of vegetation	3.394464	3.260039
	Vegetation canopy	3.47479	3.636158
	Vegetation density	1.984429	2.074757
	Vegetation average breast diameter	1.121849	1.077941
	Average age of vegetation	1.752101	1.83545
	Vegetation tree structure	1.569871	1.521129

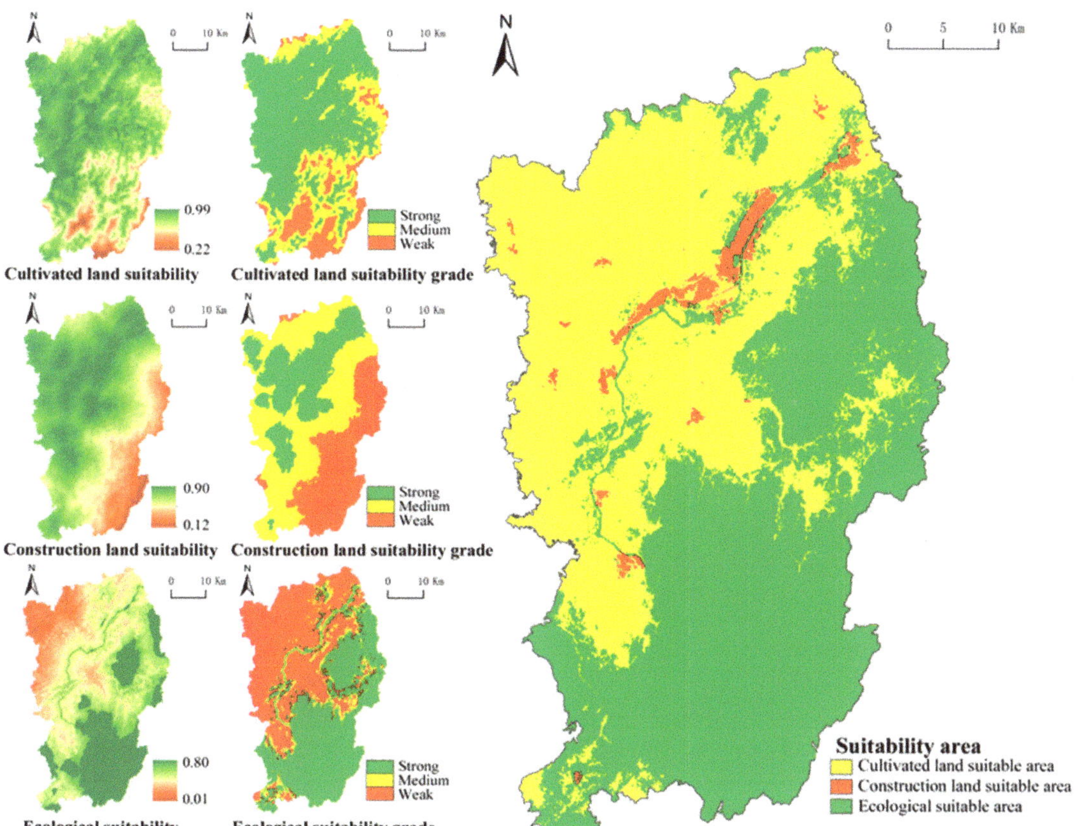

Figure 4. Delimitation of the candidate areas.

4.2. Preliminary Candidate Range for Reserves

On the basis of the specific situation of Jiangshan natural reserves and the data availability, the forestry survey unit was selected as the basic statistical unit (Figure 5) to classify or grade the geomorphological, soil, and vegetation conditions of each unit. The geomorphological conditions include the topography, slope grade, aspect, and slope position (Figure 6); the soil conditions comprise the soil thickness, soil name, and soil type (Figure 7); the vegetation conditions include the vegetation coverage, vegetation average height, canopy density, average breast diameter, density, tree age, and tree species structure (Figure 8). The values are assigned from low to high according to the grading of each feature of the resource background. The average values of the background features of PIN and POUT are calculated, Table 4 reports the statistical results. Figure 9 displays the candidate range of the protected sites for the new screening (C1).

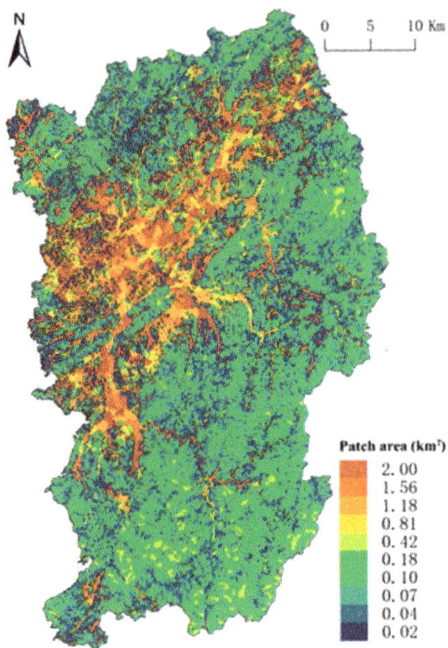

Figure 5. Basic statistical unit.

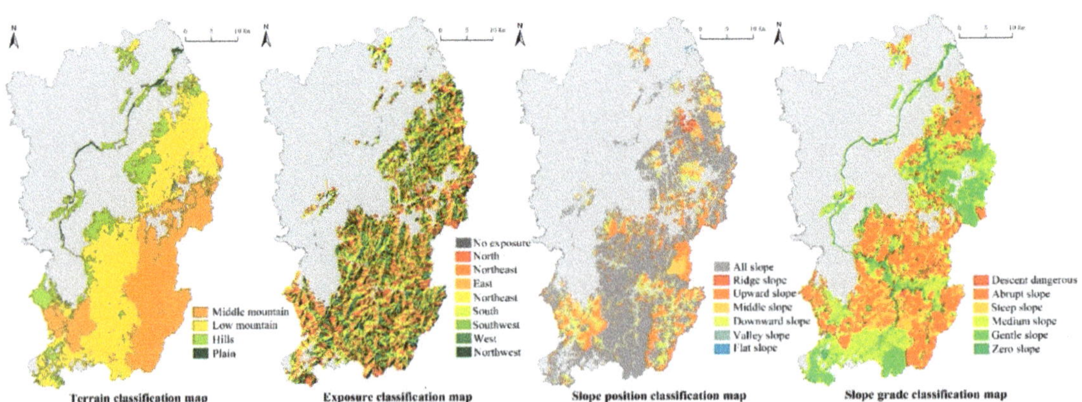

Figure 6. Geomorphological condition assessment.

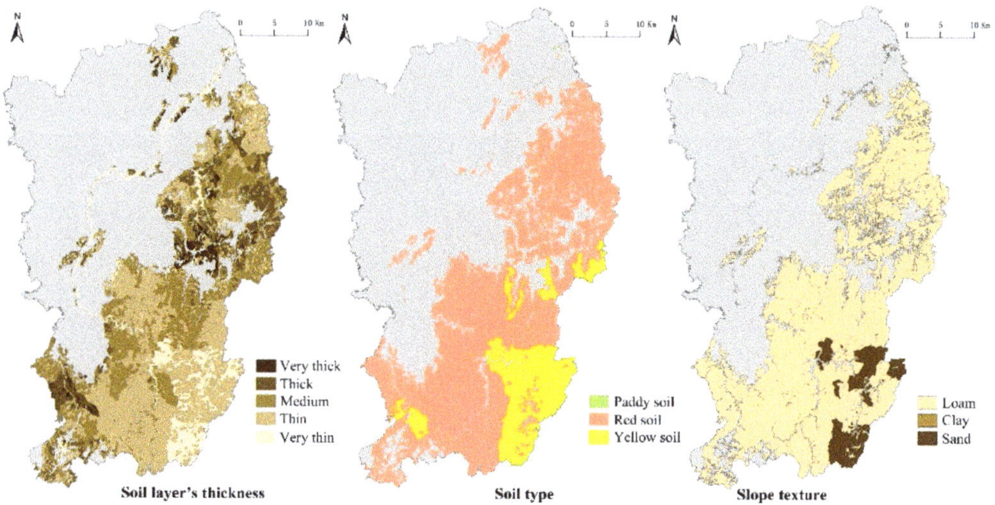

Figure 7. Soil condition assessment.

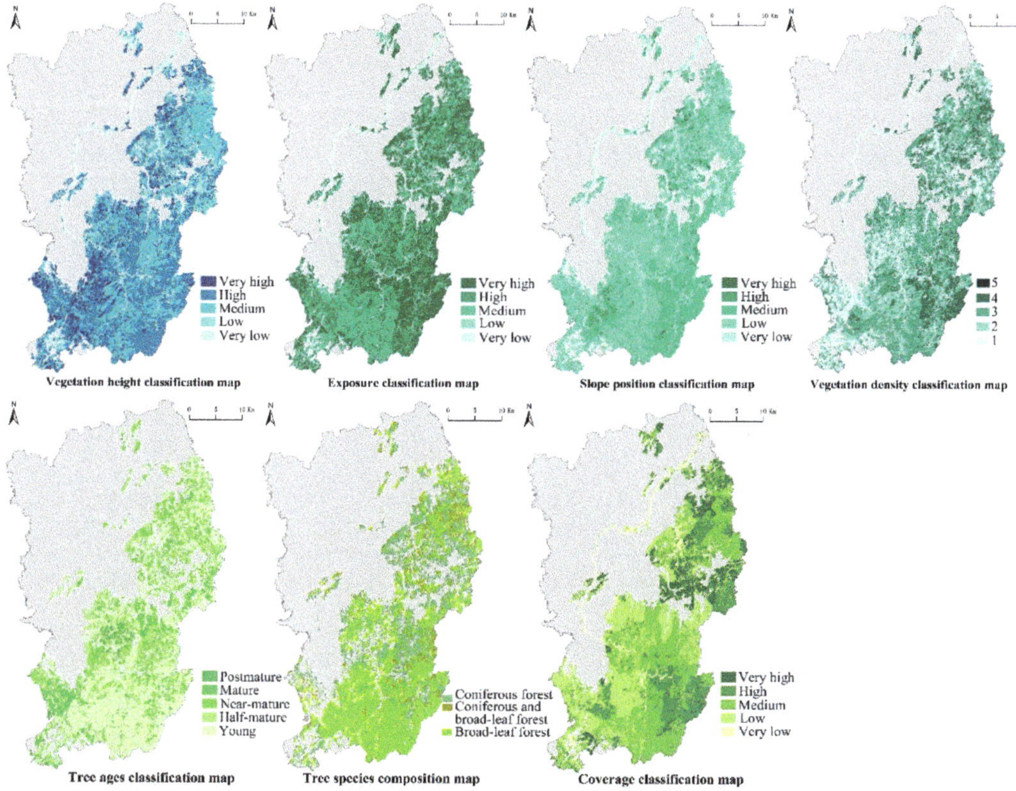

Figure 8. Vegetation condition assessment.

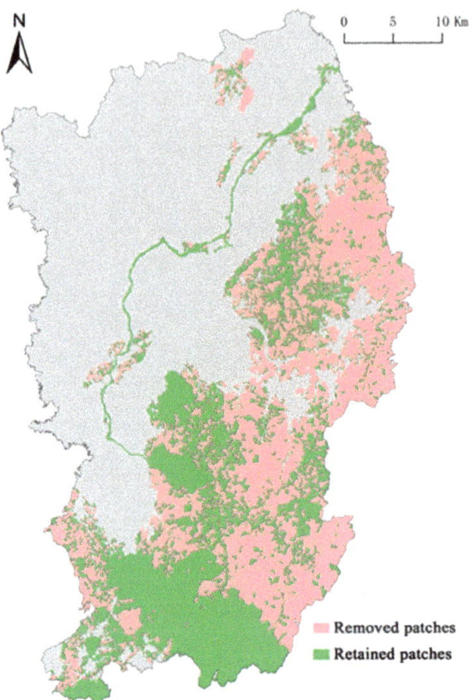

Figure 9. Reserve candidate range C1.

4.3. Results of the Secondary Optimization of the Candidate Range of Reserves

The distribution map of natural and cultural landscape resources is generated according to the general planning text of Jianglang Mountain protected area. Natural landscape resources correspond to meteorological diversity scenery, geological landscape, hydrologic scenery, biology landscape, and geological environment protection area. Meteorological diversity scenery mainly includes sun, moon, stars, snow, clouds, natural sound, and image. The geological landscape is a geological and geomorphic landscape. Hydrologic scenery mainly refers to spring, stream, lake, and lake falls. Biological landscape includes all kinds of flora and fauna landscapes. Cultural landscape resources, including the elements, such as the specialties (e.g., folk customs, crafts, and products) and historical sites and buildings in the protected areas. The spots with important ecological value and ecological security are extracted from the forest survey data (Figure 10), which include the areas with a forest coverage of greater than 65, highest protection level, complete community structure, and important water sources. After the above elements and the previously delineated protected land boundary with the candidate range C1 were merged, the new screening range C2 was obtained through intersecting the area with complete species community structure (Figure 11).

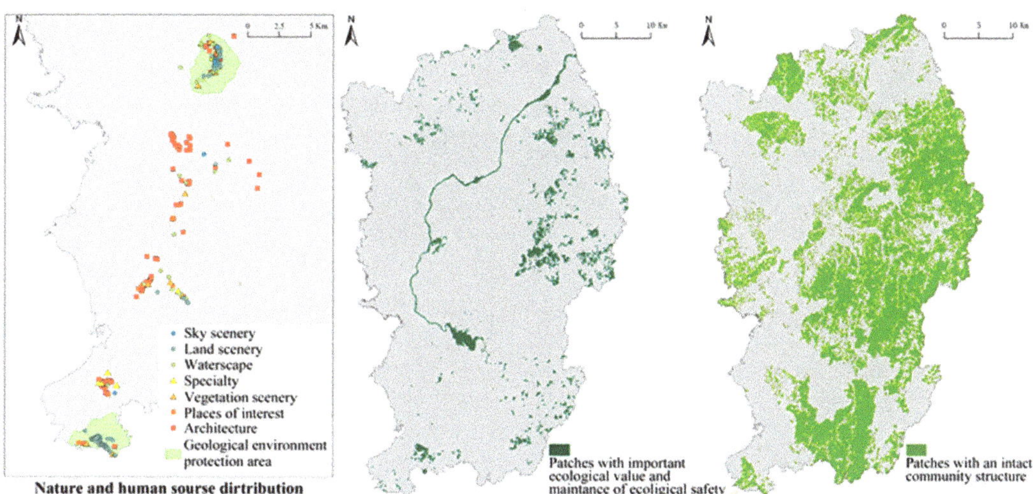

Figure 10. Analysis of the reserve elements.

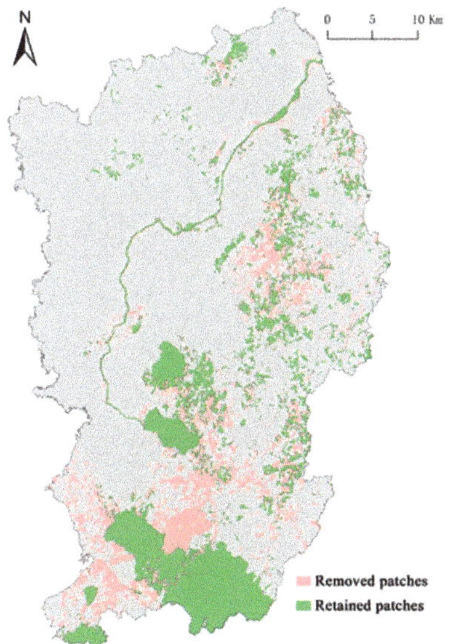

Figure 11. Reserve candidate range C2.

4.4. Tertiary Optimization to Form a New Reserve

The spots mentioned in the previous subsection were connected and coordinated with the other existing construction control conditions, and the current land-use-status information of the candidate range C2 of the reserve was assessed to remove the patches; such information includes the urban or construction town land, village land, hydraulic construction land, tea garden and orchard land, mining land, private land of forest rights, river roads, and administrative areas. Figures 12 and 13 depict the screening process of the surrounding spots of the existing reserves in Jiangshan and a sample scope of the newly formed natural protection areas. The area of the new reserve is 150.524 km^2, which is 6.682 km^2 bigger than the original one.

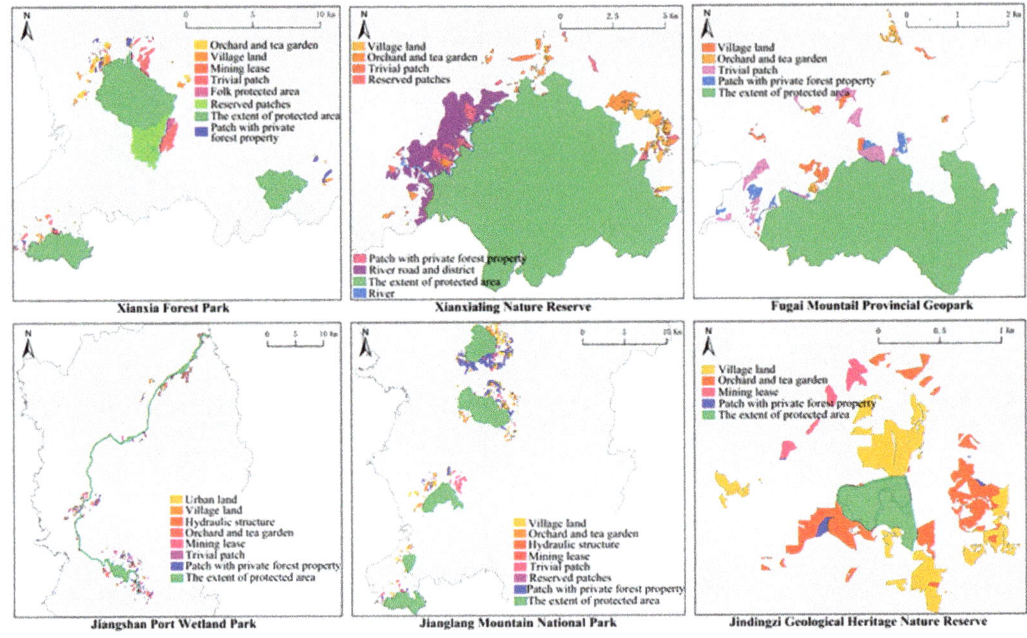

Figure 12. Screening of the spots around the reserve.

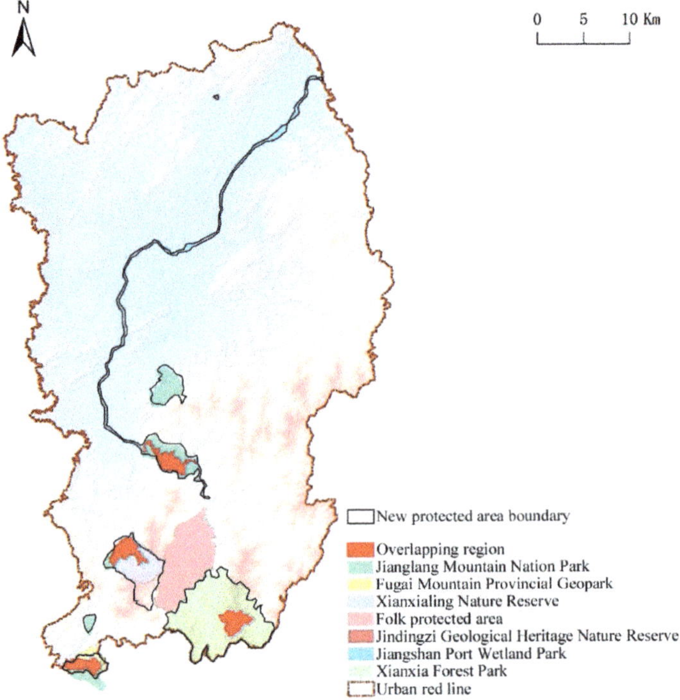

Figure 13. New delineated reserve.

It can be seen from the comparison that the original Fugai Mountain Provincial Geopark covers an area of about 9.41 km^2, which overlaps with the Fugai Mountain Provincial Geopark within the scope of Jianglang Mountain National Park. It also has high similarity in the composition of scenic resources and biological resources. Combined with the principle of giving priority to protection intensity at the same level, low levels at different levels obey high levels. The original Jiangshan Fugai Mountain Provincial Geopark should be included in the scope of Jianglang Mountain National Park and integrated with the Fugai Mountain Provincial Geopark in Jianglang Mountain National Park. The Fugai Mountain area of the original Xianxialing Natural Reserve, with an area of about 7.81 km^2, overlaps with the Fugai Mountain Provincial Geopark within the Jiangshan Fugai Mountain Provincial Geopark and the original Jiangshan Fugai Mountain Provincial Geopark in a large area, and it should be included in the Jiangshan Fugai Mountain Provincial Geopark and integrated with the Fugai Mountain Scenic spot in the Jianglang Mountain National Park.

5. Discussion

5.1. Comparison with Other Delineation Techniques for Reserves

Boundary issues play a key role in the study of natural reserves. However, few studies have discussed the boundary of natural reserves; the academic interest is mainly focused on the boundaries in the field of landscape ecology. Moreover, the systematic boundary delimitation methods for nature reserves are lacking. The study of the conservation regionalization that focuses on regional biodiversity conservation is relatively mature and includes conservation vacancy analysis, conservation priority area analysis, and ecological conservation planning.

The vacancy analysis for the biodiversity of reserves applies layer superposition and iteration methods [35] to determine the conservation gap. These approaches involve superposing the built reserves and the survey data of specific species, vegetation, and natural ecosystem type distribution or using a species distribution [36], habitat suitability [37], and other mathematical models [38] to guide the scientific layout of natural reserves. At present, numerous vacancy analyses for reserves remain confined to a part of the national key protected species or limited types of ecosystem protection, which is unfavorable for establishing an entire policy set to strengthen the comprehensive analysis. In this study, the detailed data that provided a basis for the subsequent quantitative analysis were obtained from multiple channels. Consequently, the investigation is no longer limited to the protection of a single species, because, through the overall analysis of the regional ecosystem, along with the heritage resources and construction control conditions, the small-scale and precise analysis on the boundary demarcation of the natural reserve was conducted.

The quantitative analysis of the conservation priority area was conducted on the basis of the data of biodiversity and threats [39]. The determination of the conservation priority area is generally indicated by biodiversity hotspots or species with a high indicator [40,41], because the distribution information of the existing species is mostly based on the administrative units for statistics. In addition, as the primary means of resource allocation and protection decision-making [42], the administrative unit is highly conducive to the development of the protection plan. In related studies, the administrative unit was used to determine the protection priority area [43]. However, previous reports revealed that the administrative unit lacks ecological significance, and the priority area determined on the basis of the biogeographic unit is more representative than the administrative unit. Therefore, the combination of the administrative and biogeographic units should be used to identify the protection priority area and optimize the reserve system. The division of the priority areas of landscape protection is based on the objective ecosystem vulnerability, which has a different emphasis from the evaluation standard [44], and thereby yielding spatial overlap and challenges for generating a unified zoning plan [45]. This study constructed the C-NHC framework from the national policy level to screen the preliminary candidate areas for the reserves, organically integrated the administrative and natural geographic units, and established a scientifically unified spatial evaluation standard for reserves.

The ecological zone protection planning was performed on the basis of biogeographical zoning and related protection planning. Biogeographical zoning is a widely used technique for delineating natural reserve boundaries [46]. In the existing studies, the land is divided according to the ecological relationship between adjacent ecosystems [47]. A global zoning scheme based on biological groups, such as global biota types [20], world eco-regionalization [48], and biodiversity-based biogeographic regionalization framework [21], was also proposed. The conservation value of the biodiversity in nature reserves was quantitatively evaluated from three aspects, namely, ecosystem, species diversity, and genetic germplasm resource, through overlay analysis, TWINSPAN classification, and vacancy analysis for reserves based on spatial distribution data (e.g., landform, vegetation, and natural reserve for integrated geographical regionalization of natural protection). All of these methods presented satisfactory results and patterns. Similar to the conservation priority areas, inconsistent zoning standards will lead to different geographic zoning schemes, which will affect the protection decisions. Therefore, a unified judgment standard was established in this study through the construction of a quantifiable evaluation index system, which lays the foundation for boundary demarcation and scientific management of natural reserves.

The existing studies on nature reserve boundaries involve many aspects, such as protection objectives, habitat quality, and protection strategies (Table 5) [49,50]. In this study, natural disturbance mechanisms, climate change, habitat quality, and connectivity were incorporated as delineation criteria [51,52]. However, many deficiencies remain present in natural protection zoning plans based on biodiversity data, ecosystem status [53], and environmental quality [54,55]. In practice, the phenomenon of cross-integration occurs, which is limited to a single level of ecological protection and fails to address the conflicting interest demands of different social groups, as well as the confrontation and conflict in various land use modes. This study integrated three aspects namely, natural resource background, heritage resource characteristics, and construction control conditions, to optimize the boundaries of natural reserves. On the basis of regional ecological protection, the optimized boundaries of reserves are used to alleviate the conflict between the development and protection of natural reserves, which is beneficial in solving the problem regarding the spatial overlap of reserves. Thus, a scientific demarcation of natural reserves is generated.

Table 5. Existing boundary delimitation technologies for natural reserves.

Name	Main Content	Connotation	Target	Strength	Weakness
Comprehensive geographical division of natural protection	Superposition analysis and TWINSPAN classification on the basis of the spatial distribution data of landforms, vegetation, and nature reserves.	A mathematical model and method for the quantitative evaluation of the biodiversity conservation value in terms of the ecosystem, species diversity, and genetic germplasm resources of the natural reserves.	To evaluate the effectiveness of the National Nature Reserve in protecting the natural vegetation and improving the effectiveness of the natural reserve network.	Effectively identifies the conservation values of the ecosystem and species diversity in natural reserves.	Lacks the attention to cultural heritage elements or cultural landscape heritage.

Table 5. *Cont.*

Name	Main Content	Connotation	Target	Strength	Weakness
Vacancy analysis for reserves	Layer superposition and iterative method, which are used to identify the protection vacancy of wild animals and plants, vegetation types, and land use in a certain area.	Identification of the distribution of plants, animals, and vegetation types that are not effectively protected by the reserve network.	Focus on the distribution areas (i.e., blank points) of animals and plants and their habitats that do not appear in the reserve network and promote the formation of the scientific layout of natural reserves.	Intuitive and easy to operate and conducive to the comprehensive and systematic protection of biodiversity	Requires a large number of accurate data due to the lack of authoritative data; insufficiency in practical applications.
Analysis of the conservation priority areas	Quantitative analysis through mathematical algorithms based on natural attributes, biological characteristics, connectivity, and socioeconomic costs of establishing the reserve to determine the protection objectives.	Quantitative investigation of the priority sequence of conservation in different regions on the basis of the data of biodiversity and threats.	To determine the priority areas for biodiversity conservation and guide the process of biodiversity conservation.	Contributes to the research and protection of typical ecosystems.	Large-scale, which ignores the regions that are not rich in biodiversity, faces serious threats and, thus, should be transformed to a smaller scale.
Analysis of the planning of the ecological zone protection	Distribution of biological communities, ecological relationship between adjacent ecosystems, biota, and biodiversity, which are regarded as important bases of the biogeographic division.	Construction of the relevant protection planning scheme on the basis of biogeographical zoning research.	Protect regional ecosystems, guide the construction of regional nature reserves, and provide a scientific basis for the formulation of regional biodiversity policies.	Protects the wildlife in the area and their habitats.	Poses inconsistent zoning standards, which lead to different biogeographic zoning schemes that affect biodiversity conservation decisions.

5.2. Comparison with Traditional Border Demarcation Techniques for Chinese Reserves

China has not yet established a polished national park system. All types of natural reserves have been managed through special regulations, in which most of the discussions on the boundaries of natural reserves focus on scenic spots. The establishment of scenic spots involves the source of scenery and ecological protection, continuation of the historical context, coordination of protection plans, and utilization and management of multi-layered goals. The original demarcation methods for such spots include the equidistant control method of parallel moving road and river center line, terrain method on the basis of the contour line of mountain ridges, scenic-spot control method guided by scenery source, and coordination method involving relevant planning boundaries (Table 6). However, such techniques face several disadvantages. When demarcating the boundary, the landscape resources are not specified, thereby disregarding important factors, as well as lack of organization and accuracy. In recent years, scholars have proposed an element-based spatial analysis method based on delineating the boundaries of scenic spots by using overlapping and buffer analyses. The system for boundary demarcation comprises the classification of elements and the superposition of effective organizations, which is based on the concretization of the landscape resources into elements (e.g., survey data, natural resources, human economy, facilities and basic engineering conditions, and land and other materials). At the micro-level, the boundary is drawn by using the "terrain method" and evaluated and adjusted by using the elements. The element-based demarcation of natural reserves is easy to execute, and the weight assignment of the elements involves subjective components, thereby affecting the grasp on the overall characteristics. However, given the diversity in the types of scenic spots, no universal factor-classification system has been established.

At present, the boundary delineation of nature reserves adopts the spatial analysis technology in geographic information systems (GISs) to classify lands with high precision. Compared with a pure qualitative evaluation, the quantitative processing of each evaluation index, using the GIS technology, is more objective and is equipped with an index system that can be easily applied in the study area [56]. Moreover, the relevant methods for visual landscape evaluation are based on perspective (e.g., field of view and viewing distance). Imaging methods are used to extend the boundary demarcation results to three dimensions [57]. However, the boundary delineation of natural reserves involves technical, economic, legal, and even political decision-making, while some of them can hardly be controlled by planning and design institutions. Therefore, in-depth studies on the influence of laws, regulations, and government decisions should be conducted.

The construction of the pilot areas of national parks in China should rely on the original multi-type protection lands [58]. However, the general inheritance of the original boundaries and zones is inconsistent with the protection objectives of the authenticity and integrity of the ecosystem. Consequently, the influence of the existing land-use patterns on the realization of the protection objectives and community economic development is disregarded [59]. The existing studies propose a combination of resource and environment, landscape source value, and boundary overlap assessments [60] to classify, overlay, evaluate [61], and define the boundary of the reserve. The C-NHC framework considers the influence of the policies and regulations on the boundaries of natural reserves. The proposed framework comprehensively evaluates the requirements of the ecological land, cultivated land, and urban development of the conflict spots through building a quantifiable evaluation index system and subsequently constructs a unified evaluation standard that lays a foundation for the formulation of the collaborative optimization scheme for boundary conflicts. This study performed triple optimizations of the boundary of nature reserves based on the characteristics of the resource background, heritage resources, and construction control conditions. Moreover, following the previous studies, the proposed framework conducted the following: it integrated the ecological factors; established the scenery sources, construction management conditions, and other important factors of protected areas; and supplemented the previous relatively single and one-sided boundary delimitation methods of protected areas. These conditions aimed to introduce the boundary delimitation methods that can be implemented, popularized, and replicated.

Table 6. Traditional boundary demarcation techniques for Chinese reserves.

Name	Main Content	Connotation	Target	Strength	Weakness
National Park Nature Reserve Zoning Model	Selection of the indicators in building a natural resource protection zoning index system and combining the natural resource protection zoning with the specific resource problems and protection needs of each region on the basis of ecological background, resource characteristics, and human interference.	Realization of the effective protection of natural resources and formation of a zoning model of a national park's natural resource protection that can be copied and promoted.	Identify regional resource issues and divide a reasonable resource protection spatial pattern to determine the protection goals and measures and achieve the effective protection of the natural resources.	Solves the difficult coordination problem caused by the misalignment of the existing various land types in the national park.	Focuses on the boundary division of the internal partitions without considering how to delineate the outer boundary.
National park layout analysis	Divisions of natural geography and biogeography, as well as the main functional areas, which discuss the partitions that are suitable for the layout of national parks and evaluate and screen the national parks on the basis of resource endowment, construction suitability, and management feasibility.	Establishment of the principles and characteristics for the selection of national parks selection, selection of the suitable sites for national park selection, and clarification of the overall layout of the national parks.	Provide reference for the overall layout of the national park, help build the national park system, and improve the nature protection system.	Comprehensively proposes the evaluation methods and layout plans for the candidate sites of national parks.	Difficult delivery of an objective and quantitative data analysis, because the layout analysis is qualitative in nature.
Spatial analysis method on the basis of resource elements	Completion of the systemization of the boundary demarcation through element classification and overlay analysis from the resource protection, management authority, and human behavior control levels; illustration of the terrain at the micro-level, using the topography method.	Embodiment of the landscape resources as elements and establishment of a new method framework for boundary delineation through the element-based spatial analysis method (overlap analysis + buffer zone analysis).	Demarcate the boundaries of scenic spots.	Generates a quantitative summary by using elements to simplify the boundary problem of scenic spots.	Presence of a subjective component in the weighting of the elements, which cannot grasp the overall characteristics.

Table 6. *Cont.*

Name	Main Content Connotation	Target	Strength	Weakness
Scenic source	Implementation of boundary control using landscape sources (e.g., scene, landscape, scenic spots, and landscape groups) according to the requirements of the evaluation results of the landscape resources; delimitation of the scope of influence of the landscape source and formation of the scope of the scenic spots that are separated or connected with the surrounding boundaries.	Demarcate the boundaries of scenic spots.	Possesses a certain scientific nature and plays an important role in protecting important sceneries on the basis of core values.	Difficulty in determining the radiation range (buffer size) of the scenic source; difficulty in systematically analyzing the surrounding environmental characteristics.
Topography	Delineation of the boundaries according to topographic lines, such as the ridge lines of adjacent mountains, contour lines at a certain altitude, and the border lines of the watersheds.	Demarcate the boundaries of scenic spots.	Easily implements management, sets piles, and demarcates and can effectively protect the topography, natural resources, and landscape scenery.	Lack of basis for the determination of the topographic line; difficulty in solving the scale problem, which affects the accuracy of the border demarcation; lacks relevance to the core values of the scenic spots.
Offset method	Movement of the center line of the road, center line of the river, or the shoreline of the reservoir to a parallel position relative to a certain distance to obtain the scenic area.	Demarcate the boundaries of scenic spots.	Suitable for the rough delineation of boundaries and possesses strong operability.	Lacks basis in selecting the translation subject and translation distance; lacks relevance to the core values of the scenic spots.
Coordination method	Coordination with other types of protective land boundaries, such as World Heritage and National Forest Park, including the corresponding scope or direct sharing of the boundary line; coordination with the city according to the current status of urban development.	Demarcate the boundaries of scenic spots.	Strengthens the coordination of various plans, which helps avoid conflicts in management.	Difficulty in demonstrating the boundaries of borrowing other protective land; difficulty in determining the reference factors, specific distance, and visual landscape factors in coordination with the city.

5.3. Policy Enlightenment of Optimizing Boundary for Protected Areas

5.3.1. Standardized Boundary Demarcation Technology of Protected Areas Will Boost the Construction of Protected Area Systems

The boundary delineation of protected areas is not only technical delineation, but also includes economic, legal, and even political decisions. Some factors are beyond the control of planning and design institutions. Thus, in-depth research on the impact of relevant laws, regulations and government decisions is needed. If we only focus on the boundary of the protected area itself, then understanding the complete logical relationship between the boundary generation and the boundary information expression of the protected area fully is impossible. From the influence of ideology and policy system on the construction concept of natural protected areas to the special protection planning and land-use policy formulated by government departments, inextricably logical links are found between them and the boundary information of natural protected areas.

To establish a protected area system that is universally recognized by the international community, the effectiveness of boundary planning must be protected by legislation. The introduction of the Guiding Opinions in 2019 has clarified the direction for solving outstanding problems, such as overlap, multi-head management, unscientific classification, and unreasonable scope of protected areas in China, and opened a new chapter in the modern management of protected areas. However, specific implementation measures have not been clarified yet. How are guidelines implemented? How are boundaries integrated, merged, and adjusted from the most urgent and most controversial issues in the current reform of protected areas?

The boundary delineation method proposed in this study standardizes the integration and merging modes of various protected areas and strives to ensure the effectiveness and stability of the boundaries of the protected areas from the institutional level, thereby also eliminating the differences among different departments, regions, and management systems. This method will also promote the introduction of the special legislation protection of the boundary of natural conservation areas. In the sense of political geography that is highly unified between national sovereignty and the governance of protected areas, the boundary delineation method clears institutional obstacles for the creation of a complete system of protected areas in the entire territory of China.

5.3.2. Boundary of Protected Areas Will Become an Important Part of Land Spatial Layout Optimization

The demarcation of natural reserve boundaries is a key link in the construction of natural reserve system. In the process of the rapid urbanization worldwide, natural reserve boundaries play an increasingly important role in the protection of natural resources. As a means to coordinate the protection and development of nature, realizing the protection and sustainable use of natural resources is also helpful. However, in the actual implementation process, the lack of relevant theoretical system and the limitation of technical means will cause the boundary demarcation of natural protected areas to lack scientific basis and a clear quantitative system, thus resulting in an inaccurate boundary demarcation.

Uneven spatial data distribution, weak data base, low data integrity, and insufficient information sharing are all technical barriers to the boundary demarcation of natural protected areas in the early years that affect the scientific setting and layout of natural protected areas. With the advent of the information age, the rapid development of satellite remote-sensing data and GIS and GPS technologies has laid a solid foundation for the demarcation of refined and scientific natural protection boundaries. The improvement of technological level will inevitably be accompanied by the continuous improvement of standards, and the connotation of the boundaries of protected areas has also changed from a purely natural resource control boundary to the bottom line of the national land spatial planning.

The CPC Central Committee and the State Council issued several Opinions on Establishing a Land Spatial Planning System and Supervising its Implementation, thereby requiring the special plans for the protected areas to be closely linked with relevant land spatial planning and checked against "in one map" during the compilation and review processes. The construction of natural protected land system and the land spatial planning system will be connected and coordinated in the form of space scope and realize fine management. Based on the aforementioned requirements, the research begins from stating the objectives, criteria, and indicators for the delimitation of the protected area boundaries. Subsequently, the suitability of ecological, cultivated land, and construction land for all map spots in the region are evaluated. The evaluation results lay the foundation for the formulation of the coordinated optimization plan for border conflicts and provide a scientific basis for coordinating various types of land use in the preparation of land-use planning and prevention and reduction of actual land-use conflicts. This research is of great significance for realizing the orderly layout of the three types of space, including towns, agriculture, and ecology, which is also important for the optimization of the national spatial layout.

5.3.3. Delimiting the Scientific Boundaries of Protected Areas Is the Key to Improving the Management of Protected Areas

The space superposition problem of protected areas caused by multiple international designated areas is the key to restricting the effective management of protected areas. (1) Unreasonable boundaries of protected areas will lead to the serious fragmentation of protected areas and decentralization of departmental functions, which will intensify the fragmentation and islanding of protected area. (2) The staggered boundaries of natural protected areas also inevitably lead to the fuzzy management target boundaries of protected areas, which results in the unclear positioning of all kinds of protected areas in the management and causes difficulty in carrying out targeted management. (3) The unclear boundaries between the powers and responsibilities of various departments in the multiple international designated area will cause problems, such as indistinct management objectives, unclear responsibilities, and disordered management policies.

The boundary of natural protected areas is mainly used to define the authority of management. A reasonable boundary must contain the core value of natural resources and must not cause difficult coordination problems due to the large scope to serve the work of conservation management, rational utilization, and planning and implementation better. The boundary demarcation of protected areas, as a means of space control, can promote the improvement of the management level of protected areas from two aspects: one is to define the protection objectives, and the other is to balance the interests of multiple parties. (1) To clarify the protection objective, determining the protection object, the urgency, the main means, and the effect of protection and converting the protection objective into quantitative index are necessary. This study built a three-dimension (e.g., ecology, cultivated land, and construction) suitability evaluation index system that considers the ecological value, social economic security demand, and survival and judged the relationship between protection targets on the basis of defining a single protection target. The boundary demarcation based on the above principles is of great significance for the effective identification of protection targets. (2) The ecosystem service value is the core of the protected area. Therefore, this study took the natural background as the core analysis element and then analyzes the influence of soft factors, such as heritage resource characteristic and existing construction control conditions on the designation of protected area, in detail. The coordination of stakeholders at the boundary can be achieved by integrating natural resource management into the social selection framework based on the understanding of the key characteristics of protected areas and the consideration of respect for indigenous and traditional knowledge. The control measures will be implemented in space to achieve the unification of the physical and control boundaries.

The protected areas' boundary, which is based on the coordination of resource utilization and ecological protected areas, provides a basic guarantee for the establishment of institutional standards. In the practice of protected area construction, boundary demarcation should not be regarded as the establishment of geographical boundary. The boundary of protected area should be adjusted in time according to the change in economic development and protection target, and the boundary of protected area should be transformed into the control problem of coordinating the land-use mode, scale, and intensity of stakeholders to clarify the boundary of protected land as the basis of implementing the spatial control of protected land.

6. Summary and Conclusions

With the continuous improvement of international protection awareness, the number of protection organizations has dramatically increased, and they have promoted various reserve-naming methods under different objectives, purposes, and management requirements; thus, the phenomenon of overlapping reserves has prominently increased. Therefore, the demarcation of the boundaries of scientific and reasonable reserves has become the key to the effective management and sustainable development of natural reserves. The delineation of the boundaries of natural reserves provides a scientific basis for the formulation of regional ecological protective policies. As an effective tool for managing reserves, the boundaries serve as a bridge to alleviate the current conflicts between the development and protection of nature reserves, as well as a key factor and guarantee that play multiple functions in the implementation of an effective management.

Rebuilding the natural reserve system by using national parks as the main body is a crucial exploration in the construction of an ecological civilization system in China. On this basis, this study conducted an in-depth investigation of the technical methods for delineating the spatial boundaries of natural reserves. Taking Jiangshan Nature Reserve as a case, the C-NHC framework was constructed, and the boundaries of the new reserves were delineated. First, the preliminary candidate areas of the reserve were selected. Second, a comprehensive vector evaluation was performed on the basis of the background characteristics of the resources, heritage resource characteristics, and construction management conditions. Third, resource background evaluation and initial aggregation optimization were executed, the existing elements of the reserve were extracted, and the resource-based clustering results were intersected and optimized. Lastly, the existing construction management conditions were connected and coordinated to form a scientific, reasonable, and clear boundary of the nature reserves. Through the quantitative evaluation of the conflict patches and the triple optimization of the boundary of the reserve, this study has changed the status quo of the spatial overlap of the reserve, which is considerably important in the improvement of the planning and management of natural reserves and establishment of a unified management authority.

This study utilized geographic information technology, and the protection area boundaries were optimized on the basis of three dimensions (i.e., resource background characteristics, heritage resource characteristics, and construction management conditions). The newly delineated boundaries play a key role in coordinating the construction and development of Jiangshan and the ecological environment protection, restoring and improving the service functions of the regional natural ecosystems, and ensuring the sustainable use of resources. This study will exert an extensive influence on the in-depth analysis of the spatial pattern of natural reserves in the future, improvement of the planning and management, and overall promotion of the construction of natural reserves. This technical framework can provide a useful reference for the boundary delimitation of nature reserves in China and even the world. This study also has some limitations. We did not provide a method for selecting the appropriate restrictive suitability neighborhood influence factors and characteristics of the heritage resources in different study areas. When our methods are used in different study areas, researchers are still required to select the elements to participate in the study, a process that is vulnerable to great subjectivity.

Although China has formulated numerous planning and management schemes, the current natural reserve system remains limited, given the prominent overlapping phenomenon, which hinders the scientific construction and management of the natural reserves. As China continues to promote a "national park-based nature reserve system", natural protection zones that can precisely reflect the characteristics of the regional natural resources and construction conditions should be developed to provide a basis for the construction and management of the natural protection land. To solve the serious problems and contradictions in the management of the natural reserves in the country, the proposed boundary delimitation method has re-integrated all types of natural reserves and established a scientific spatial pattern of natural reserves. We believe that the construction management conditions have the greatest impact on the final results. The proposed method can enhance the management system; solve relevant problems (e.g., multi-head management); and promote a highly systematic, integrated, and collaborative natural reserve system.

Author Contributions: Conceptualization, H.G.; methodology, H.G.; software, Y.W.; validation, Y.W.; formal analysis, Y.W.; investigation, Y.W.; resources, H.G.; data curation, H.G.; writing—original draft preparation, Y.W.; writing—review and editing, Y.L.; visualization, Y.W.; supervision, Y.D.; project administration, H.G.; funding acquisition, H.G. All authors have read and agreed to the published version of the manuscript.

Funding: This work was supported by the Social Science Fund of Zhejiang Province (21NDJC034YB), and Centre for Balance Architecture, Zhejiang University (2021-KYY-669000-0014).

Institutional Review Board Statement: Not applicable.

Informed Consent Statement: Not applicable.

Data Availability Statement: Not applicable.

Acknowledgments: The authors wish to thank anonymous reviewers for their valuable comments and suggestions for improving the quality of this paper.

Conflicts of Interest: The authors declare no conflict of interest.

Appendix A

Table A1. Discriminant matrix of suitable land use.

Type Code	Land Suitability Combination			Appropriately	Land Use Adjustment Instructions
	Suitability of Construction Land	Suitability of Cultivated Land	Suitability of Ecological Land		
1	High	High	High	Ecological land, cultivated land	The three types of land are suitable. When conflicting with ecological land, the principle of natural reserves should be followed to maintain the original ecological land. When there is no conflict of ecological land, considering that cultivated land also has certain ecological service value, the original cultivated land should be maintained.
2	High	High	Low	Construction land and cultivated land	Both construction land and agricultural land are suitable areas. Due to the obvious comparative advantages of construction land and the low urgency of ecological protection, it is more likely that cultivated land will be converted to construction land. Consider conversion to construction land. When there is no conflict in regard to construction land, maintain or convert to cultivated land.
3	High	Medium	High	Ecological land, construction land	The suitability of ecological land and construction land is very strong, because construction land has obvious comparative advantages and the possibility of expansion of construction land is high, but the principle of nature protection should be considered, so maintenance or conversion to ecological land should be considered. It can be converted into construction land when there is no conflict in regard to ecological land.

Table A1. *Cont.*

Type Code	Land Suitability Combination			Appropriately	Land Use Adjustment Instructions
	Suitability of Construction Land	Suitability of Cultivated Land	Suitability of Ecological Land		
4	High	Medium	Low	Construction land and cultivated land	Construction land has obvious advantages and high expansion potential; thus, it can be considered to be converted to construction land. It is maintained or converted to cultivated land when there is no conflict in regard to construction land.
5	High	Low	High	Ecological land, construction land	The suitability of ecological land and construction land is very strong, because the comparative advantage of construction land is obvious; the possibility of construction land expansion is high, but the principle of natural protection should be considered, so it should be considered to maintain or transform into ecological land; it can be considered to transform into construction land when there is no conflict in regard to ecological land.
6	High	Low	Low	Construction land and original land	Construction land has obvious advantages, and the possibility of expansion is high. It can be considered to be converted into construction land; when there is no conflict in regard to construction land, you can consider maintaining the original land type.
7	Medium	High	High	Ecological land, cultivated land	The comparative advantages of construction land and ecological land are obvious, but following the principle of ecological priority, it can be considered to be converted into ecological land; when there is no conflict in regard to ecological land, it can be considered to be maintained or converted into cultivated land.
8	Medium	High	Low	Cultivated land, construction land	Cultivated land has obvious advantages, and the original cultivated land should be maintained; when there is no conflict of cultivated land, it can be considered to be converted into construction land.
9	Medium	Medium	High	Ecological land, cultivated land	The ecological land maintenance capability is strong, and the original ecological land should be maintained; when there is no ecological land conflict, the conversion into cultivated land should be considered according to the principle of priority of cultivated land protection.
10	Medium	Medium	Low	Cultivated land, construction land	The maintenance capacity of ecological land is weak, and the suitability of construction land and cultivated land is equivalent, but the principle of priority of cultivated land protection should be followed, and the original cultivated land should be kept from encroaching as much as possible; when there is no conflict in regard to cultivated land, it can be considered to be converted into construction land.

Table A1. *Cont.*

Type Code	Land Suitability Combination			Appropriately	Land Use Adjustment Instructions
	Suitability of Construction Land	**Suitability of Cultivated Land**	**Suitability of Ecological Land**		
11	Medium	Low	High	Ecological land, construction land	Ecological land has obvious advantages; when there is no conflict of ecological land, it can be considered to be converted to construction land.
12	Medium	Low	Low	Construction land and original land	Due to the comparative advantage of construction land, the possibility of maintaining or converting land into construction land is high; when there is no conflict between construction land, consider maintaining the original land type.
13	Low	High	High	Fierce conflict	Due to the conflict between agricultural land and ecological land, the possibility of expansion of construction land is low; due to the comparative advantage of agricultural land output rate and the shortage of agricultural land resources in the region, the possibility of conversion of suitable unused agricultural land to agricultural land is high, but the possibility of conversion is determined by the comparison of two kinds of policies: agricultural land and ecological land.
14	Low	High	Low	Weak conflict	Cultivated land has obvious advantages and should be maintained or reclaimed as cultivated land.
15	Low	Medium	High	Ecological land, cultivated land	Ecological land has obvious advantages. The original ecological land should be maintained, or the land consolidation should be considered as ecological land; when there is no ecological land conflict, it should be maintained or converted into cultivated land for type 8 or 9.
16	Low	Medium	Low	Arable land, original land	The suitability of cultivated land is high, and the land should be maintained or converted into cultivated land; when there is no conflict of cultivated land, the original land type should be maintained.
17	Low	Low	High	Ecological land, original land type	Ecological land has obvious advantages, so we should maintain the original ecological land or consider returning farmland or land consolidation to ecological land; when there is no conflict in regard to ecological land, maintain the original land type.
18	Low	Low	Low	Current status	It is more likely to maintain the status quo of land use.

Table A2. Ecological Suitability Evaluation Index.

Target	Guidelines	Factor		Element
Evaluation of the importance of ecological function and competitiveness	Water conservation, soil and water conservation area, sand prevention, and sand fixation area	Natural conditions	Vegetation factor	Vegetation coverage
				Vegetation cover type
			Terrain factor	Slope
				Slope position
				Slope length
			Soil factor	Soil texture
				Soil thickness
			Natural location factor	Distance from river
				Distance from lakes and reservoirs
				Distance from pit
				Distance from existing delineated reserve and ecological red line
		Landscape pattern	Plaque characteristics	Plaque size
				Plaque shape index
			Plaque aggregation degree	Degree of aggregation
				Separation
	Biodiversity Reserve	Natural conditions	Resource status	Surface cover type
				Water network density
				Vegetation coverage
			Species distribution	Species diversity
				Species rarity
				Species distribution concentration
				Habitat nature
		Landscape pattern	Plaque characteristics	Plaque size
				Plaque shape index
			Plaque aggregation degree	Plaque fragmentation
				Patchy landscape diversity
		Network connectivity		Plaque centrality
				Median Index
				Correlation length index

Table A3. Cultivated Land Suitability Evaluation Index.

Target	Guidelines	Factor	Element
Cultivated land suitability evaluation	Restrictive factors	Planning factors	Whether it is in a nature reserve
			Is it located in the forest park
			Whether it is located in a national scenic spot
			Whether it is located in the primary and secondary water source protection zone
			Whether it is located in the returned forest area
		Natural conditions	Whether the slope is greater than 15°
	Suitability factor	Farming conditions	Slope
			Soil organic matter content
			Surface soil thickness
		Location factor	Distance from road
			Distance from river
			Distance from the reservoir
			Distance to nearest village
		Planning factors	Whether the current land is cultivated land
			Cultivated land conversion cost
		Geometric Features	Patch size
			Patch shape index
	Influencing factors of neighborhood		The largest area in the buffer zone
			The largest perimeter in the buffer zone
			Proportion of cultivated land area in the buffer zone
			Patch density of cultivated land in the buffer zone
			Plaque aggregation degree of buffer farmland

Table A4. Construction Land Suitability Evaluation Index.

Target	Guidelines	Factor	Element
Construction land suitability evaluation	Restrictive factors	Planning factors	Whether it is in a nature reserve
			Whether it is located in the forest park
			Whether it is located in a national scenic spot
			Whether it is located in the primary and secondary water source protection zone
			Whether it is located in the basic farmland protection area
		Natural conditions	Whether it is an area with frequent natural disasters
	Suitability factor	Urban construction conditions	Slope
			Elevation
			Terrain relief
		Location factor	Distance from road
			Distance from river
			Distance from the reservoir
			Distance from main city
		Planning factors	Whether the current land is construction land
			Construction and development costs
		Geometric Features	Patch size
			Patch shape index
	Influencing factors of neighborhood		The largest area in the buffer zone
			The largest perimeter in the buffer zone
			Proportion of existing construction land area in the buffer zone
			Plaque density of construction land in the buffer zone
			Plaque aggregation degree of buffer construction land

References

1. DeFries, R.S.; Foley, J.A.; Asner, G.P. Land-Use Choices: Balancing Human Needs and Ecosystem Function. *Front. Ecol. Environ.* **2004**, *2*, 249–257. [CrossRef]
2. Nogués-Bravo, D.; Araújo, M.B.; Romdal, T.; Rahbek, C. Scale Effects and Human Impact on the Elevational Species Richness Gradients. *Nature* **2008**, *453*, 216–219. [CrossRef] [PubMed]
3. Orme, C.D.L.; Davies, R.G.; Burgess, M.; Eigenbrod, F.; Pickup, N.; Olson, V.A.; Webster, A.J.; Ding, T.-S.; Rasmussen, P.C.; Ridgely, R.S. Global Hotspots of Species Richness Are Not Congruent with Endemism or Threat. *Nature* **2005**, *436*, 1016–1019. [CrossRef]
4. Liu, Y.B.; Li, R.D.; Song, X.F. Analysis of Coupling Degrees of Urbanization and Ecological Environment in China. *J. Nat. Res.* **2005**, *20*, 105–112.

5. Grimm, N.B.; Foster, D.; Groffman, P.; Grove, J.M.; Hopkinson, C.S.; Nadelhoffer, K.J.; Pataki, D.E.; Peters, D.P. The Changing Landscape: Ecosystem Responses to Urbanization and Pollution across Climatic and Societal Gradients. *Front. Ecol. Environ.* **2008**, *6*, 264–272. [CrossRef]
6. Schulze, E.-D.; Mooney, H.A. *Biodiversity and Ecosystem Function*; Springer Science & Business Media: Cham, Switzerland, 2012.
7. Balmford, A.; Green, R.E.; Jenkins, M. Measuring the Changing State of Nature. *Trends Ecol. Evol.* **2003**, *18*, 326–330. [CrossRef]
8. Groombridge, B.; Jenkins, M.D. *Global Biodiversity: Earth's Living Resources in the 21st Century*; World Conservation Press: Cambridge, UK, 2000.
9. Pinheiro, J. *Parks for Biodiversity Policy Guidance Based on Experience in ACP Countries*; IUCN: Gland, Switzerland, 1999.
10. Dudley, N. *Guidelines for Applying Protected Area Management Categories*; IUCN: Gland, Switzerland, 2008.
11. Deguignet, M.; Juffe-Bignoli, D.; Harrison, J.; MacSharry, B.; Burgess, N.D.; Kingston, N. *United Nations List of Protected Areas*; UNEP-WCMC: Cambridge, UK, 2014.
12. Borrini-Feyerabend, G.; Bueno, P.; Hay-Edie, T.; Lang, B.; Rastogi, A.; Sandwith, T. A Primer on Governance for Protected and Conserved Areas. In *Proceedings of the Stream on Enhacing Diversity and Quality of Governance, 2014 IUCN World Parks Congress*; IUCN: Gland, Switzerland, 2014.
13. Finke, G. Linking Landscapes: Exploring the Relationships between World Heritage Cultural Landscapes and IUCN Protected Areas. In *IUCN World Heritage Study*; IUCN: Gland, Switzerland, 2013; Volume 11.
14. Osipova, E.; Shadie, P.; Zwahlen, C.; Osti, M.; Shi, Y.; Kormos, C.; Bertzky, B.; Murai, M.; Van Merm, R.; Badman, T. *IUCN World Heritage Outlook 2: A Conservation Assessment of All Natural World Heritage Sites*; IUCN: Gland, Switzerland, 2017; Volume 92.
15. Zhao, Z.C.; Peng, L.; Yang, R. Reconstruction of Protected Area System in the Context of the Establishment of National Park System in China. *Chin. Landsc. Archit.* **2016**, *7*, 11–18.
16. Jim, C.Y.; Xu, S.S. Recent Protected-Area Designation in China: An Evaluation of Administrative and Statutory Procedures. *Geogr. J.* **2004**, *170*, 39–50. [CrossRef]
17. Chen, S.-B.; Jiang, G.-M.; Ouyang, Z.-Y.; XU, W.-H.; Xiao, Y. Relative Importance of Water, Energy, and Heterogeneity in Determining Regional Pteridophyte and Seed Plant Richness in China. *J. Syst. Evol.* **2011**, *49*, 95–107. [CrossRef]
18. Arponen, A.; Heikkinen, R.K.; Thomas, C.D.; Moilanen, A. The Value of Biodiversity in Reserve Selection: Representation, Species Weighting, and Benefit Functions. *Conserv. Biol.* **2005**, *19*, 2009–2014. [CrossRef]
19. Tang, X. Analysis of the Current Situation of China′ s Nature Reserve Network and a Draft Plan for Its Optimization. *Biodivers. Sci.* **2005**, *13*, 81. [CrossRef]
20. Prentice, I.C.; Cramer, W.; Harrison, S.P.; Leemans, R.; Monserud, R.A.; Solomon, A.M. Special Paper: A Global Biome Model Based on Plant Physiology and Dominance, Soil Properties and Climate. *J. Biogeogr.* **1992**, *19*, 117–134. [CrossRef]
21. Kreft, H.; Jetz, W. A Framework for Delineating Biogeographical Regions Based on Species Distributions. *J. Biogeogr.* **2010**, *37*, 2029–2053. [CrossRef]
22. Alberts, H.C.; Hazen, H.D. Maintaining Authenticity and Integrity at Cultural World Heritage Sites. *Geogr. Rev.* **2010**, *100*, 56–73. [CrossRef]
23. Unesco. Recommendation Concerning the Preservation of Cultural Property Endangered by Public or Private Works 19 November 1968. Available online: http://portal.unesco.org/en/ev.php-URL_ID=13085&URL_DO=DO_TOPIC&URL_SECTION=201.html (accessed on 1 December 2021).
24. Gullino, P.; Larcher, F. Integrity in UNESCO World Heritage Sites. A Comparative Study for Rural Landscapes. *J. Cult. Herit.* **2013**, *14*, 389–395. [CrossRef]
25. Jokilehto, J. Considerations on Authenticity and Integrity in World Heritage Context. *City Time* **2006**, *2*, 1.
26. Wells, M.; Brandon, K.; Hannah, L. *People and Parks: Linking Protected Area Management with Local Communities*; World Bank: Washington, DC, USA, 1992.
27. Margules, C.R.; Pressey, R.L. Systematic Conservation Planning. *Nature* **2000**, *405*, 243–253. [CrossRef] [PubMed]
28. Jones, K.R.; Venter, O.; Fuller, R.A.; Allan, J.R.; Maxwell, S.L.; Negret, P.J.; Watson, J.E. One-Third of Global Protected Land Is under Intense Human Pressure. *Science* **2018**, *360*, 788–791. [CrossRef] [PubMed]
29. Sannigrahi, S.; Chakraborti, S.; Joshi, P.K.; Keesstra, S.; Sen, S.; Paul, S.K.; Kreuter, U.; Sutton, P.C.; Jha, S.; Dang, K.B. Ecosystem Service Value Assessment of a Natural Reserve Region for Strengthening Protection and Conservation. *J. Environ. Manag.* **2019**, *244*, 208–227. [CrossRef] [PubMed]
30. Al-Harbi, K.M.A.-S. Application of the AHP in Project Management. *Int. J. Proj. Manag.* **2001**, *19*, 19–27. [CrossRef]
31. Bunruamkaew, K.; Murayama, Y. Land Use and Natural Resources Planning for Sustainable Ecotourism Using GIS in Surat Thani, Thailand. *Sustainability* **2012**, *4*, 412–429. [CrossRef]
32. Yu, K.; Guo, G.-D.; Li, J.; Lin, S. Quantum Algorithms for Similarity Measurement Based on Euclidean Distance. *Int. J. Theor. Phys.* **2020**, *59*, 3134–3144. [CrossRef]
33. Silverman, H. Heritage and Authenticity. In *The Palgrave Handbook of Contemporary Heritage Research*; Springer: Cham, Switzerland, 2015; pp. 69–88.
34. Xu, J.; Chen, L.; Lu, Y.; Fu, B. Local People's Perceptions as Decision Support for Protected Area Management in Wolong Biosphere Reserve, China. *J. Environ. Manag.* **2006**, *78*, 362–372. [CrossRef] [PubMed]
35. Jennings, M.D. Gap Analysis: Concepts, Methods, and Recent Results. *Landsc. Ecol.* **2000**, *15*, 5–20. [CrossRef]

36. Catullo, G.; Masi, M.; Falcucci, A.; Maiorano, L.; Rondinini, C.; Boitani, L. A Gap Analysis of Southeast Asian Mammals Based on Habitat Suitability Models. *Biol. Conserv.* **2008**, *141*, 2730–2744. [CrossRef]
37. Hopton, M.E.; Mayer, A.L. Using Self-Organizing Maps to Explore Patterns in Species Richness and Protection. *Biodivers. Conserv.* **2006**, *15*, 4477–4494. [CrossRef]
38. Peterson, A.T.; Kluza, D.A. New Distributional Modelling Approaches for Gap Analysis. In *Animal Conservation Forum*; Cambridge University Press: Cambridge, UK, 2003; Volume 6, pp. 47–54.
39. Groves, C.R.; Jensen, D.B.; Valutis, L.L.; Redford, K.H.; Shaffer, M.L.; Scott, J.M.; Baumgartner, J.V.; Higgins, J.V.; Beck, M.W.; Anderson, M.G. Planning for Biodiversity Conservation: Putting Conservation Science into Practice: A Seven-Step Framework for Developing Regional Plans to Conserve Biological Diversity, Based upon Principles of Conservation Biology and Ecology, Is Being Used Extensively by the Nature Conservancy to Identify Priority Areas for Conservation. *BioScience* **2002**, *52*, 499–512.
40. Brum, F.T.; Graham, C.H.; Costa, G.C.; Hedges, S.B.; Penone, C.; Radeloff, V.C.; Rondinini, C.; Loyola, R.; Davidson, A.D. Global Priorities for Conservation across Multiple Dimensions of Mammalian Diversity. *Proc. Natl. Acad. Sci. USA* **2017**, *114*, 7641–7646. [CrossRef]
41. Dobson, A.P.; Rodriguez, J.P.; Roberts, W.M.; Wilcove, D.S. Geographic Distribution of Endangered Species in the United States. *Science* **1997**, *275*, 550–553. [CrossRef]
42. Moilanen, A.; Arponen, A. Administrative Regions in Conservation: Balancing Local Priorities with Regional to Global Preferences in Spatial Planning. *Biol. Conserv.* **2011**, *144*, 1719–1725. [CrossRef]
43. Yang, F.; Hu, J.; Wu, R. Combining Endangered Plants and Animals as Surrogates to Identify Priority Conservation Areas in Yunnan, China. *Sci. Rep.* **2016**, *6*, 30753. [CrossRef]
44. Pimm, S.L.; Lawton, J.H. Planning for Biodiversity. *Science* **1998**, *279*, 2068–2069. [CrossRef]
45. Brooks, T.M.; Mittermeier, R.A.; Da Fonseca, G.A.; Gerlach, J.; Hoffmann, M.; Lamoreux, J.F.; Mittermeier, C.G.; Pilgrim, J.D.; Rodrigues, A.S. Global Biodiversity Conservation Priorities. *Science* **2006**, *313*, 58–61. [CrossRef]
46. Udvardy, M.D.; Udvardy, M.D.F. *A Classification of the Biogeographical Provinces of the World*; International Union for Conservation of Nature and Natural Resources Morges: Gland, Switzerland, 1975; Volume 8.
47. Bailey, R.G.; Hogg, H.C. A World Ecoregions Map for Resource Reporting. *Environ. Conserv.* **1986**, *13*, 195–202. [CrossRef]
48. Schultz, J. *The Ecozones of the World: The Ecological Divisions of the Geosphere*; Springer Science & Business Media: Cham, Switzerland, 2005.
49. Kujala, H.; Moilanen, A.; Araujo, M.B.; Cabeza, M. Conservation Planning with Uncertain Climate Change Projections. *PLoS ONE* **2013**, *8*, e53315.
50. Moilanen, A.; Laitila, J.; Vaahtoranta, T.; Dicks, L.V.; Sutherland, W.J. Structured Analysis of Conservation Strategies Applied to Temporary Conservation. *Biol. Conserv.* **2014**, *170*, 188–197. [CrossRef]
51. Hodgson, J.A.; Moilanen, A.; Wintle, B.A.; Thomas, C.D. Habitat Area, Quality and Connectivity: Striking the Balance for Efficient Conservation. *J. Appl. Ecol.* **2011**, *48*, 148–152. [CrossRef]
52. Leroux, S.J.; Rayfield, B. Methods and Tools for Addressing Natural Disturbance Dynamics in Conservation Planning for Wilderness Areas. *Divers. Distrib.* **2014**, *20*, 258–271. [CrossRef]
53. Sharafi, S.M.; Moilanen, A.; White, M.; Burgman, M. Integrating Environmental Gap Analysis with Spatial Conservation Prioritization: A Case Study from Victoria, Australia. *J. Environ. Manag.* **2012**, *112*, 240–251. [CrossRef]
54. Jenkins, C.N.; Joppa, L. Expansion of the Global Terrestrial Protected Area System. *Biol. Conserv.* **2009**, *142*, 2166–2174. [CrossRef]
55. Cabeza, M.; Arponen, A.; Jäättelä, L.; Kujala, H.; Van Teeffelen, A.; Hanski, I. Conservation Planning with Insects at Three Different Spatial Scales. *Ecography* **2010**, *33*, 54–63. [CrossRef]
56. Rui, H.Y.Y. Study on Boundary Cognizance of Famous Scenic Sites. *Chin. Landsc. Archit.* **2011**, *27*, 56–60.
57. Theberge, J.B. Guidelines to Drawing Ecologically Sound Boundaries for National Parks and Nature Reserves. *Environ. Manag.* **1989**, *13*, 695–702. [CrossRef]
58. Zhou, D.Q.; Edward Grumbine, R. National Parks in China: Experiments with Protecting Nature and Human Livelihoods in Yunnan Province, Peoples' Republic of China (PRC). *Biol. Conserv.* **2011**, *144*, 1314–1321. [CrossRef]
59. Saviano, M.; Di Nauta, P.; Montella, M.M.; Sciarelli, F. Managing Protected Areas as Cultural Landscapes: The Case of the Alta Murgia National Park in Italy. *Land Use Policy* **2018**, *76*, 290–299. [CrossRef]
60. Sriarkarin, S.; Lee, C.-H. Integrating Multiple Attributes for Sustainable Development in a National Park. *Tour. Manag. Perspect.* **2018**, *28*, 113–125. [CrossRef]
61. Habtemariam, B.T.; Fang, Q. Zoning for a Multiple-Use Marine Protected Area Using Spatial Multi-Criteria Analysis: The Case of the Sheik Seid Marine National Park in Eritrea. *Mar. Policy* **2016**, *63*, 135–143. [CrossRef]

Article

The Evolution of Ecological Space in an Urban Agglomeration Based on a Suitability Evaluation and Cellular Automata Simulation

Yipu Chen, Bohong Zheng and Runjiao Liu *

School of Architecture and Art, Central South University, Changsha 410017, China; chenyipu542@163.com (Y.C.); zhengbohong@csu.edu.cn (B.Z.)
* Correspondence: 218042@csu.edu.cn

Abstract: Changing and reconstructing the ecological space of urban agglomerations is inevitable for ecological conservation and a scientific problem that needs urgent attention from geography, ecology, and urban and rural planning. Using ArcGIS and other software for data processing, this study established a spatial attribute database, constructed a land use conversion matrix of the Changsha-Zhuzhou-Xiangtan (CZX) urban agglomeration's ecological space, and quantitatively analyzed the main changes in ecological land. Using a trained cellular automata model with predicted land use in 2035 as the threshold value, the simulation research was presented by creating two simulation scenarios for the spatial distribution of land use by 2035 in the "Green Heart" area of the CZX urban agglomeration. The simulation results were compared, and the constraining role of land use suitability evaluation on ecological space evolution was analyzed. This study found that the total area of ecological space in the Green Heart area saw a rapid reduction, and it predicted that, by 2035, the total area of the CZX Green Heart area will have decreased. Comparing the two simulation scenarios proved the hypothesis that zoning ecological space reconstruction based on a land suitability evaluation can effectively protect ecological space and ensure ecological network functions are harnessed.

Keywords: ecological space; suitability evaluation; Changsha-Zhuhai-Xiangtan urban agglomeration; cellular automata simulation

Citation: Chen, Y.; Zheng, B.; Liu, R. The Evolution of Ecological Space in an Urban Agglomeration Based on a Suitability Evaluation and Cellular Automata Simulation. *Sustainability* **2022**, *14*, 7455. https://doi.org/10.3390/su14127455

Academic Editors: Qingsong He, Jiayu Wu, Chen Zeng and Linzi Zheng

Received: 25 April 2022
Accepted: 15 June 2022
Published: 18 June 2022

1. Introduction

The requirements for economic growth and urban diversity have exerted an impact on increasing pollution and declining environmental quality [1]. As metropolitan regions and urban agglomerations emerge and the spatial patterns evolve, the multi-level correlation and local impact of rapid urbanization have already become one of the most prominent issues in sustainable development worldwide. A series of development and construction activities has contributed to greater erosion of the ecological space. Every metropolitan region or urban agglomeration is confronted with the conflict between ecological conservation and resource development [2,3]. It is a major challenge posed to countries around the world to protect the ecological space between and around cities, rationally optimize the ecological space of urban agglomerations, and build a favorable regional ecological security pattern [4].

Ecological space provides urban and rural ecosystem services, and is indispensable to ensure urban and rural ecological security and improve the quality of people's life [5]. Since the 1980s, clear, unified conclusions or research methods have not been achieved in research on ecological land as the most basic material carrier of ecological space elements. Other countries' ecological land categorizations tend to regard land as a whole and emphasize the natural attributes of land [6]. Many scholars have subsequently studied the connotations of regional ecological land from various perspectives [7–11]. Drawing on the concept and classification standards of Long et al. [11], this study defined regional ecological space as land space, excluding artificially hardened surfaces, with self-adjustment and recovery capabilities, to which other ecosystem services can provide direct or indirect environmental regulation and biological support, including arable land, woodland, grassland, and waterbodies.

Since ecological land is the core foundation that supports ecological space, there must be a guide through effective planning approaches. The protection and control of ecological space has also become the key to national land space planning. There have been attempts at this in major cities across the world from the 19th century to date. For example, in 1876, Frederick Law Olmsted designed the "Emerald Necklace", an urban park and open-space system, which was the first example of resolving urban issues by establishing a way to connect parkways and park green spaces [12,13]. To prevent urban sprawl from erosion to rural and suburban ecological land, the "Green Belt Policy" [14] was launched in the Greater London Plan; actions to protect the "Green Heart" in the Netherlands were proposed in five national spatial plans towards the disorderly expanding Randstad [15]; the urban agglomerations in the Pearl River Delta in Guangdong, China have effective connections to more than 200 forest parks, nature reserves, scenic spots, country parks, etc. through the established greenways [16].

From the perspectives of landscape ecology and geography, most of the current researches on ecological land used geographic information systems (GIS) and remote sensing (RS) to analyze the laws of ecological land change and proposed plans and strategies towards ecological land optimization; in addition, starting from the ecological security pattern, these researches investigated the relationship between ecological security and ecosystem services [17,18]. In recent years, many scholars have begun to use qualitative and quantitative methods, comprehensive models, etc. [19] to investigate the evaluation of ecological land use suitability and ecological service value. Land use suitability evaluation, also known as land ecological suitability evaluation, was first proposed by Ian McHarg who defined land suitability as "the inherent suitability of the land for some specific, persistent uses. This land is determined by such characteristics as hydrology, geography, topography, geology, biology, sociology, etc. [20]." From a variety of methods for land suitability analysis, McHarg's overlay analysis is the most widely used. The sophisticated and popular geographic information system (GIS) has effectively overcome such issues as opacity of multi-map overlay drawings and limited weight of elements and breadth of evaluation factors [21]. Some scholars have based green space classifications and green space system layouts on land-use suitability evaluations, and investigated the temporal and spatial changes in ecological land at different scales from the perspective of data reconstruction [22]. Additionally, with the rapid development of computer technology, it will become a new research trend to use mathematical methods such as CA, ant colony algorithm, grey system theory and Agent for scenario simulations and evaluations of ecological land [23–25].

As for the research on ecological land in China, the existing studies mainly concentrated on the economically developed coastal areas, such as the Beijing-Tianjin-Hebei region [26], Tianjin Binhai New Area [27], and the city of Shenzhen [28]. In contrast, other regional ecological spaces, such as inland urban agglomerations, have received limited attention from scholars. The Green Heart project in the Changsha-Zhuzhou-Xiangtan (CZX) urban agglomeration is currently the only large-scale Green Heart area in an urban agglomeration in Chinas; with an area three times that of the Green Heart area in the Randstad urban agglomeration in the Netherlands (approximately 150 km^2), it is the largest urban agglomeration in the world [29]. Over the years, the rapid expansion of construction land in Changsha, Zhuzhou, and Xiangtan has posed major threats to the ecological environment in the Green Heart area, where the ecological land has been invaded continuously. The protection plans for the Green Heart area developed by the government departments to protect its ecological environment are human-oriented land management schemes using primarily qualitative analytical methods. These protection plans are thus too subjective to fully reflect the objective laws of conversion between ecological land and construction land, making them less relevant and effective in protecting the ecological land in the Green Heart area. It is urgent for us to update the original planning techniques. Therefore, in addition to the conventional qualitative analytical methods, more objective quantitative analytical methods were introduced. Specifically, GIS was applied for spatial overlay analysis of remote sensing image data in different years, with a summary of the characteristics about the changing quantity and space of ecological land in the Green Heart area. Meanwhile, using an ecological land conversion matrix, the study analyzed the main direction of flow of ecological land in the CZX Green Heart area. Based on a land-use suitability evaluation as well as using the cellular automata (CA) model for simulation prediction of ecological space change characteristics of the CZX Green Heart area in 2035 under the different scenarios, the study also leveraged a comprehensive comparison and evaluation of the simulation results to investigate the key factors affecting the ecological land change in the Green Heart area. In this way, more relevant and feasible ecological space protection measures can be developed to provide a strong guarantee for the sustainable development and ecological land protection of the CZX urban agglomeration.

2. Study Area Overview and Data Sources

2.1. Study Area Overview

The concept for the CZX urban agglomeration Green Heart area was formally proposed in 2003. As shown in Figure 1, the Overall Plan for the Ecological Green Heart Area of the Changsha-Zhuzhou-Xiangtan Urban Agglomeration formulated in 2010 stated that the project would encompass 17 towns and townships, 4 sub-district offices, 124 administrative villages (neighborhood committees), and 638 settlements across the urban agglomeration's three cities, covering a total area of 522.87 km^2. The Green Heart area has a subtropical monsoon climate, with hot summers and cold winters, and four distinct seasons. It contains many rivers and lakes and thus has abundant water resources. However, due to poor overall circulation and uneven rainfall distribution, floods, and droughts frequently occur in summer [30].

Figure 1. Map of the Green Heart Area.

2.2. Data Sources

Land-use data for the CZX urban agglomeration for the five years of 1980, 1990, 2000, 2010, and 2017 published by the Geographical Information Monitoring Cloud Platform were used as the research data, with a resolution of 500 m × 500 m.

China's national land-use data product released by the Geographical Information Monitoring Cloud Platform uses Landsat TM/ETM/OLI remote sensing images as the main data source. After image fusion, geometric correction, image enhancement and stitching, the national land-use types are divided into six first-order categories, 25 s-order categories, and some third-order categories (as shown in Table 1). With a relatively mature classification system, this data product has been approved by Chinese competent authorities and widely used by local government agencies. To maintain a scientific and consistent study, the land-use classification standard of the platform was applied in the present study instead of using a new classification standard.

Table 1. Land use categories (http://www.dsac.cn/DataProduct/Detail/200804, accessed on 12 February 2022).

First-Order Categories		Second-Order Categories	
No.	**Name**	**No.**	**Name**
1	Arable land	11 12	Paddy field Dryland
2	Forestland	21 22 23 24	Woodland Shrubland Sparse woodland Other woodland
3	Grassland	31 32 33	High-coverage grassland Moderate-coverage grassland Low-coverage grassland
4	Waterbody	41 42 43 44 45 46	Canal Lake Reservoir Permanent glacier Tidal flat Flood land
5	Urban and rural, industrial and mining, residential land	51 52 53	Urban land Rural settlement Other construction land
6	Unused land	61 62 63 64 65 66 67	Sandy land Desert Saline-alkaline land Marshland Bare land Exposed rock Other

3. Methodology

3.1. Research Approach

Figure 2 shows the research approach of this study. The first step was to collect relevant data from field inspections, use ArcGIS software to process them, establish a spatial attribute database, and use 500 m × 500 m grids as the basic units to evaluate land suitability. The second step involved using ArcGIS software tools (Esri Corporation, Redlands, CA, USA) to conduct an overlay analysis of five-year ecological spatial thematic maps. More specifically, a series of dynamic evolution conversion matrices of ecological space for four periods (including 1980–1990, 1990–2000, 2000–2010, and 2010–2017) were constructed. Third, the threshold values predicted by a trained CA model was used to simulate the spatial distribution of land use in the Green Heart area in 2035. After that, this study established different simulation scenarios, compared the simulation results, and analyzed the constraining effect of land suitability evaluation on the evolution of ecological space.

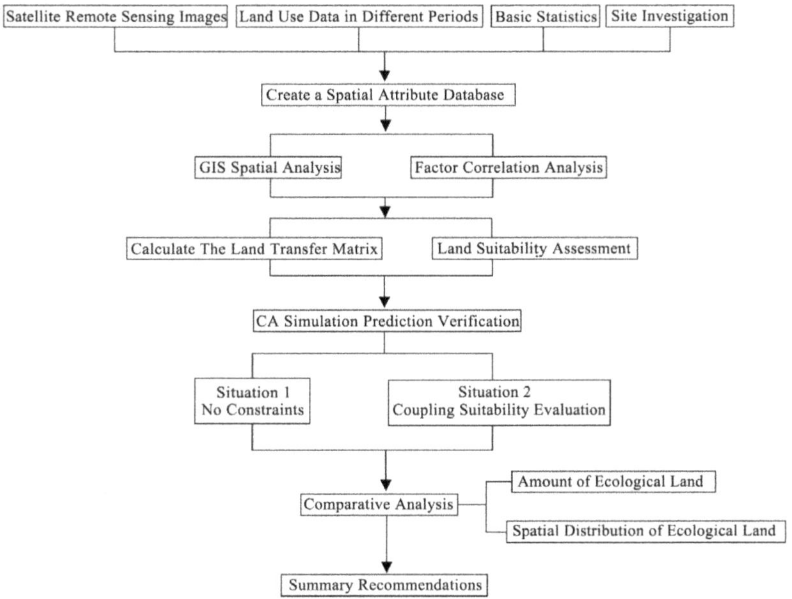

Figure 2. Research Process.

3.2. Ecological Land Use Conversion Matrix

A land-use conversion matrix was used to describe quantitatively the circulation state of a system in system analysis. Moreover, a land-use matrix can be used to directly reflect the trends and quantities of land circulated in a specific period. Based on its principles and methods, a series of ecological land conversion matrices that reflect regional ecological changes were established, including a conversion-in matrix and a conversion-out matrix. The conversion-in matrix represents the conversion from non-ecological land to ecological land, reflecting an increase and the source of ecological land. The conversion-out matrix represents the conversion from ecological land to non-ecological land, reflecting a decrease in and the direction of the flow of ecological land. This is calculated as follows:

$$A_{ij} = \begin{bmatrix} A_{11} & \cdots & A_{1n} \\ \vdots & \ddots & \vdots \\ A_{n1} & \cdots & A_{nn} \end{bmatrix} \quad (1)$$

where n is the total number of ecological land types, i and j are the pre-conversion and post-conversion land categories, respectively, and A_{ij} is the area of converted land.

3.3. Dynamic Measurement of Ecological Space

The total area and dynamic measurement of ecological space are the main indicators for measuring the characteristics of ecological space. The dynamic measurement of ecological space (R_d) refers to the speed of change of ecological space in a certain period and is used to describe the dynamic change of ecological space [31]. The calculation formula is as follows:

$$R_{\rm d} = (U_b - U_a) \div U_a \times \frac{1}{T} \times 100\% \quad (2)$$

where U_a and U_b are the total areas (km^2) of ecological space at the beginning and end of the study period (T), respectively.

3.4. Evaluating Land Suitability

The following principles were adopted for the indicators of this evaluation process: (1) factors with a dominant role; (2) factors with obvious differences and critical values; (3) relatively stable evaluation factors; (4) measurable factors; and (5) relatively independent factors. This study used an expert scoring method to determine the importance of comparison and preliminary scoring of different indicators and then employed the analytic hierarchy process software (YAAHP) to construct an index system and evaluation index judgment matrix. After that, the weights of the evaluation factors were subsequently obtained through the consistency test. Data standardization and uniform value ranges revealed that the estimated values used for the assumed factors were dimensionless. The evaluation values were 0–10. The higher the evaluation value, the better the suitability and the greater the benefit to development and construction.

Referring to the stipulations in the Food and Agriculture Organization's *A Framework for Land Evaluation* [32], a land-use evaluation was carried out on all aspects of the evaluation units based on the suitability of land use, the strength of restrictions, and the locational benefits to obtain an overall score for each grid. After data processing, the final evaluation results were determined. For each delineated spatial unit, the standardized score for each factor was multiplied by the weight, and was added together to determine the corresponding overall evaluation value.

$$K = \sum_{i=1}^{n} x^i a^i \quad (3)$$

where K represents the comprehensive score of each evaluation unit, i represents the i-th evaluation factor, x^i is the value of the i-th factor, and a^i represents the weight of the i-th factor.

In this study, the main influencing factors were divided into terrain, transportation, urban and rural construction, and land use obstruction factors. For terrain factors, the three most representative indicators, namely the slope, elevation, and water system, were selected. Generally, the greater the slope and the higher the elevation, the poorer the land-use suitability and the lower the score [33]. Different slopes have different water and heat conditions and physical and chemical soil properties due to different solar irradiance and sunshine hours. As a result of different incident angles of the sun, hillside slopes receive different solar radiation; thus, their temperature and other ecological factors are constantly changing. Altitude is one of the most relevant factors for changes in mountain topography. The transportation factors were divided into two categories: external transportation and internal transportation. Internal transportation is an important locational benefit, whereas external transportation determines the isolation effect. Accordingly, the two categories were separated. Urban and rural construction factors mainly consider the impact of current construction on the land and inherent locational benefits [33].

These factors consider the impact of different land-use types and current construction intensity. The closer the current centralized construction land is to concentrated construction land and the higher the construction intensity, the better the construction conditions and the higher the score. The land-use obstruction factors were divided into the ecological conservation indicator and the natural and human resources impact indicator. Natural and human resources include many important natural and human landscapes, such as nature reserves, forest parks, and scenic spots, which have good economic attractiveness and development advantages. However, if they are too close, it impacts and undermines related resources. The ecological conservation indicator refers to the ecological sensitivity of different vegetation zones in the Green Heart area and divides them into low-sensitivity areas, ordinary-sensitivity areas, moderate-sensitivity areas, high-sensitivity areas, and extremely high-sensitivity areas [34]. The greater the ecological sensitivity, the more fragile the ecological environment, and the worse the construction conditions, and vice versa.

3.5. Cellular Automata Model

3.5.1. Operating Principle

The CA model is a network dynamics model proposed by Von Neumann in the 1940s to simulate discrete network interactions and causal relationships in time and space through simple local operations. It is also a type of parallel computing model [35]. The standard CA model determines the conversion of the state of the central cell based on the state of its neighboring cells, which is expressed in the following mathematical formula [36]:

$$S^{t+1}(x,y) = f(S^t(x,y), N)$$

The entire cell space is the distribution of state set S at time t in a one-dimensional space. In the above formula, S^{t+1} represents the state of the cell at the next moment, S^t the state of the cell at time t, N the overall state of the cells neighboring the target point, and f the conversion rule function. Not only does the expansion of land use depend on the expansion of the state, but it also controls the simulation process by adding some constraints to the CA model, thereby producing a sustainable and reasonable layout of land-use expansion.

Based on the suitability evaluation results, the range of alternative construction land was determined, and a CA model was constructed by taking the total amounts of actual construction land in 2018 and 2019 as the threshold values for the expansion of land in the two simulations before and after those dates. Taking the grid covering current construction land as the initial state, several rounds of cell state change iterations (state change from non-construction land to construction land) were performed until the total amount of construction land reached the total construction scale in the specific year, at which point the simulation ended.

The details of how the CA model works at different stages and the logical decision rules are illustrated below as well as in Figures 3 and 4.

Step 1 used the 3 × 3 Moore model as the cellular neighborhood model and obtained all non-prohibited grid attribute tables with five attribute values as the T_0 state of the land boundary.

Step 2 input the T_0 state into the recursive function of the CA model and calculated the new evaluation value of each non-construction land grid based on the conversion rule, replacing the original evaluation value in the T_0 state. The conversion rules are as follows:

$$value_{new} = 0.4 * \sum_{rid} value_{rid} * type_{rid} + 0.6 * value_{before} \quad (4)$$

The new suitability evaluation value of the central grid was calculated as the sum of the products of value and type of the surrounding adjacent grids and its own original suitability evaluation value.

Step 3 adjusted the internal parameter v of the CA and selected different growth rates (non-construction land grids were sorted, from large to small in order of their new evaluation values, the percentage was set, and the part of the grid with larger evaluation values became construction land, that is, the type value in the T_0 state was changed from 0 to 1).

Step 4 recursed once to arrive at the land boundary growth result in the TT_1 state.

Step 5 counted the total number of construction land grids and determined whether the upper limit of construction land for the preset period had been reached. When it was reached, the process proceeded to the next step; otherwise, it returned to Step 2 and input T_1 to continue the recursion.

Step 6 output land-use boundary growth result T_n, where n is the number of recursions, both the value and type changed, and finally, land-use growth results under different growth rates were obtained.

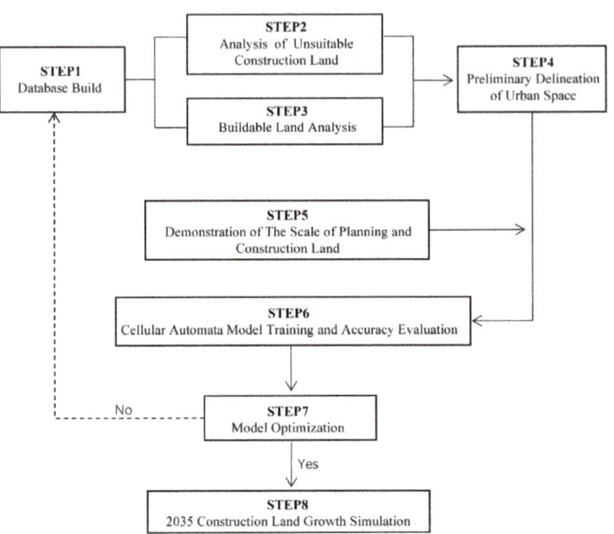

Figure 3. Technology Road Map of Cellular Automata Model Simulation.

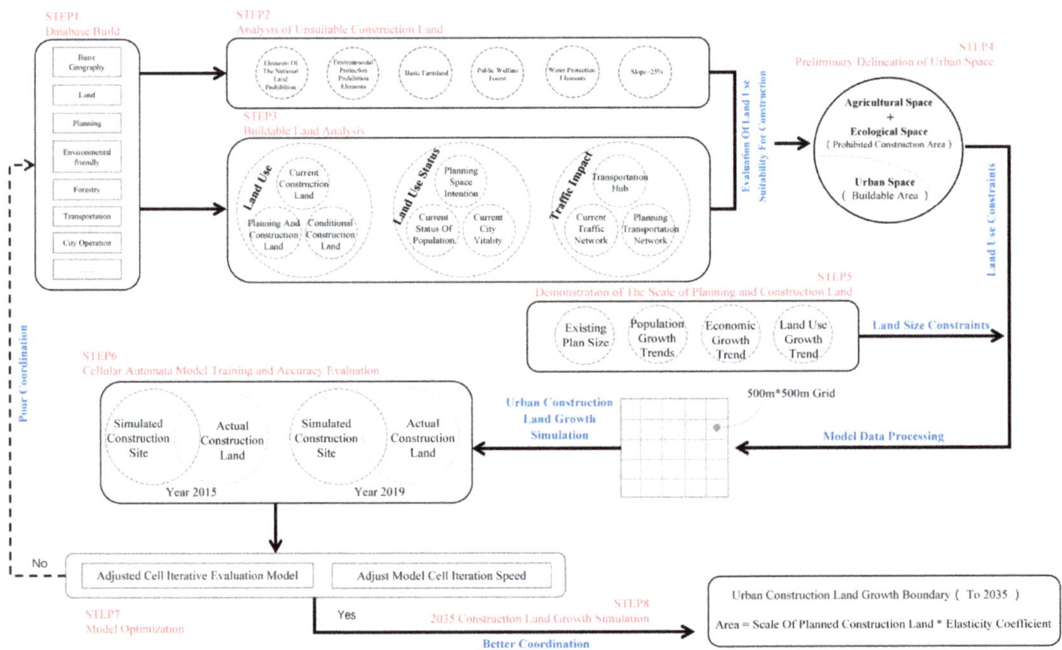

Figure 4. Flow Chart of the Cellular Automata Model Simulation Process.

3.5.2. Parameter Optimization

In the CA model, the iteration velocity parameter v was adjusted to obtain multiple sets of results. Under the models with different parameters, the number of cell transitions from non-construction land to construction land had a strong positive correlation with the average land suitability evaluation value of the cell set with the same number of changes. This demonstrated that the higher the average land suitability evaluation value, the more frequent cell state changes. As shown in Figure 5, the effect of the experimental data using 4 sets of different iteration velocity parameters was significantly lower than six and eight sets of parameters. Therefore, in order to obtain a better simulation result, this study decided to use the inflection point value of the accuracy evaluation result in the experiment of eight sets of iteration velocity parameters, as the optimal parameter value.

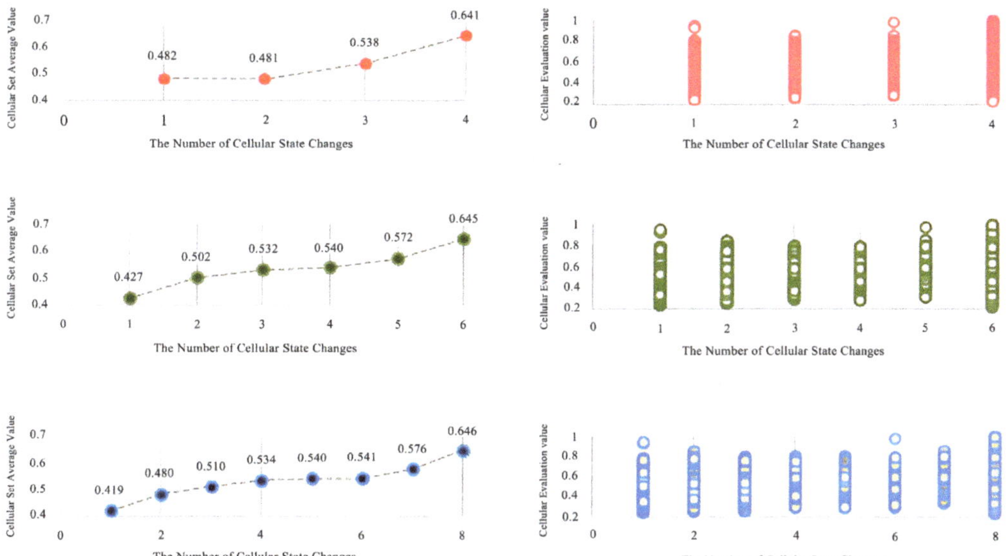

Figure 5. Comparison of Model under Different Parameters.

4. Results

4.1. Land-Use Suitability Evaluation Results

The second-order land suitability evaluation factors given in Table 2 above were used in the GIS database with Euclidean distance and then reclassified. Subsequently, all elements were unified into standardized value ranges according to the evaluation criteria to obtain single-factor evaluation results for all the factors (Figure 6). Finally, the normalized score of each factor was multiplied by its previously obtained weight and added together to obtain the final overall evaluation result. As shown in Table 3 and Figure 7, based on the overall suitability evaluation results, the overall score for the CZX Green Heart area was 0.1043–0.5514, which can be divided into five intervals. The total area of land with scores of 0.1043–0.1367 was 91.08 km^2, accounting for 17.42% of the total area; 188.10 km^2 of land had scores of 0.1367–0.1465, accounting for 35.98% of the total; 169.62 km^2 of land had scores of 0.1465–0.1537, accounting for 32.44% of the total; 48.06 km^2 of land had scores of 0.1537–0.1727, accounting for 9.19% of the total; and 26.02 km^2 of land had scores of 0.1727–0.5514, accounting for 4.98% of the total land area.

Table 2. Statistical table of land adaptability evaluation factors in the Green Heart area.

First-Order Factor	Weight	Second-Order Factor	Weight	Classification Standards	Normalized Value Range
Terrain factors	0.3068	Slope indicator	0.1655	0–2°	10
				2–7°	7
				7–15°	4
				15–25°	1
				>25°	0
		Elevation indicator	0.0911	35–55 m	10
				55–75 m	7
				75–100 m	4
				100–150 m	1
				>150 m and <35 m	0
		Water system indicator	0.0501	<50 m	0
				50–100 m	7
				100–500 m	10
				500–1000 m	4
				>1000 m	1
Transportation factors	0.1752	External transportation indicator	0.0584	<100 m	0
				100–200 m	1
				200–500 m	4
				500–1000 m	10
				>1000 m	7
		Internal transportation indicator	0.1168	<200 m	10
				200–500 m	7
				500–1000 m	4
				1000–2000 m	1
				>2000 m	0
Urban-rural construction factor	0.1346	Current construction area impact indicator	0.1346	<100 m	10
				100–200 m	7
				200–500 m	4
				500–1000 m	1
				>1000 m	0
Land use obstruction factors	0.3835	Ecological conservation index	0.2876	low-sensitivity area	10
				ordinary-sensitivity area	7
				moderate-sensitivity area	4
				high-sensitivity area	1
				extremely high-sensitivity area	0
		Natural and human resources impact indicator	0.0959	<100 m	0
				100–300 m	4
				300–500 m	7
				500–1000 m	10
				>1000 m	1

Table 3. Land area of the various score ranges from the evaluation of land suitability for the CZX Green Heart area.

Score Range	Area (km^2)	Percentage
0.1043–0.1367	91.08	17.42%
0.1367–0.1465	188.10	35.98%
0.1465–0.1537	169.62	32.44%
0.1537–0.1727	48.04	9.19%
0.1727–0.5514	26.02	4.98%
Total	522.87	100.00%

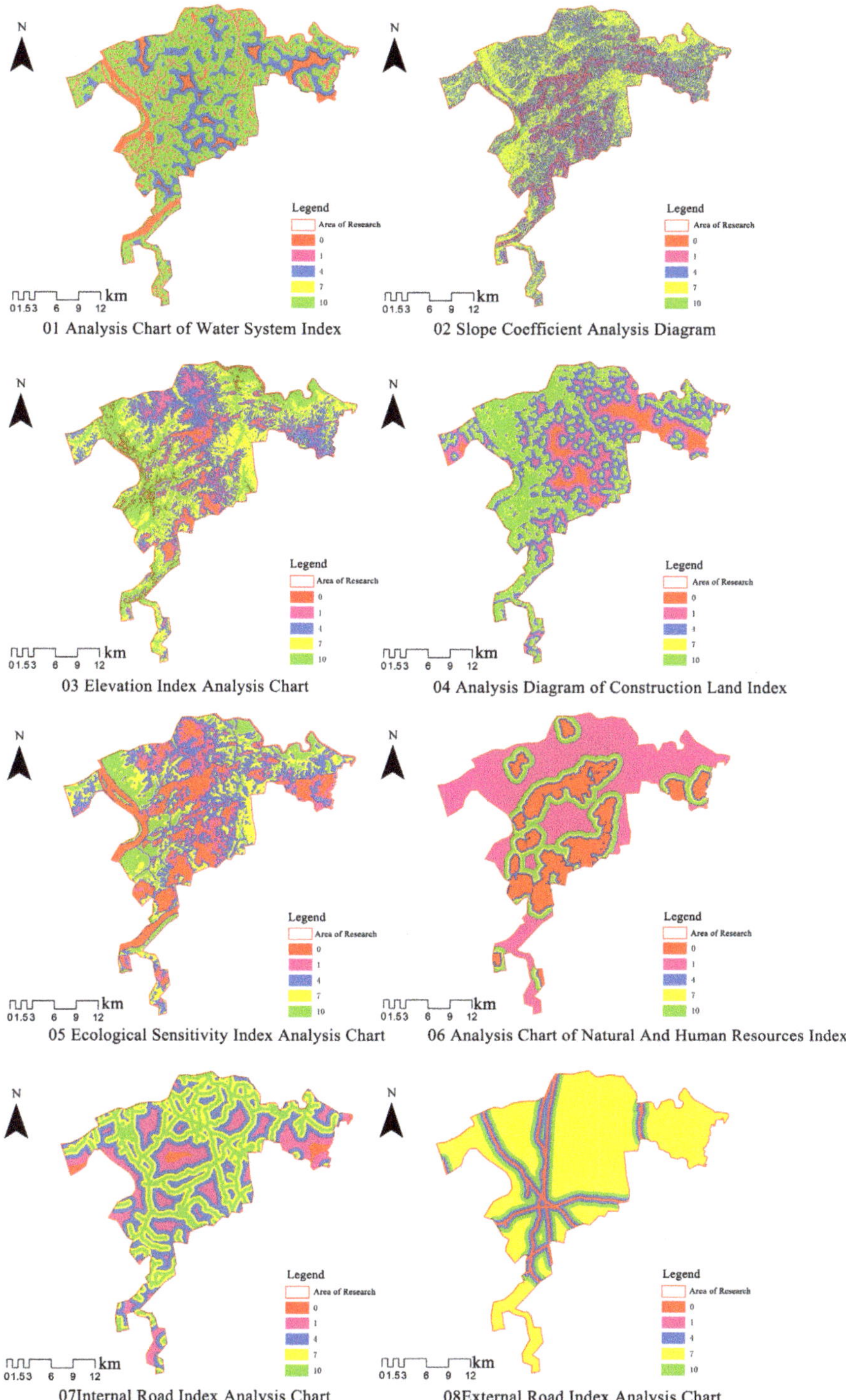

Figure 6. Analysis of factors for evaluating land suitability in the CZX Green Heart area.

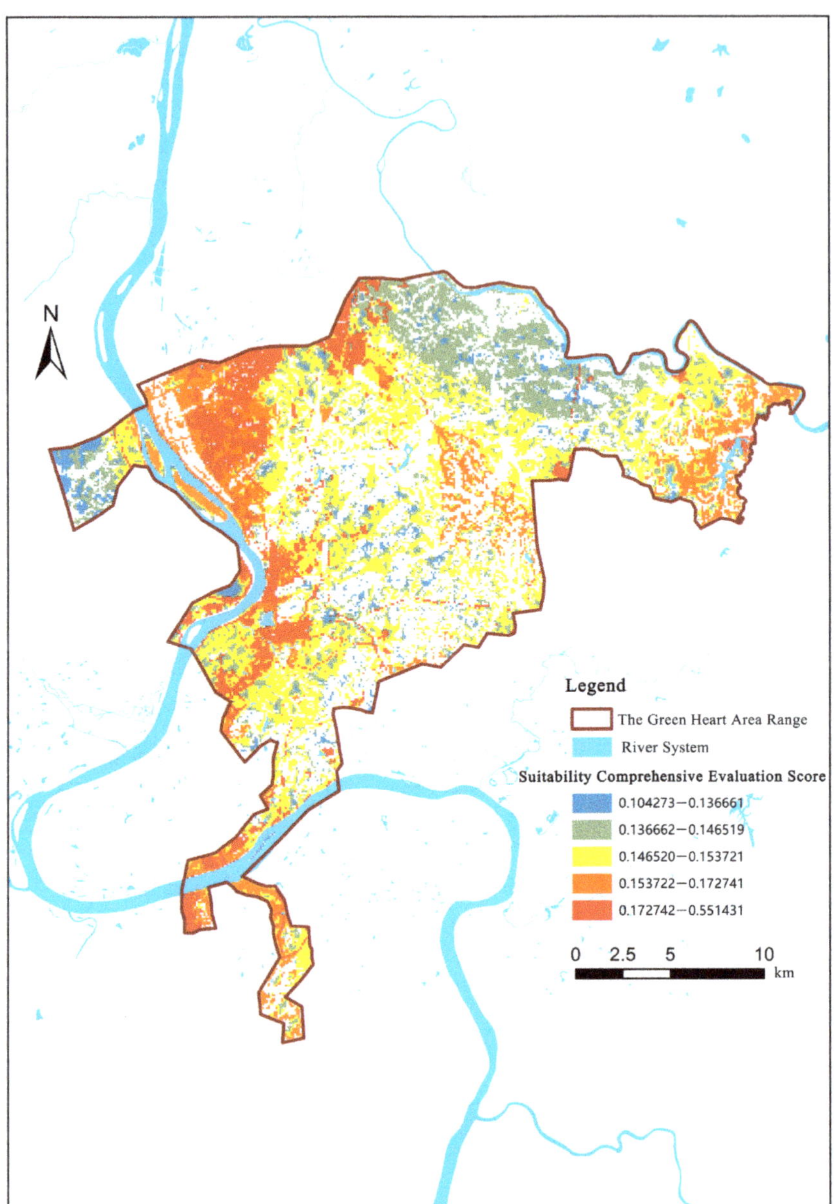

Figure 7. Ecological space suitability results for the CZX Green Heart area.

4.2. Summary of Current Land-Use Conversion Characteristics

Tables 4–6 show that between 1980 and 2017, the conversion-in area of ecological space in the study area was only 0.76 km^2, the conversion-out area of ecological space was 49.81 km^2, and the mutual conversion area of ecological space was 9.64 km^2. The conversion-out area of ecological space was far greater than the conversion-in area; thus, the total area of ecological space continuously shrank.

Table 4. Land conversion matrix in the CZX Green Heart area 1980–2017 (unit: km^2).

1980 / 2017	Grass-Land	Urban and Rural Construction Land	Arable Land	Forest-Land	Water-Body	Unused Land	2017 Total
Grassland	4.50			0.04			4.54
Urban and rural construction land	0.50	800	22.08	26.22	1.00		57.81
Arable land		0.01	160.93	1.95	0.79		163.70
Forestland	0.54		3.14	262.91	0.06	0.75	267.39
Waterbody			2.29	0.81	24.09		27.20
Unused land			1.25	1.00			2.25
1980 total	5.54	8.01	189.44	293.18	25.95	0.75	522.87

Table 5. Land conversion in the CZX Green Heart area 1980–2017 (unit: km^2).

Initial Categories	Conversion Categories					
	Grass-Land	Urban and Rural Construction Land	Arable Land	Forestland	Water Body	Unused Land
Grassland		0.50	0	0.54	0	0
Urban and rural construction land	0		0.01	0.25	0	0
Arable land	0	22.08		2.89	2.29	1.25
Forestland	0.04	26.22	1.95		0.81	1.00
Waterbody	0	1.00	0.79	0.06		0
Unused land	0	0	0	0.75	0	

Table 6. Total area, change, and dynamic measurement of ecological space in the CZX Green Heart area 1980–2017.

Year	Total Area (km^2)	Change in Area (km^2)	Dynamic Measurement (%)
1980	514.11		
1990	511.12	−2.99	−0.0581
2000	507.14	−3.98	−0.0779
2010	489.63	−17.50	−0.3451
2017	464.30	−25.33	−0.5174

The trend shown in Tables 4 and 5 and Figure 8 indicates that the total area of ecological space in the study area decreased annually by 1.84 km^2, decreasing by 49.81 km^2 between 1980 and 2017 in total. This change can be divided into two stages. The period from 1980 to 2000 was a period of slow reduction. During this period, the total area of ecological space decreased relatively slowly, with a total reduction of 6.97 km^2, an average annual decrease of less than 0.35 km^2. The period from 2000 to 2017 was a period of rapid reduction. During this period, the total area of ecological space declined by 46.82 km^2, with an average annual rate of 2.75 km^2, nearly eight-fold higher than that of the previous 20 years. With regard to dynamic measurement, the degree of change increased sharply from 2000, and it was highest after 2010, reaching −0.5174%. The reduction in ecological land was concentrated in the Xiangjiang River area, mainly for new urban construction land and township construction land.

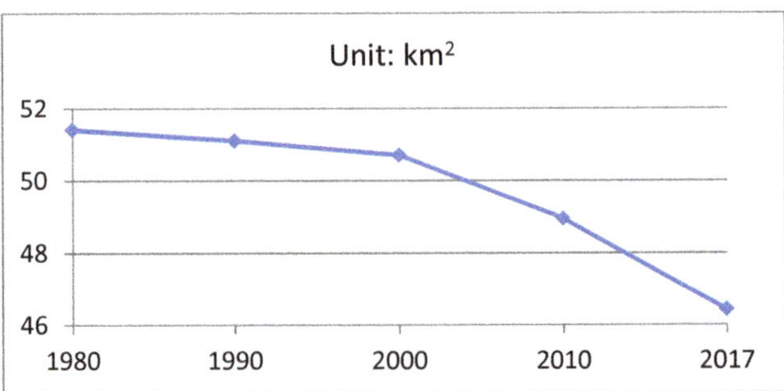

Figure 8. Change in ecological space of the CZX Green Heart area 1980–2017.

From 1980 to 2017, there were notable differences in the conversion to non-ecological uses of the various land-use types, although most of the land was used as urban and rural construction land after conversion. Arable land and forestland were the main land-use types to be converted, followed by grassland and water bodies, with 26.22 km^2 of forestland and 22.08 km^2 of arable land converted (Figure 9), accounting for 52.65% and 44.33% of the total converted land, respectively. The study results indicate that the conversion of ecological space to construction land is highly common, mainly for urban development and transportation infrastructure.

There are three main reasons for this. One is intensive human activities in the CZX metropolitan area. To meet human working and living requirements, the size of the central urban area has expanded continuously, which has inevitably led to the conversion of ecological space into urban and rural construction land. Between 2000 and 2010, the integration process of the three cities of the CZX agglomeration continued to develop, greatly affecting the Green Heart area. From 2010 to 2017, due to the government's policies to further encourage the integrated development of the three cities, their integration gathered pace. Governments at all levels in the three cities seized the development opportunities and accelerated the development trend for their own benefit. The Green Heart area, located in the center of the three cities of Changsha, Zhuzhou, and Xiangtan, which occupies a large area of land, has become a bonus area of the integrated development process. Second, the local economic benefits derived from construction land are far greater than those from ecological land, especially forestland. As the Chinese state has strengthened efforts to protect arable land, forestland has become the primary choice for conversion to construction land, which is a fundamental reason why forestland is the most converted land type. Third, policies to guide and protect the Green Heart area conflict with the urban master planning of the three municipal governments. In this study, the GIS software was applied for stitching, overlay, and comparison of Green Heart planning [37] and the construction land plan in the urban master planning of Changsha, Zhuzhou, and Xiangtan [38–40]. It revealed many inconsistencies between the Green Heart planning and the urban master planning, leaving much of the ecological land in the Green Heart area occupied for construction (as shown in Figure 10). This was mainly because the protection plans for the CZX Green Heart area were prepared by the Hunan Provincial People's Government, who aimed to take account of the coordinated development of the CZX urban agglomeration as a whole, regarding the Green Heart area as the ecological isolation area of the urban agglomeration while keeping up with the trajectory of ecological conservation; the main objective for the local governments was development-oriented, as local governments treated the Green Heart area as a backup area for urban development and construction. Meanwhile, most of the construction land arranged by the three cities within the Green Heart conflicts with the protection plans for the Green Heart area, making it difficult to obtain the approval of the government at a higher level (the provincial government). As a result, many construction projects never started.

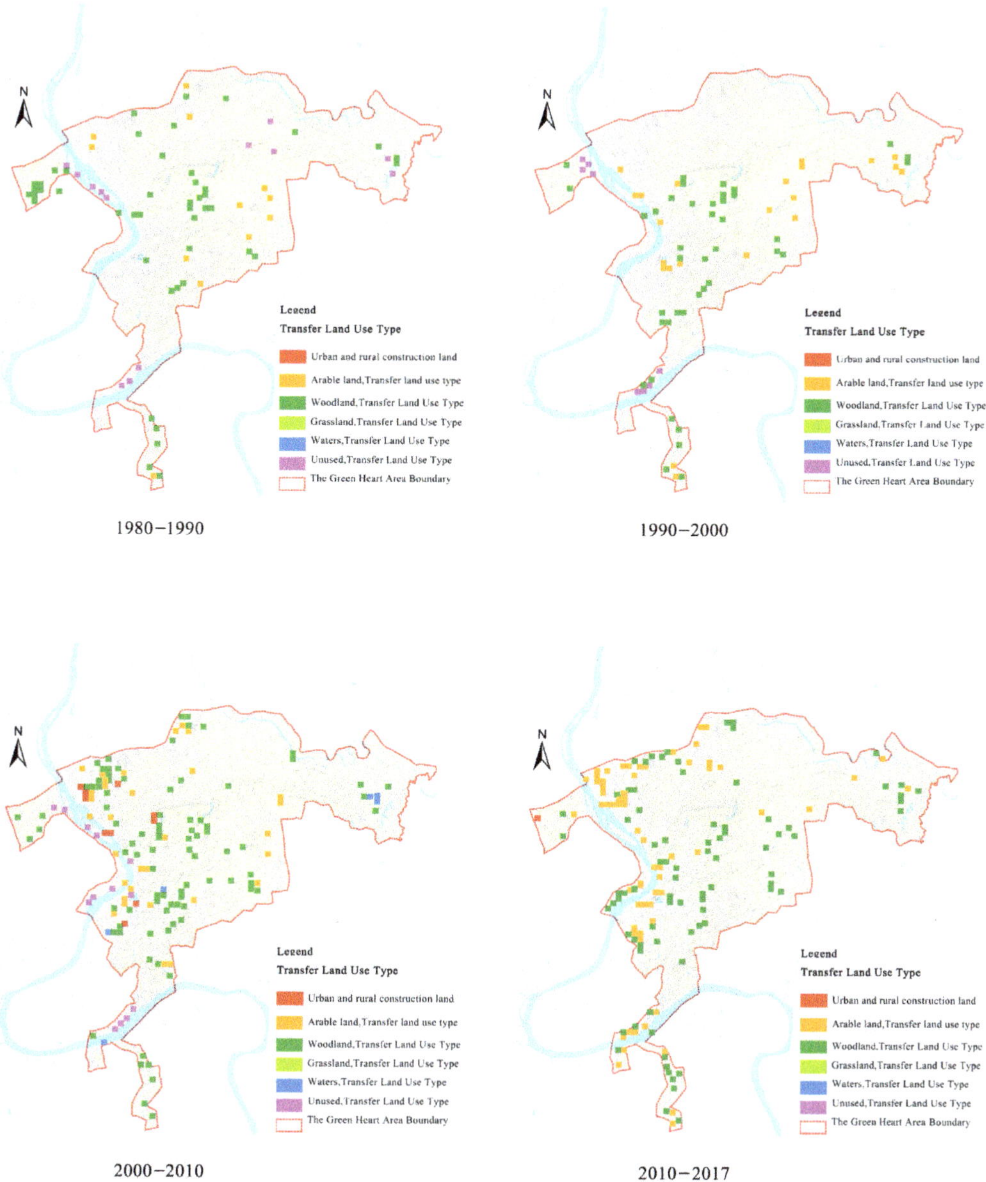

Figure 9. Map of the conversion of ecological space in the CZX Green Heart area 1980–2017.

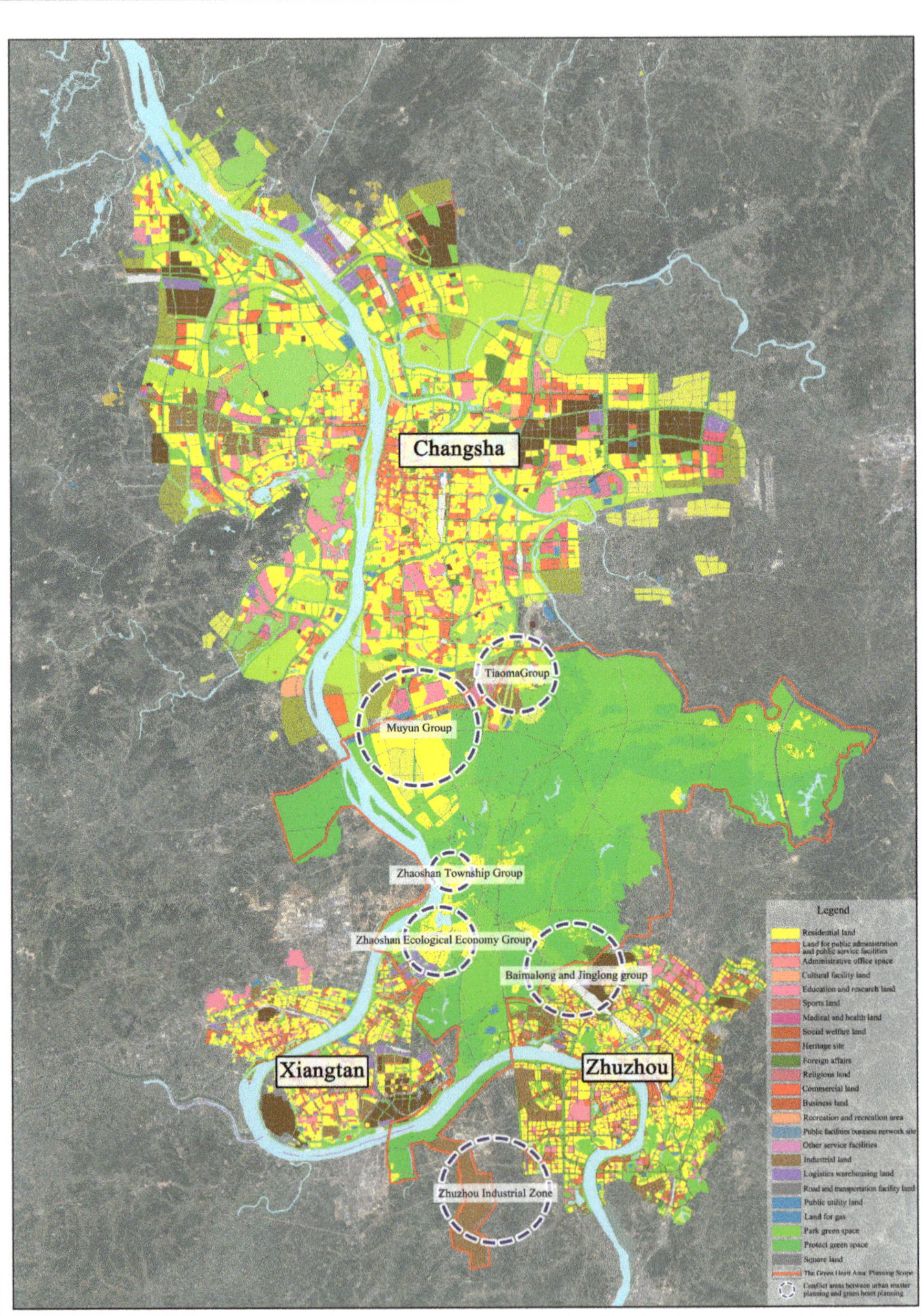

Figure 10. Comparison of Conflicts between Master Planning of Changsha, Zhuzhou, and Xiangtan and Planned Range of Green Heart [37–40].

4.3. Simulation Accuracy Analysis

For a comprehensive research, this study combined the point-by-point comparison method and Kappa coefficient for calculation to verify the accuracy of the simulation experiment. The point-by-point comparison method focuses on the consistency of each cell scale, whereas the overall morphological method pays more attention to the consistency of the overall layout. From the comparison results, the point-by-point comparison accuracy simulated in the model experiment reached 89.1%, and the Kappa coefficient was 0.59, which indicates that the model achieved good results. Table 7 shows the confusion matrix for the evaluation of the simulation accuracy. In addition, the Moran's I index, which reflects the overall shape of the city, was calculated, and the value was 0.477 in 2019, the value of actual urban land in 2018 was 0.479, and the value of the simulated urban land in 2018 was 0.481. This indicates that the overall form produced by the simulation was highly similar to the actual form.

Table 7. Confusion matrix for evaluating simulation accuracy.

		2018 Simulation			2019 Simulation		
		Not Converted	**Converted**	**Precision**	**Not Converted**	**Converted**	**Precision**
Actual land use	Not converted	1375	128	91.5%	1275	161	88.8%
	Converted	98	491	83.3%	98	558	85.0%
	Total accuracy			89.1%			87.6%
	Kappa			0.59			0.56

4.4. Simulation Prediction of Future Development

Based on the number and pattern of ecological green land in the CZX urban agglomeration in 2017, this study fitted the construction land data over the years to predict construction land in 2035. It used the trained CA model to obtain the predicted value of the land in 2035 as the threshold and simulated the spatial distribution of land use in the Green Heart area in 2035, creating two simulation scenarios to obtain the corresponding spatial evolution patterns. Then, the land use simulation results were compared under the two scenarios to verify whether the zoning reconstruction strategy based on the suitability evaluation results could play a protective role in the development of ecological space. Scenario 1: No constraints were set. Under the condition of no control, the guiding and controlling role of the government was minimized. Under this scenario, the only constraints were terrain factors, whereas the market and capital were dominant factors in urban development, and economic development was the only driving force of expansion. Scenario 2: The coupling suitability evaluation results were used as constraints. From the perspective of ecological conservation, the CZX Green Heart area was divided into the following five zones: prohibited construction, strictly limited construction, generally restricted construction, basically suitable for construction, and suitable for construction [41] (Table 8 and Figure 11). Based on the zoning in the model, this study then increased corresponding constraints and predicted the future development of land use.

Table 8. Reconstruction and Zoning of Land in the CZX Green Heart Area.

Score Range	Zoning Classification
0.1043–0.1367	Prohibited construction
0.1367–0.1465	Strictly limited construction
0.1465–0.1537	Generally restricted construction
0.1537–0.1727	Basically suitable for construction
0.1727–0.5514	Suitable for construction

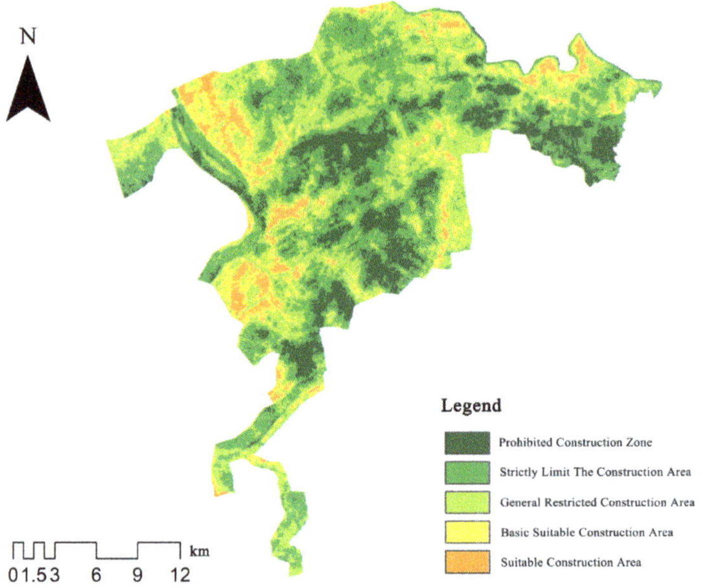

Figure 11. Map of ecological space reconstruction in the CZX Green Heart area [33].

Based on the conversion matrix of land-use types in the Green Heart area from 1980 to 2017 (Table 4), a conversion probability matrix of land-use types was calculated for 1980–2017, and a land-use conversion matrix was calculated for 2017–2035 based on the areas of land-use types in 2017 (Table 9).

Table 9. Land-use conversion probability matrix for the CZX Green Heart area 2017–2035.

Land Use Type		2017					
		Grassland	Urban and Rural Construction Land	Arable Land	Forestland	Water	Unused Land
1980	Grassland	0.813					
	Urban and rural construction land	0.09	0.998	0.117	0.089	0.039	
	Arable land		0.002	0.849	0.008	0.031	
	Forestland	0.097		0.017	0.897	0.002	1
	Water			0.012	0.003	0.929	
	Unused land			0.005	0.003		

Figure 12 and Table 10 show that the total area of ecological space in the CZX Green Heart area will decrease by 2035 under Scenario 1 and Scenario 2, with reductions of 136.22 km^2 and 114.44 km^2, respectively. This includes decreases of 36.33 km^2 (Scenario 1) and 30.77 km^2 (Scenario 2) of forestland; a decrease of 0.11 km^2 of grassland under both scenarios; decreases of 96.78 km^2 and 81 km^2 of arable land; and decreases of 3 km^2 and 2.56 km^2 of waterbodies. Evaluating from the land-use conversion matrix (Table 11) and the results of Simulation Scenario 1, by 2035, nearly 7 km^2 of arable land and more than 30 km^2 of forestland will have been converted to other ecological space, which indicates that occupation pressures on forest spaces will remain high. The main areas where ecological space is occupied are distributed through the west of the Green Heart area and east of the Xiang River. Ecological spaces in Zhuzhou and Xiangtan, in contrast, will not change significantly. According to the analysis of the dynamic measurement of single land uses (Table 12), under the two scenarios from 2017 to 2035, the dynamic measurement of forestland under Scenario 1 is most negative, reaching −0.75%, followed by water bodies, reaching −0.61%. Under Scenario 2, the dynamic measurement of water bodies is most negative, reaching −0.52%, followed by grassland, at −0.13%.

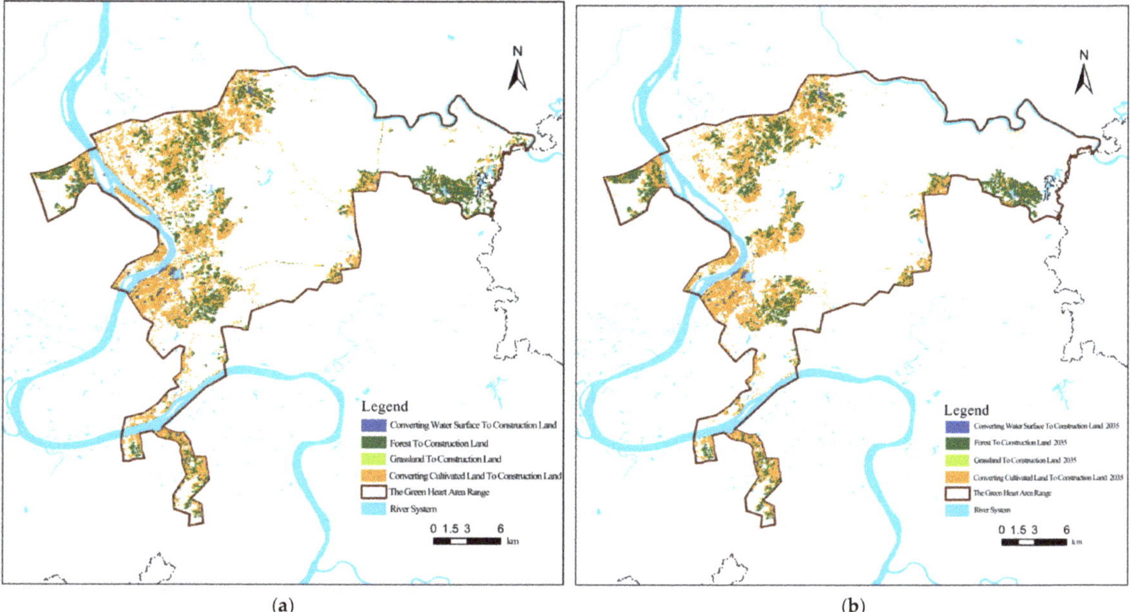

(a) (b)

Figure 12. Comparison of ecological space simulation results for the Green Heart area in 2035. (**a**) Scenario 1 simulation prediction. (**b**) Scenario 2 simulation prediction.

Table 10. Comparison of predicted ecological space in the CZX Green Heart area (2017 and 2035).

Year	Forestland	Grassland	Waterbody	Arable Land
2017	267.39	4.54	27.20	163.70
2035 Scenario 1	231.06	4.43	24.20	156.92
2035 Scenario 2	264.32	4.43	24.64	162.33

Table 11. Land-use conversion matrix for the CZX Green Heart area 2017–2035 (unit: km^2).

	Land Use Type	2035				
		Grassland	Urban and Rural Construction Land	Arable Land	Forestland	Water
2017	Grassland	3.69				
	Arable land		0.33	138.98	1.31	5.07
	Forestland	25.94		4.55	239.85	0.53
	Water			0.33	0.08	25.27

Table 12. Dynamic measurements of ecological space change in the CZX Green Heart area under the two scenarios (2017–2035).

Land USE type	Scenario 1 Dynamic Measurement	Scenario 2
Grassland	−0.0013	−0.0013
Arable land	−0.0023	−0.0005
Forestland	−0.0075	−0.0006
Waterbody	−0.0061	−0.0052

In terms of the pattern of ecological space, it does not change greatly compared with 2017. Forestland still accounts for the majority and is relatively concentrated. Comparing the two simulation prediction scenarios showed that the reduction in the area of ecological space after setting ecological reconstruction constraints in Scenario 2 was much smaller than that in Scenario 1, with the change in forestland kept to within approximately 3 km^2 and the change in arable land within 1.5 km^2. Therefore, zoning of ecological space reconstruction based on the land suitability evaluation can effectively protect ecological space and ensure that ecological network functions are harnessed effectively. There is an overall increase in the area of construction land; under Scenario 1, the main areas at risk of urbanization in the CZX urban agglomeration are concentrated in the suburbs of Changsha, Zhuzhou, and Xiangtan, as well as the junctions of the three cities, with each city cluster becoming increasingly integrated. There is a notable trend of urban integration caused by the rapid expansion of city clusters in all directions along major transportation arteries [4]. Scenario 2 is better at maintaining the connectivity of natural ecosystems, preventing disorderly urban sprawl, and shaping a positive pattern of urban-rural development.

5. Discussion

5.1. Characteristics of Ecological Land Change

Between 1980 and 2017, the area of land converted from ecological space to non-ecological space in the CZX Green Heart area was much larger than the growth in ecological space, reaching 49.81 km^2. Between 2000 and 2017, there was a rapid reduction in ecological space, totaling 46.82 km^2, giving an average annual rate of more than 2.75 km^2, nearly eight-fold higher than the previous 20 years. In terms of actual land use after 2000, urban expansion was the main reason for reduced ecological land. With regard to dynamic measurement, the degree of change increased sharply from 2000, and it was highest after 2010, reaching −0.5174%. In terms of land type, the area of converted forestland was the largest, reaching 26.22 km^2, whereas the area of converted arable land was 22.08 km^2, accounting for 52.65% and 44.33% of all converted land, respectively.

By 2035, the total area of ecological space in the CZX Green Heart area is predicted to decrease under both Scenario 1 and Scenario 2. The reduction in the area of ecological space is much smaller under Scenario 2 than that under Scenario 1. The change in forestland is kept within approximately 3 km^2, and the change in arable land is kept within 1.5 km^2. Therefore, the zoning of ecological space reconstruction based on the land suitability evaluation can effectively protect ecological space and ensure that ecological network functions are harnessed effectively.

5.2. Ecological Space Protection Strategy

Previous studies have shown that excessive fragmentation of patches of ecological space is a cause of the deterioration in the overall ecological environment. The elements of ecological space vary in the Green Heart area. While establishing the entire Green Heart space system, these elements must be integrated into a complete system to improve the complex ecosystem, ensure the ecological security of the urban agglomeration, and promote the transformation of industrial ecology. From its inception since 1970s, the Dutch defragmentation program (Meerjarenprogramma Ontsnippering, MJPO) has laid a solid foundation for the construction of a nature network. This policy also set quantitative growth targets and proposed the elimination of obstacles to the connectivity from transport infrastructure to habitat over the years. In 1990, the Dutch government launched the National Ecological Network program to connect individual ecological units into a continuous and complete ecological space subject to zoning control [42]. Although zoning control was also adopted for Amsterdamse Bos, or the Amsterdam Forest, which lies in two municipalities, the Amsterdam Forest was zoned based on a combination of considerations, including the distribution of people residing in the cities and surrounding areas, recreational needs, use intensity, and surveys of background natural resources [43]. The Randstad area has seen constant changes in its urban development structure in the five spatial plans, developing from the initial concept of a single "Green Heart" to a delta "blue-green network", and finally a mature urban network and open space system. At the same time, the core concept of "Green Heart" also follows the requirements of the higher-level planning [44]. Therefore, the approach to protection of the "Green Heart" area in the Randstad is shifting from "absolute protection" to a central open space based on the Green Heart. Similarly, "absolute protection" does not meet the development needs of the CZX urban agglomeration. The present study was directed towards finding a balance between regional development and Green Heart protection.

In view of the above experiences, the study took the perspective of ecological conservation, and based on the comprehensive score results of the suitability evaluation, zoned the CZX ecological Green Heart area for spatial control. This offered the Green Heart area greater room for development and adjustment while protecting its ecological space. In addition, the following guidelines were proposed for the spatial planning of metropolitan regions or urban agglomerations:

(1) Focus on the network construction of ecological spatial planning. With reference to the experience of establishing a regional ecological network system, create connections for elements within the ecological space, provide habitats for wild animals and plants, and consider policies for restoration of habitat fragmentation due to infrastructure, with an emphasis on ecological connections between urban agglomerations.

(2) Focus on spatial differences and diversity. Depending on the characteristics and the nature of different spaces, set development strategies and goals, improve the natural and cultural quality of the spaces, meet different usage requirements, and promote the quality of life for urban residents.

(3) Encourage cooperation among non-governmental organizations and engagement from the public. The prospect of regional development is related to the whole population, involving different authorities and individual participants. Healthy and orderly development in the region can be achieved in the future only with concerted efforts.

(4) Strictly implement policies of ecological conservation and reasonably delineate redlines for ecological conservation. Redlines for ecological conservation refer to the areas with special important ecological functions whose protection must be strictly enforced. The delineation of redlines for ecological conservation is a rigid demand for maintaining ecological security, an important measure to improve ecosystem service functions and build a complete ecological security pattern, and an essential guarantee to contain the destruction of the ecological environment [45].

6. Conclusions

Using ArcGIS and other software for data processing, this study established a spatial attribute database for suitability evaluation, and quantitatively analyzed the main changes in ecological land in Changsha-Zhuzhou-Xiangtan (CZX) Green Heart area by constructing a land use conversion matrix of the ecological space. On this basis, a trained cellular automata (CA) model was used to simulate and predict the spatial distribution of land use by 2035 in the ecological "Green Heart" area under different scenarios. The simulation results were compared, analyzing the constraining role of land use suitability evaluation on ecological space evolution. To sum up, the study had the following findings:

1. For the suitability evaluation, four categories and seven sub-categories of land-use suitability factors were selected, and an index evaluation system was constructed with a combination of the analytic hierarchy process and evaluation. Based on the land-use evaluation, a final comprehensive evaluation result was obtained using weighted indicators, which allowed this study to divide the CZX Green Heart area into five categories: prohibited construction, strictly limited construction, generally restricted construction, basically suitable for construction, and suitable for construction zones. This provides a reliable method for regulating the use of ecological space in urban agglomerations.

2. The land-use conversion matrix quantitatively reflected the main characteristics of land use changes in the CZX urban agglomeration's ecological space. It was a useful

tool for summarizing the main driving factors of ecological land change, thereby further investigating the degree to which the driving factors impacted and damaged the local ecological environment.

3. The comparison of the two results predicted by the trained CA model produced sound results. On the one hand, the CA model combined the data obtained at the local small scale with the neighborhood conversion rules, and it used computers to simulate the dynamic characteristics of the system at a large scale [46]; on the other hand, the CA model effectively simulated and revealed ecological processes through the given special conditions; in addition, its data structure facilitated high integration with systems such as GIS [47].

However, with limited access to data resources and simulation techniques, the present study has shortcomings and requires improvements. This study used different software to separate GIS and CA model, generating potential errors in data conversion between software. Based on GIS with programming function, future research can further develop models to reduce errors in an integrated study. The CA model also has some limitations, including the following: (1) it uses cellular automata for iteration, which consumes much computing time and is impacted by the size of the surrounding neighborhood, especially when the object of study is on a large scale; (2) the temporal and spatial resolution in an ecological spatial pattern analysis has a great influence on the results, so the simulation results are highly dependent on the resolution; (3) the main principle of the CA model is a cell learning the state of the adjacent cells, from which the entire region is deduced, so the focus is more on the interaction between cells in a spatial neighborhood and less on the influence of macroscopic factors on spatial processes [25]. In order to solve the above issues, attempts can be made to improve the cellular automata. On the one hand, its termination conditions can be defined to reduce the number of iterations of the cellular automata and improve its operating efficiency. On the other hand, more accurate multi-level vector data can be used; the construction of a spatial unit system that is hierarchical, nested, and recursive for multi-level urban planning would allow the multi-level spatial control and conduction effects to be applied to the vector cells at different levels to realize the linkage control of land conversion simulation in the whole process and in all aspects [48].

Therefore, besides further in-depth discussions on the socio-economic driving factors affecting ecological space changes, the future study will continue to use updated simulation techniques to analyze the impact of ecological space reconstruction on regional ecological functions, especially the ecological effects of ecological space reconstruction. More typical regions will be selected for comparative simulation experiments, so as to increase the scientific and practical significance of the relevant findings.

Author Contributions: Conceptualization, Y.C. and B.Z.; Data curation, Y.C.; Formal analysis, R.L.; Funding acquisition, R.L.; Investigation, Y.C. and B.Z.; Methodology, Y.C. and B.Z.; Project administration, R.L.; Resources, Y.C.; Software, Y.C. and R.L.; Supervision, B.Z. and R.L.; Validation, Y.C.; Visualization, Y.C.; Writing—original draft, Y.C.; Writing—review & editing, R.L. All authors have read and agreed to the published version of the manuscript.

Funding: This research was funded by National Nature Science Foundation of China, grant number 52008397. The APC was funded by National Nature Science Foundation of China, grant number 52008397.

Institutional Review Board Statement: Not applicable.

Informed Consent Statement: Not applicable.

Data Availability Statement: Not applicable.

Conflicts of Interest: The authors declare no conflict of interest.

References

1. Cervero, R.; Murakami, J. Rail and Property Development in Hong Kong: Experiences and Extensions. *Urban Stud.* **2009**, *46*, 19–43. [CrossRef]
2. Taylor, P.J.; Derudder, B. *World City Network: A Global Urban Analysis*, 2nd ed.; Routledge: London, UK, 2016.
3. Foley, J.A.; Defries, R.; Asner, G.P.; Barford, C.; Bonan, G.; Carpenter, S.R.; Chapin, F.S.; Coe, M.T.; Daily, G.C.; Gibbs, H.K.; et al. Global consequences of land use. *Science* **2005**, *309*, 570–574. [CrossRef] [PubMed]
4. Li, Z.H.; Yu, Y.; Chen, L.; Gan, J.J. Study on ecological space optimization of Chang Zhu Tan urban agglomeration. *Cent. South Univ. For. Technol.* **2019**, *13*, 33–39.
5. Wang, F.Y.; Wang, K.Y.; Chen, T.; Li, P. Research progress and Prospect of urban ecological space. *Prog. Geogr.* **2017**, *36*, 207–218.
6. Tan, Y.Z.; Zhao, Y.; Cao, Y.; He, J. Research Progress on regional ecological land classification in China. *China Land Sci.* **2016**, *30*, 28–36.
7. Ngom, R.; Gosselin, P.; Blais, C. Reduction of disparities in access to green spaces: Their geographic insertion and recreational functions matter. *Appl. Geogr.* **2016**, *66*, 35–51. [CrossRef]
8. Neuenschwander, N.; Hayek, U.W.; Grêt-Regamey, A. Integrating an urban green space typology into procedural 3D visualization for collaborative planning. *Comput. Environ. Urban Syst.* **2014**, *48*, 99–110. [CrossRef]
9. Nutsford, D.; Pearson, A.L.; Kingham, S.; Reitsma, F. Residential exposure to visible blue space (but not green space) associated with lower psychological distress in a capital city. *Health Place* **2016**, *39*, 70–78. [CrossRef]
10. Yu, F.; Li, X.B.; Zhang, L.J.; Xu, W.H.; Fu, R.; Wang, H. Study of ecological land in China: Conception, classification, and spatial-temporal pattern. *Acta Ecol. Sin.* **2015**, *35*, 4931–4943.
11. Long, H.L.; Liu, Y.Q.; Li, T.T.; Wang, J.; Liu, A.X. A primary study on ecological land use classification. *Ecol. Environ. Sci.* **2015**, *24*, 1–7.
12. Erickson, D.L. The Relationship of Historic City Form and Contemporary Greenway Implementation: A Comparison of Milwaukee, Wisconsin (USA) and Ottawa, Ontario (Canada). *Landsc. Urban Plan.* **2004**, *68*, 199–221. [CrossRef]
13. Charles, E.L. *Greenways for America*; The Johns Hopkins University Press: Baltimore, MD, USA, 1995; pp. 1–10.
14. Amati, M.; Yokohari, M. Temporal changes and local variations in the functions of London's green belt. *J. Landsc. Urban Plan.* **2004**, *75*, 125–142. [CrossRef]
15. Kühn, M. Greenbelt and Green Heart: Separating and Integrating Landscapes in European City Regions. *Landsc. Urban Plan.* **2003**, *64*, 19–27. [CrossRef]
16. Wu, Z.L. The international experience and enlightenment of urban eco-function zone planning: In reference to the Greater London and the Randstad. *Int. Urban Plan.* **2015**, *30*, 95–100.
17. Chen, S.; Liu, Y.X.; Peng, L.H. Dynamics of urban ecological space evolution and policy responses: A case study of Nanjing City. *Acta Ecol. Sin.* **2008**, *28*, 2270–2278.
18. Guan, X.K.; Zhang, F.R.; Wang, X.L.; Zhao, F.H.; Jiang, G.H. Study on spatial evolution and layout optimization of ecological land in Beijing. *Areal Res. Dev.* **2013**, *32*, 119–124.
19. Yu, B.B. Ecological effects of new-type urbanization in China. *Renew. Sustain. Energy Rev.* **2021**, *135*, 110239. [CrossRef]
20. Mcharg, I.L. *Design with Nature, Garden City*; Doubleday: New York, NY, USA, 1969.
21. Lathrop, R.G., Jr.; Bognar, J.A. Applying GIS and landscape ecological principles to evaluate land conservation alternatives. *Landsc. Urban Plan.* **1998**, *41*, 27–41. [CrossRef]
22. An, G.Q.; Qin, X.M.; Xu, X.X.; Chen, L.F.; Xu, W.H.; Xu, Y.Y. Ecological land use change and evaluation of driving factors in Shandong province. *Chin. J. Agric. Resour. Reg. Plan.* **2020**, *41*, 45–54.
23. Tao, Y.H.; Wang, H.J.; Zhang, B.; Zeng, H.R.; Sun, J. Cellular automata modeling and urban expansion simulation based on agent and artificial neural network. *Geogr. Geo-Inf. Sci.* **2022**, *38*, 79–85.
24. Xing, R.S. *Simulation of Ecological Spatial Layout Change and Habitat Quality Assessment Based on ANN CA Markov Model—A Case Study of Wanzhou District*; Chongqing Technology and Business University: Chongqing, China, 2020.
25. Chen, Y.L. *Study on Ecological Spatial Change and Reconstruction of Chang Zhu Tan Urban Agglomeration*; Hunan Normal University: Changsha, China, 2018.
26. Wang, S.H.; Huang, L.; Xu, X.L.; Li, J.H. Temporal and spatial differentiation of ecological space and ecological carrying state of mega urban agglomeration. *Acta Geogr. Sin.* **2022**, *77*, 164–181.
27. Ma, Z.X.; Du, J.M.; Sun, Y.L.; Wang, Z.L. Study on ecological land use evaluation in Tianjin Binhai New Area. *Areal Res. Dev.* **2016**, *35*, 109–112.
28. Chen, L.X.; Hong, W.Y.; Ao, Z.H. Discussion on comprehensive and fine management of ecological space in Shenzhen. *Planners* **2018**, *34*, 46–51.
29. Zhang, X.C.; Long, D.; Bian, F. The protection of "Green Heart" in Randstad: The construction of regional coordination and the innovation of spatial planning. *Int. Urban Plan.* **2015**, *30*, 57–65.
30. Zhu, Y.Y. *Study on Ecological Compensation in the Green Heart Area of Chang Zhu Tan Urban Agglomeration—Taking Ecological Public Welfare Forest as an Example*; Hunan Normal University: Changsha, China, 2015.
31. Wang, H.M. *Study on the Evolution of Ecological Land Use Pattern in the Core Area of Chang Zhu Tan*; Hunan University of Technology: Zhuzhou, China, 2016.

32. Food and Agriculture Organization of the United Nations. FESLM: An International Framework for Evaluating Sustainable Land Management. 1993. Available online: https://www.doc88.com/p-6961965229433.html?r=1 (accessed on 17 February 2022).
33. Chen, Y.P.; Zheng, B.H.; Wang, H. Study on the evolution and reconstruction of green core ecological space of Urban Agglomeration Based on the suitability of land and space development. *Resour. Environ. Yangtze Basin* **2022**, *31*, 551–562.
34. Huang, X.Y.; Zhao, X.M.; Guo, X.; Jiang, Y.F.; Lai, X.H. The natural ecological spatial management zoning based on ecosystem service function and ecological sensitivity. *Acta Ecol. Sin.* **2020**, *40*, 1065–1076.
35. Von Neumann, J. The general and logical theory of automata. In *Systems Research for Behavioral Sciencesystems Research*; Routledge: New York, NY, USA, 2017; pp. 97–107.
36. Li, X.; Anthony, G.O.Y. Modelling sustainable urban development by the integration of constrained cellular automata and GIS. *Int. J. Geogr. Inf. Sci.* **2000**, *14*, 131–152. [CrossRef]
37. Master Planning for Ecological Green Heart Area of Changsha Zhuzhou Xiangtan Urban Agglomeration. 2021. Available online: https://www.hunan.gov.cn/ (accessed on 21 February 2022).
38. Master Planning for Changsha City. 2021. Available online: http://zygh.changsha.gov.cn/ (accessed on 21 February 2022).
39. Master Planning for Zhuzhou City. 2021. Available online: http://gtzyj.zhuzhou.gov.cn/ (accessed on 21 February 2022).
40. Master Planning for Xiangtan City. 2021. Available online: http://zygh.xiangtan.gov.cn/ (accessed on 21 February 2022).
41. Gu, C.L.; Ma, T.; Yuan, X.H.; Zhang, X.M.; Wang, X. Ecological protection and development of green land in the Chang Zhu Tan urban agglomeration. *Resour. Environ. Yangtze Basin* **2010**, *19*, 1124–1131.
42. Hootsmans, M.; Kampf, H.; Bos, P. *Ecological Networks: Experiences in the Netherlands: "A Joint Responsibility for Connectivity"*; Ministry of Agriculture, Nature and Food Quality: The Hague, The Netherlands, 2004.
43. Liu, M.X. *Research on the Planning of Mega Urban Green Space*; South China University of Technology: Guangzhou, China, 2018.
44. Deng, H.T. *A Study on Green Heart Urban Open Space in the Randstad Region, The Netherlands*; Southeast University: Nanjing, China, 2019.
45. Hu, M.S.; Ye, C.S.; Lu, L. Study on ecological space and ecological protection red line delimitation in Nanchang City. *Ecol. Environ. Sci.* **2021**, *30*, 631–643.
46. Wu, J.G. *Landscape Ecology: Pattern, Process, Scale and Hierarchy*; Higher Education Press: Beijing, China, 2000.
47. Zhang, X.F.; Cui, H.W. Spatio-temporal analysis and modeling based on the integration of GIS and CA model. *Image Graph.* **2000**, *5A*, 1012–1018.
48. Sun, Y.Z.; Yang, J.; Song, S.Y.; Zhu, J.; Dai, J.J. Multi-level vector cellular automata modeling and land use change simulation. *Acta Geogr. Sin.* **2020**, *75*, 2164–2178.

Article

A Two-Stage Fuzzy Optimization Model for Urban Land Use: A Case Study of Chongzhou City

Jinjiang Yao [1], Bingkui Qiu [2], Min Zhou [3], Aiping Deng [3,*] and Siqi Li [3]

[1] Science and Technology Innovation and Public Management Research Center of Shanghai, Fudan University, Shanghai 200433, China; jinjiangyao@126.com

[2] Department of Tourism Management, Jin Zhong University, Jinzhong 033619, China; qbk@jzxy.edu.cn

[3] College of Public Administration, Huazhong University of Science and Technology, Wuhan 430074, China; shijieshandian00@163.com (M.Z.); withdoca@163.com (S.L.)

* Correspondence: DAPAipingDeng@outlook.com; Tel./Fax: +86-27-87543047

Citation: Yao, J.; Qiu, B.; Zhou, M.; Deng, A.; Li, S. A Two-Stage Fuzzy Optimization Model for Urban Land Use: A Case Study of Chongzhou City. *Sustainability* **2021**, *13*, 13961. https://doi.org/10.3390/su132413961

Academic Editors: Marc A. Rosen, Miguel Amado and George D. Bathrellos

Received: 16 October 2021
Accepted: 15 December 2021
Published: 17 December 2021

Publisher's Note: MDPI stays neutral with regard to jurisdictional claims in published maps and institutional affiliations.

Abstract: Under the background of New-type Urbanization, with the continuous advancement of urbanization and the all-round development of cities, all kinds of demands are also rising. In the case of demand, it is difficult to quickly adjust from the land supply side and to guide the optimization of the structure and layout of land use is one of the methods to achieve this based on the current situation and shortage of urban land use structure and spatial arrangement. Because of the complexity, uncertainty and dynamics of the land use system, it is necessary to use an uncertain model to accurately describe and propose the approximate optimal solution, so this study analyzes the influencing mechanism of land use and optimize the land use structure under uncertainties by using a Bayesian network and fuzzy mathematical programming. Based on the results of the two stages of analysis, the cellular automata simulation is completed. The framework is applied to Chongzhou city in western China. The results indicated that the optimal land space for cultivated land is in the middle and the south based on the joint influence probability of arable land and urban construction land. The conversion probability of the area near the east is low, and the joint impact probability of construction land in all areas is generally similar except for the western protection area. After the optimization of the fuzzy planning, the optimal construction land scale is 69.42 km^2. Under the condition that the cultivated land's red line is guaranteed, there is still 98.87 km^2 of space for the increase in cultivated land. It is found through simulation that the increase in construction land would occur in the central and western parts of Chongzhou, which may be caused by the urban siphon effect. According to Monte Carlo verification, when the conversion probability exceeds 50%, the cultivated land could be turned into urban construction land, with an accuracy of 91.99%. Therefore, this proposed framework is helpful to understand the process of land use and provides a reference for making scientific and reasonable territorial spatial planning and guiding land use practice under uncertainties.

Keywords: two-stage land use; Bayesian network; fuzzy mathematical programming; Chongzhou city; uncertainty

1. Introduction

Land-use systems are complex, diverse and large, and the academic community tries to narrow the boundaries of land-use systems from the perspective of scenario setting to facilitate the use of various models and methods for research [1]. In terms of land use simulation, different land use systems, methods and models derive different driving mechanisms [2–4] containing various assumptions in the simulation process, which provide new ideas and angles for exploring the uncertainty in land use system and form our understanding of land use system today. Although there are systematic differences in scenario-based land use simulation and modeling, uncertainty does not change with the model methods and scenarios set [5], namely, changes caused by changing model parameters and modifying a hypothesis do not change the uncertainty as well. Reducing uncertainty in land use forecasts is desirable, and some technical efforts can be made, but it is problematic to determine an "accurate" model for a specific purpose. Compared with other systems, the complexity and hugeness of land use systems do not allow repeated proofreading and collection of data that occurred in the past, which makes it more difficult to test the validity of simulation. Reducing such known uncertainties is a common problem faced by experts and scholars. What academics need to pay attention to is how to improve their understanding of the laws of land use systems and how to use these uncertainties to guide land use. It is a current problem that needs to be solved urgently, and it is also the way used to guide practice in a timely manner. Moreover, with the continuous advancement of urbanization and the all-round development of cities, all kinds of demands are also rising, so it is not feasible to reduce the demand in a short period of time [6]. Therefore, in the case of demand, it is difficult to quickly adjust from the land supply side, and to guide and play the role of land in the supply side reform elements may be one of the solutions.

With the development of land use research, more and more scholars have been inspired to capture uncertainty. The uncertainty of spatial factors will affect land use decision making, and incorporating uncertainty into the optimization of land use structure has gradually become one of the important modeling directions of the future [7,8]. Land use structure optimization and spatial optimization together provide an important theoretical basis for the formulation of territorial spatial planning. Territorial spatial planning is an important guide for land use practice, and planning often has problems with unclear planning intentions and inaccurate spatial grasp at the driving level of land use changes [9]. This is due to the inappropriate perception by the framers of uncertainties and laws in land use. In addition to emphasizing the situational and social construction in space [10], planning should also objectively describe and measure the actual existence and summarize it, including correctly understanding the law of land use and identifying various uncertainties [5]. Uncertainty is inherent. Although the tools used to study land use change may never know all the information, some uncertain methods are still expected to be applied to express and capture some of the uncertainties. Some representative studies are as follows, Verstegen et al. (2012) constructed a PCraster land use change model, which integrated the functions of visualization, simulation and uncertainty analysis. Van et al. (2012) simulated land supply with known drivers of land use, and then analyzed uncertainty using the Monte Carlo method. Meyer et al. (2012) used Bayesian network to help assess the uncertainties brought about by stakeholder knowledge modeling. Different from the uncertainty of spatial influencing factors, the uncertainty in the optimization of land use structure is often ignored. Most studies used ordinary planning to balance the land use demand of various sectors [11–15]. However, it is necessary to consider the uncertainty in the optimization of the land use structure. In order to perfect the research on this aspect, fuzzy mathematical programming is applied to it. Moreover, the uncertainty in the land use system has not been widely studied, and there are few relevant data. Some studies that take into account this problem are also always described as random distributions [16], which means that some research choses to avoid uncertainty and generally define the sources of uncertainty. The related research only analyzed the uncertainty of the influencing factors, but seldom applied the results obtained by the uncertainty method to the spatial simulation.

Mathematical models are widely used in optimization research in various fields to solve the problems of multi-constraint conditions and multi-objective optimal configuration. In the field of land use, mathematical planning has long been applied. The mathematical programming model is the optimal allocation scheme determined by mathematical programming theory [17–27]. For example, integer programming [28,29] is used to solve the optimization problem of integer variables. Dynamic programming [30] is used to solve the optimal problem of decision-making processes. Linear programming [31,32] is used to solve the one-dimensional optimal allocation problem. Nonlinear programming [33–35] is used to solve multi-dimensional allocation optimization problems. Among them, the use of integer programming needs to limit the variable to integers, so the scope of application is not wide enough. Dynamic programming can realize the optimization of the dynamic decision-making process of linkage, but there may be a multi-dimensional solution dilemma. Generally, the time cost is high. Linear programming is simple to solve and more applicable than other kinds of programming, but only for simple linear programming problems. Fuzzy programming, random programming, chance-constrained programming, parameter programming, etc., are all linear programming methods. Stochastic programming is mainly used for stochastic processes, which can be solved by intelligent models. Parameter planning is applicable to exploring the influence of parameter changes on the final results and has a specific scope of application. Both types of planning are scenario specific. Fuzzy programming and chance-constrained programming have a relatively wide range of applications. They can express the general decision-making process and can be solved even when constraints are violated and the value of variables is not clear. The results can be used as a reference for decision makers. The applicability of nonlinear programming is not as broad as that of linear programming because of its multi-dimensional characteristics, but nonlinear programming has better performance for some complex problems.

In view of the shortcomings of the above models, the purpose of this paper is to fully consider the uncertain factors in the land use system and optimize the land use quantity and the spatial layout by combining cellular automata model so as to provide quantitative support for decision makers. Based on the complexity, uncertainty and dynamic characteristics of land use systems, the modeling method of capturing uncertainty is used to find more spatial information from the perspective of uncertainty, and the function of multiple spatial impact factors was constructed to depict the law of land use impact. To a certain extent, this expands the depth of the analysis of influencing factors, provides certain theoretical significance for the study of land use development and further deepens the cognition of the uncertainty in the process of land use. Under the background of the new urbanization strategy proposed by the Chinese government to pay more attention to the coordinated development of population, economy, society and ecological environment [36], based on the current situation and shortage of urban land use structure and spatial arrangement, combined with regional development policies, starting from the reform of land supply side structure, using the method of Bayesian network and fuzzy mathematical programming, this paper makes an in-depth analysis of the influencing mechanism of land use and the optimization of land use structure in two stages. It is helpful to provide theoretical support for the formulation of territorial space planning. The optimization result aiming at economic carbon efficiency is combined as well to provide a certain reference for the sustainable development of urban land use. China's New-type Urbanization and agriculture-oriented cities will be used as case demonstration.

2. Methodology

2.1. Two-Stage Land Use Optimization Model

Figure 1 shows a two-stage land use optimization framework under uncertain conditions. This study divides land use analysis into two stages. The first stage is to analyze land use influencing factors through a Bayesian network. It performs well in dealing with uncertainties and can reflect a more real world. The second stage is to optimize land use through fuzzy mathematical programming. In addition to economic and social constraints, this stage also involves environmental constraints marked by carbon emissions. In order to link the two stages, cellular automata are used to finally realize the analysis results of the two stages, and the neighborhood influence is added.

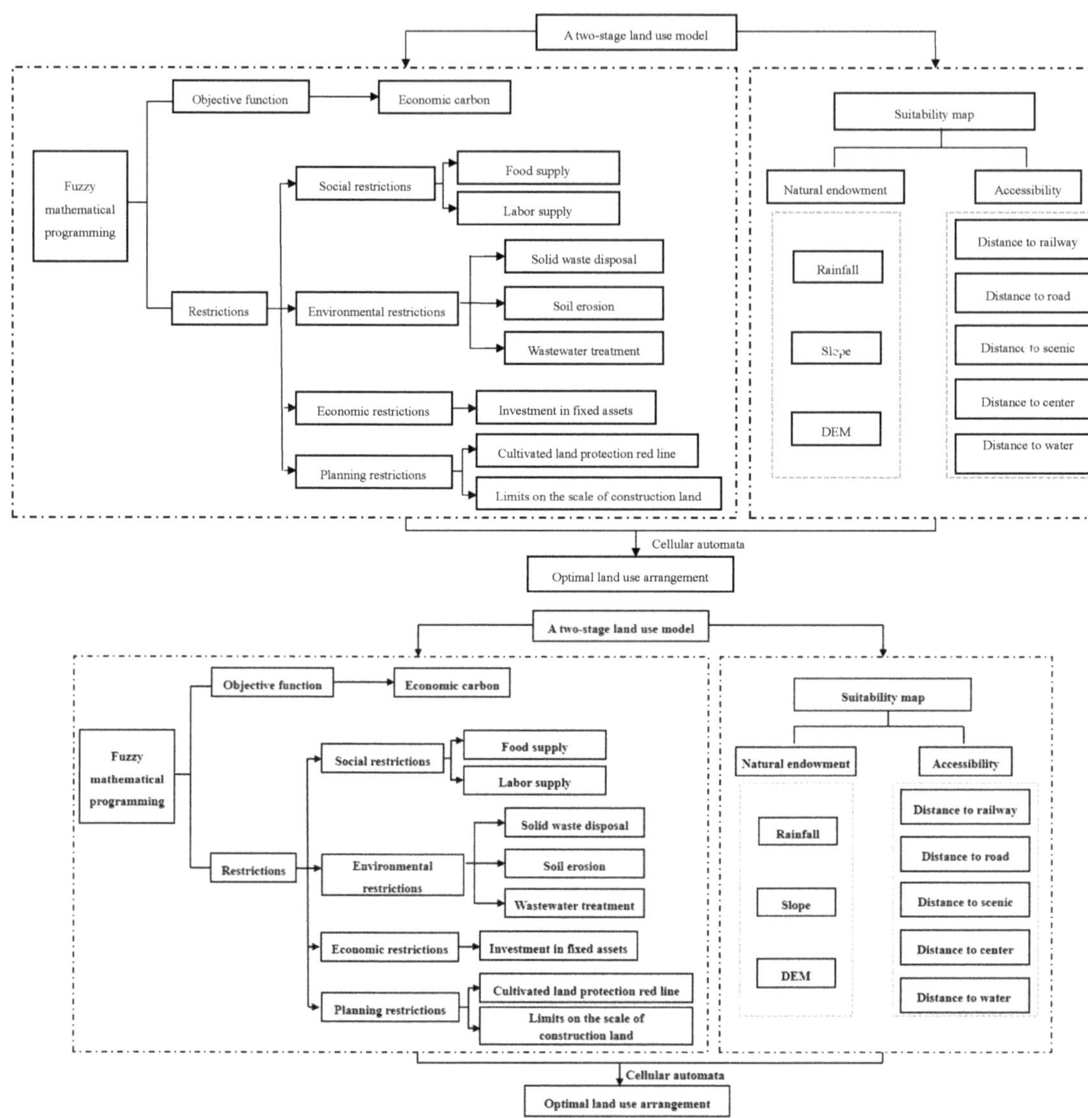

Figure 1. Two-stage optimization framework.

2.1.1. Bayesian Network for Land Use

For the occurrence or non-occurrence of land use events, there are usually only two probabilities of 1 and 0, that is, either occurrence or non-occurrence. Few studies have considered the occurrence probability of land use events, and Bayesian network has the ability to solve this problem. Bayesian network is particularly flexible, and new information can often be found during modeling to improve the performance and applicability of the network [37]. It is a way to deal with uncertainty [12] and provides more information for interpretation of the evaluation results. Bayesian network clearly shows the complex relationship between land use change, its driving factors and its change process with its good graphical description method. Although the influence of each uncertain factor on the estimation results can be measured by sensitivity analysis, it requires the estimation of complex partial derivatives. Bayesian network not only does not need complex differentiation, but it is closer to the reduction of the real situation, in which the actual process is probabilistic in nature [38]. A node X points to another node Y, where X is the parent of Y. All parent nodes can be used as a set $X = \{x_1, x_2, \ldots x_k\}$, child nodes can also be represented by the set $Y = \{y_1, y_2, \ldots y_k\}$, represented as a directed acyclic graph, which is the advanced form of Markov chain. The analysis of land use suitability can also be completed by Bayesian network. The Bayesian formula is shown in Formula (1):

$$P(X|Y) = \frac{P(Y|X) \cdot P(X)}{P(Y)} \tag{1}$$

In Formula (1), $P(X|Y)$ represents the probability that the parent node X is in a certain state under the condition of child node Y, that is, the posterior probability. $P(Y|X)$ is the same. $P(X)$ is the prior probability of X without regard to Y, and $P(Y)$ is the same thing. According to Bayesian theory, in land use analysis, it is usually assumed that all factors are independent of each other, so their relationship can be expressed by the joint probability Formula (2):

$$P(x_1, x_2, \ldots x_k) = P(x_k|x_1, x_2, \ldots x_{k-1}) \cdots P(x_2|x_1)P(x_1) \tag{2}$$

The construction of a Bayesian network involves qualitative and quantitative steps. The first step is to formulate the node and its network structure, and the second is to calculate the parameters of each node. It is found that policy and socio-economic factors show complex and close links under the influence of natural factors such as resource endowment and spatial geographical characteristics [4]. At the same time, the design and implementation of land planning will be affected by local socio-economic factors and external processes [6]. This study extracts candidate variables from spatial data, constructs the network structure according to common sense, expert knowledge and statistical correlation between variables and avoids creating parameter sets with excessive data as much as possible [39]. Therefore, in terms of spatial factor analysis, historical state, elevation, slope, accessibility, natural endowment and other factors are taken into consideration in the construction of the Bayesian network, as shown in Figures 2 and 3. First is the linear features of the impact of land use, railway effects on the city of the population, material and information exchanges with the outside world, belonging to the clear on the larger influence on urban development tool, so it leads to the improvement of the region and all aspects of development, both in cultivated land activities and urban economic activities, and can produce a good liquidity. In addition to assuming the connection with the city, the highway also serves as a bridge for communication within the city, facilitating the effective connection and interaction of internal elements. So linear features can lead to strip development around them. Secondly, point features have a certain radiating effect on urban construction, such as scenic spots and commercial centers. These point features are active centers of population and capital in cities, which attract population and investment, are easier to form intensive land use and have more potential to lead development. Moreover, natural endowments are vital to both urban and agricultural development. For agriculture, water supply will directly have an important impact on crop quality and yield. Rice, the main food crop in Chongzhou City, will have its yield affected by the lack of water resources. At the same time, water is an important resource guarantee for urban construction. Sufficient water supply will support urban life and production and, vice versa, will limit urban sustainable development. In order to improve the quality of water resources, provide a good resource endowment for urban life and production and improve the ecological and landscape conditions of the city, Chongzhou City has implemented measures such as the 'water control ten' river (lake) chief system, black and odorous water treatment and comprehensive water environment management. Therefore, this shows that the current agricultural development and urban construction are both limited by water and dependent on water.

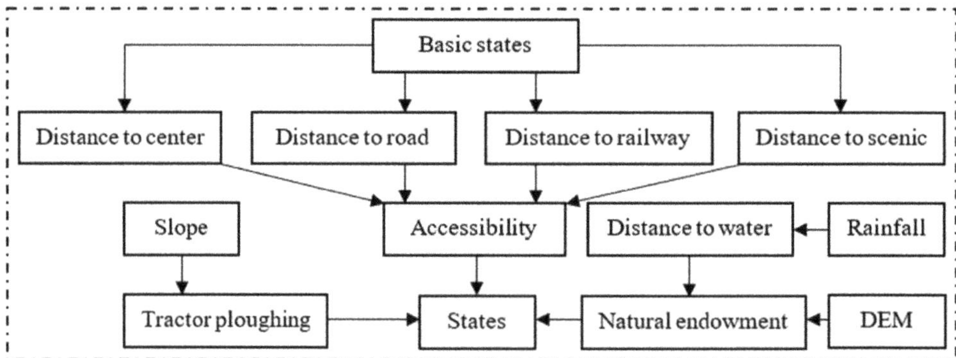

Figure 2. Bayesian networks for cultivated land.

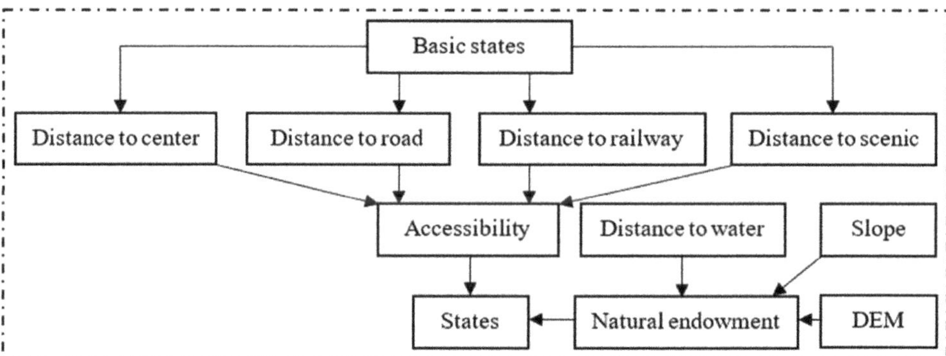

Figure 3. Bayesian network of urban construction land.

Bayesian network can help to search various laws in land use systems, and Gaussian radial basis functions can help to express these laws. The various influencing factors of land use change have different forces, so the attraction of each factor to different land types is also different. For example, the closer the area is to the city center, the easier it is to develop into urban construction land. The attraction of the city center is higher than that of the county center and the town center. The land close to the highway is easier to form strips of urban construction land on both sides of the highway, and the higher the grade of the highway, the easier it is. Similarly, rivers have a huge impact on urban development, and urban construction land is often gathered along the banks. Therefore, there are differences in the influence ranges of different influencing factors. It is generally believed that there is a certain rule with distance. In order to express the difference of forces generated by distance, a Gaussian radial basis function is adopted, as shown in Formula (3):

$$P_k(x_k|m) = exp\left(-\frac{1}{2\sigma_k^2}(x_k - \overline{x}_k)^2\right) \tag{3}$$

where $P_k(x_k|m)$ represents the probability of developing into a certain land type, and σ represents the standard deviation of the influencing factor variable. The larger the value of σ, the wider the influence range of the variable, while the smaller the value, the more concentrated the influence range. x_k represents the value of the variable, and x_k represents the central value, that is, the variable value most likely to develop into the target land type, at which time the probability of $P_k(x_k|m)$ reaches the maximum.

2.1.2. Land Use Optimization Mathematical Programming Model

Next, we estimated the uncertainty in the land use system is crucial for investigating land use policies, evaluated urban risk response strategies and quantified the impact of land use change on climate and the environment. Uncertainty involves a wide range, and this study uses a small entry point to express one aspect of the uncertainty, namely, the uncertainty of the target conditions for economic carbon efficiency, as shown in Formula (4):

$$Max\ Z = \frac{\sum_{i=1}^{n} x_i \cdot UB_i}{\sum_{i=1}^{n} x_i \cdot WC_i + \sum_{i=1}^{n} x_i \cdot PC_i + \sum_{i=1}^{n} x_i \cdot EC_i + \sum_{i=1}^{n} x_i \cdot SC_i + AC} \tag{4}$$

In Formula (4), x_i represents the area of each type of land; Max means the maximum economic benefit; and UB_i represents the level of economic benefit generated per unit land area (kg/km^2). In this study, the gross output of primary industry, secondary industry and tertiary industry represents the gross output of cultivated land and urban construction land. UB_i is the ratio of the gross output of each type of land to the area of each type of land before optimization. Min represents the minimization of carbon emission. WC_i represents the carbon emissions (kg/km^2) generated by wastewater quantified on all kinds of land, which can be obtained by the ratio of the amount of cultivated land wastewater and urban wastewater to the area of all kinds of land obtained from statistical data. PC_i is the carbon emission from human body on various types of land (kg/km^2). EC_i represents the carbon emission from energy consumption of various types of land (kg/km^2). SC_i represents the carbon emissions from solid waste generated from various types of land (kg/km^2). AC stands for carbon emissions from livestock and poultry farming. In order to more clearly understand the relationship between economic efficiency and carbon emissions, this research integrates economic benefit targets and carbon emission targets to explore the economic carbon efficiency, that is, how much economic benefit each unit of carbon emission produces and whether it is achieved between the sacrifice of the climate and the economic benefit balance. Therefore, the objective function is optimized.

(1) Environmental constraints

Since both agricultural and urban development have an impact on the soil and water environment, it is necessary to consider the status of soil. In the Formula (5), r represents soil and water loss rate (%). A_r represents the area of loss (km^2). Both farmland operation and urban production and living will produce sewage. If sewage is discharged into the environment without treatment, it will not only pollute the water environment but also cause excessive heavy metals and poisons in the soil, which will increase the burden of the ecological environment and make it unbearable, affecting the environmental sustainability. In the Formula (5), G_i represents the sewage discharge coefficient (ton/km^2) generated on a certain type of land, which can be obtained by the ratio of sewage discharge to the area of each type of land before optimization. P_G denotes sewage treatment plant capacity (tons). Both agricultural production and urban production and life need to consume a large number of solid materials and produce a large amount of solid waste. Considering that the rural solid waste is treated on-site, which is not easy to count, only the solid waste generated in the city is considered. However, the processing capacity of urban solid waste is limited, so certain constraints must be met. In Formula (5), USW_i represents the discharge coefficient of solid waste generated on a certain type of land (ton/km^2), which can be obtained by the ratio of the amount of solid waste generated by the city to the area of urban construction land before optimization. PSW stands for solid waste disposal capacity (tons).

$$\begin{aligned} &\sum_{i=1}^{n} r \cdot x_i \lesssim A_r \\ &\sum_{i=1}^{n} G_i \cdot x_i \lesssim P_G \\ &\sum_{i=1}^{n} G_i \cdot x_i \lesssim P_G \end{aligned} \tag{5}$$

(2) Economic constraints

The fixed asset investment required for agricultural production includes mechanical farming, electromechanical irrigation, mechanical seeding, mechanical harvesting, transportation roads and production facilities, etc. For Sichuan Province, with many mountains and hills, it belongs to an area with a high level of mechanization, so the investment ratio of fixed assets is also large. The fixed asset investment in urban construction has been the main part of the total investment. In Formula (6), $DSFAI$ represents the total investment (CNY 100 million); $SFAI_i$ represents the investment per unit land area (CNY 100 million). In this study, the fixed asset investment in primary industry, secondary industry and tertiary industry represents the fixed asset investment in cultivated land and urban construction land respectively. $ISFAI$ represents import investment (CNY 100 million). $OSFAI$ represents export investment (CNY 100 million). According to Formula (6), the per capita annual ration consumption is about 135 kg:

$$\sum_{i=1}^{n}(SFAI_i \cdot x_i) + ISFAI - OSFAI \gtrsim DSFAI \tag{6}$$

(3) Social constraints

Agricultural products are one of its main production targets in the key county-level city for agriculture and ecological protection, so the balance of the food supply is particularly important. In Formula (7), US_i represents the amount of grain produced per unit land area (kg/km^2), which can be obtained by the ratio of grain yield to cultivated land area before optimization. D stands for grain demand (kg). Agricultural production and urban secondary and tertiary industries both need labor. In Formula (7), UL_i represents the labor force per unit land area (person/km^2). In this study, the employees in the primary industry, the secondary industry and the tertiary industry represent the labor force on cultivated land and urban construction land, and the ratio of them to each type of land area before optimization is taken as the value of UL_i. TL stands for total labor force (people).

$$\begin{array}{c} \sum_{i=1}^{n} x_i \cdot US_i \gtrsim D \\ \sum_{i=1}^{n} (UL_i \cdot x_i) \lesssim TL \end{array} \tag{7}$$

(4) Programming constraint

The planning conditions mainly include the restriction of cultivated land protection red line and the restriction of construction land area. In Formula (8), FC represents the minimum cultivated land protected area (km^2). FB refers to the maximum area of construction land (km^2) A 338 km^2 farmland protection red line has been set, accounting for 31.62% of the total area.

$$\begin{array}{l} x_i \geq FC,\ i = 1 \\ x_i \leq FB,\ i = 2 \end{array} \tag{8}$$

(5) Other Constraints

$$x_i \geq 0,\ i = 1,2 \tag{9}$$

In Formula (9), x_1 represents cultivated land. x_2 represents urban construction land.

(6) Carbon effect estimation

Before estimating the carbon effect, it is necessary to define the organizational boundary. The organizational boundary set in this study is the carbon effect within the range of cultivated land production activities and urban consumption, which mainly includes carbon sources, carbon emissions generated by mechanical farming and urban production and living energy consumption in cultivated land activities, carbon emissions from farmland and municipal wastewater treatment, carbon emissions from municipal solid waste treatment, carbon emissions produced by the human body and carbon emissions from livestock and poultry farming. In terms of carbon sinks, cultivated land, shrub land, woodland and water are all included.

2.1.3. Cellular Automata Simulation

Cellular automata determines cellular state through transformation rules [40]. The improved transformation rules not only include the influence of cellular state and neighborhood action at the previous moment but also include quantified social, economic, spatial and other factors:

$$S_{t+1} = f(S_t, N, \ldots) \tag{10}$$

In Formula (10), S refers to the set of all possible states, f is the transition rule, N is the neighborhood of the cell, t moment is the current moment and $t+1$ represents the next moment. In fact, cellular automata take into account spatial explicit weighting factors in suitability, rather than completely random allocation. Each cell's land type is considered the primary type, and that cell is masked, while all other land types are considered the "claim" type and compete for that cell. The "claim" type of cell that influences the possibility of the future land type of the principal cell must be both inherently suitable and close to the principal cell. This study adopts a neighborhood range of 3 × 3, with a kernel as shown in Figure 4.

0	1	0
1	1	1
0	1	0

Figure 4. Cellular automata kernel.

This study defines three basic scenarios affecting land use change and the basic settings for each scenario. The first situation is the role of neighborhood. If there is an urbanized area around, the central cell is also inclined to develop into an urbanized cell under the influence of the neighborhood state, which is manifested as the concentration of the urbanized area. In the second case, the protected area is regarded as the forbidden area for development, and it will not be developed into an urban area no matter whether it has the conditions for the development of urban area. Third case: In the case of limited city size, in order to ensure the balanced improvement of various functions of the city, it is not feasible to consider only a single goal. Therefore, the consideration of the red line of cultivated land protection and the control of construction land scale is added. This paper mainly studies two types of land with fierce competition, namely, cultivated land and urban construction land.

3. Case Study

Chongzhou is located in southwestern China, covering an area of 1089 km^2, located between 30°30′~30°53′ N and 103°07′~103°49′ E (Figure 5.). In terms of geographical conditions, the terrain of Chongzhou is high in the northwest and low in the southeast. The west is mostly high mountains, and the terrain is steep. High mountains account for 38.4% of the total area. It has mild climatic conditions, abundant agricultural conditions, more rainfall and less sunshine. In terms of history and culture, Chongzhou is one of the cradles of ancient Shu civilization and the birthplace of agricultural civilization. In terms of economy and society, the GDP of Chongzhou in 2017 was CNY 30.04 billion, and the proportion of primary, secondary and tertiary industries was 12.1:49.5:38.4. The permanent resident population is 664,800, and the number of people employed in the primary, secondary and tertiary industries is 110,500, 282,800 and 149,600, respectively. Chongzhou is located in the west of Chengdu City, in the middle and upper reaches of Minjiang River, and is an important water source and ecological conservation area of Chengdu Plain. According to the strategy of "advancing to the east, expanding to the south, controlling to the west, transforming to the north, and providing excellent" in the General Plan of Chengdu City (2016–2035), Chongzhou is mainly responsible for the development of green industries and providing ecological services. Meanwhile, it strictly implements the ecological red line control, with cultivated land and woodland occupying the main position. According to existing data, generally speaking, cultivated land and urban construction land are mainly distributed in the central and eastern regions. Agriculture and urban construction will form a competitive relationship in terms of terrain. Thus, elevation and slope both constitute the main influencing factors of agricultural development and urban construction. In addition to the above linear features, namely, point features, elevation and slope conditions, for agricultural development, there are also natural conditions such as rainfall, soil productivity and heat that have a certain impact on the characteristics and productivity of agricultural products in the region. Taking into account the accessibility of data, this study selects rainfall as a representative to participate in the construction of the Bayesian network. In order to test the performance of the model, the Bayesian network modeling was replicated in Chongzhou ecological service zones, agricultural protection zones and major urban areas within Chengdu. On the one hand, natural endowment drives the stepwise development of land use. On the other hand, as one of the key pilot cities of New-type Urbanization in western China, it is still in the stage of urban expansion with highly developed social and economic activities, so it is suitable for this model. The general situation of Chongzhou is shown in Figure 6.

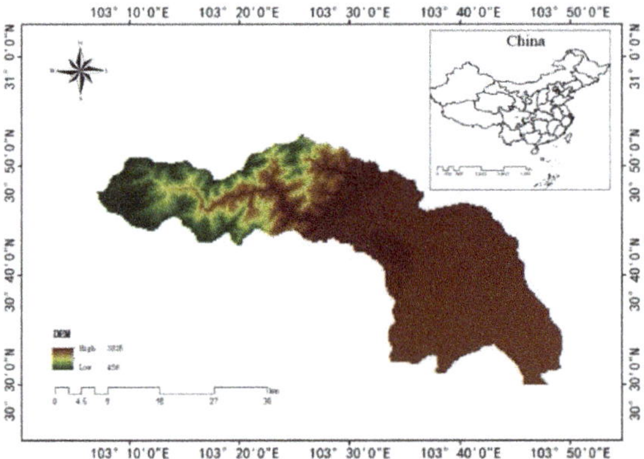

Figure 5. A location map of Chongzhou.

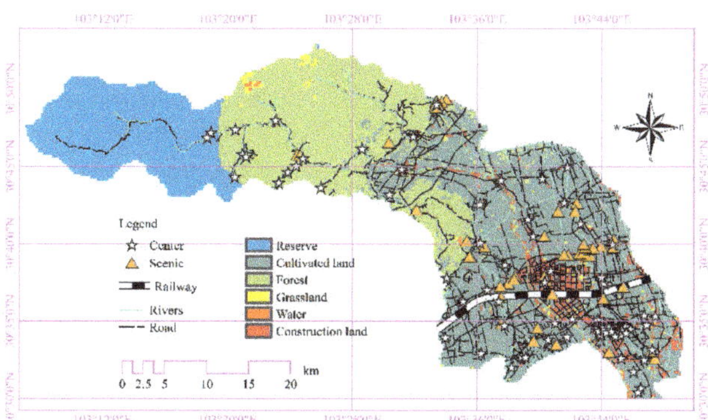

Figure 6. Overview of the study area.

The data mainly came from Chengdu Yearbook 2018, Chongzhou Yearbook 2018 and Sichuan Statistical Yearbook 2018, which recorded the economic and social data in 2017. The elevation and spatial land use data are from the geographical science and resources institute of the Chinese Academy of Sciences and Tsinghua University; China's present situation of land use remote sensing monitoring database 30 m global land use data (http://data.ess.tsinghua.edu.cn/fromglc2017v1.html accessed on 6 March 2019) and spatial rainfall data were collected from National Tibetan Plateau Scientific Data Center [41,42]; and spatial data such as roads, water systems, attractions and commercial centers are derived from OpenStreetMap database. According to China's industrial structure and the status quo of land use, combined with the classification standard of China's land use status in 2017, this study divides land types into cultivated land, woodland, shrubby land, water area, protected area and urban construction land. The data on livestock and poultry rearing quantities, feeding cycles, electricity use, fuel use and management costs are from Chongzhou Yearbook 2018, Sichuan Statistical Yearbook 2018, Sichuan Survey Yearbook 2018, National Cost-Benefit Compilation of Agricultural Products 2018 and statistical data of Ministry of Agriculture and Rural Affairs, PRC (http://zdscxx.moa.gov.cn:8080/nyb/pc/search.jsp accessed on 14 December 2021). Nitrogen emissions, CO_2, CH_4 and N_2O emissions factor from the 2006 national greenhouse gas inventory guide, the provincial greenhouse gas inventory preparation guide (2011) and the United Nations Food and Agriculture Organization (http://www.fao.org/faostat/zh/#data/GM accessed on 16 July 2021). Data on the CO_2-eq emission factors of standard coal are from China Energy Statistical Yearbook 2018, and the CO_2-eq emission factor of electric power data are from literature. [43,44]. To avoid double counting, the organization boundaries of this study are beef cattle, pigs, mutton sheep and meat poultry, as well as the production area, feed area and management area of breeding in Chongzhou [45,46], as shown in Figure 7.

The classification of land use, the sign of land use conversion and the carbon emission form of the study are explained as follows: (1) Biomass storage and carbon sequestration of agricultural products are not considered in this study. (2) Because of the forest and herbs, there are many different crops and the waters of the biomass, and this paper summarizes land use cover change by the change of land type. The vegetation on the land's net ecosystem productivity (NEP) is also changed, and combined with the geographical position and climate characteristics, they determine the net ecosystem productivity of different land types. The average carbon sequestration capacity of different land types is also calculated. (3) The carbon emission form studied in this paper is CO_2, that is, CO_2 generated by various land use activities.

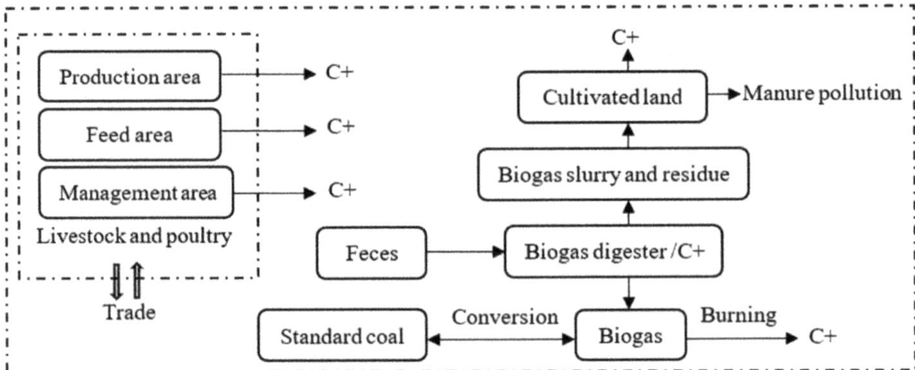

Figure 7. Organizational boundaries of this study.

4. Results and Discussion

4.1. Influencing Factors of Land Use: Bayesian Network

A Bayesian network is constructed and completed in Netica software (www.norsys.com, accessed on 16 July 2021), using data from railways, highways, scenic spots, commercial centers, slopes, water area, rainfall, elevation and other data processed by ArcGIS. Considering that the area of the research area is the size of a county, this research created 500 random sample points to train the Bayesian network with the sample data. Since the creation of random points results in the absence of values in the attributes of some points, 4 points were eliminated, and the total sample size became 496. The coordinate system of all kinds of spatial data was unified as GCS_WGS_1984, and all data processing was carried out in this coordinate system. The vector spatial data were converted to raster data through the element-to-raster operation, and the pixel size was 250 × 250 m. By generating the nearest neighbor table and extracting the point of arrival, all kinds of spatial attributes were linked with land use, and the attribute characteristics of each grid cell were obtained. Statistical spatial attribute data were the input type of the Netica software. Modules are constructed in Netica, and each module was divided into 10 continuous interval states. The initial probability of each module is equal so as to conduct initial exploration, determine parent nodes and child nodes according to the model design and form a directed acyclic graph.

Figure 8 shows the probability map of influencing factors of cultivated land, and Figure 8 shows the probability map of influencing factors of urban construction land. DEM is the height, Slope is the slope, Accessibility is the accessibility to all kinds of lines and points, and so on. A state refers to protected areas, B state refers to cultivated land, C state refers to woodland, D state represents shrub land, E state represents water area and F state represents urban construction land. Among them, the data of accessibility, natural endowment and farming are only statistical data without spatial data, so the preliminary determination is based on the expert knowledge, so that the follow-up research can be carried out smoothly.

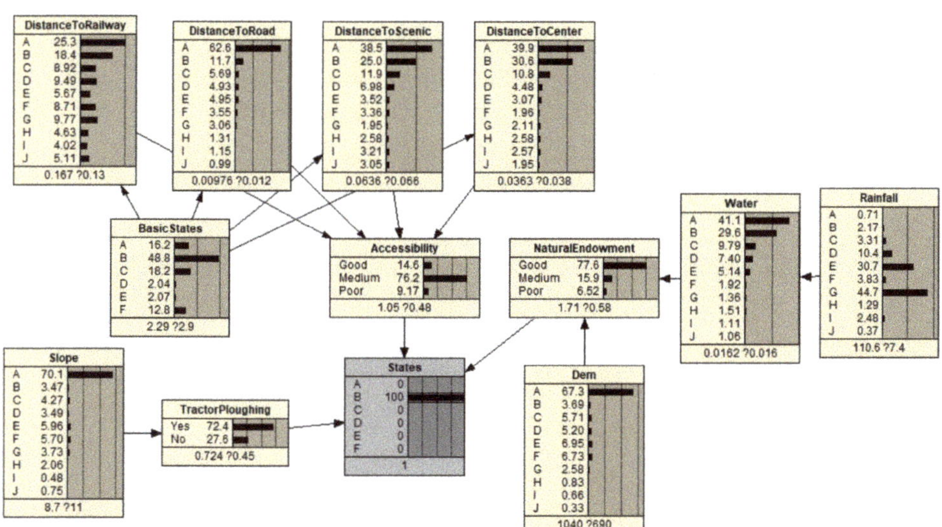

Figure 8. Probability diagram of factors affecting cultivated land: "DistanceToRailway" means the distance to the railway, "DistanceToRoad" means the distance to the road, "DistanceToScenic" means the distance to the scenic spot, "DistanceToCenter" means the distance to the commercial center, "NaturalEndowment" refers to natural abilities or qualities, "Tractorploughing" refers to mechanized farming or not, "BasicStates" refers to initial land use status and "States" refers to simulated land use status.

According to the results of the Bayesian network in Figure 8, the linear ground features have the same characteristics for cultivated land and urban construction land, that is, the closer the land is to railways, highways, scenic spots and commercial centers, the easier it is to develop into cultivated land, and the trend of decline is from near to far. Due to the large amount of data, all the data of the 496 sample points were not included in this paper, so the sample data were explained in words. The first is the influence brought by the railway. As an important means of transportation to other places, the railway has a wider scope of influence than other linear ground objects. In Figure 8, the states of the "Distancetorailway" node are divided into 10, each state represents a continuous interval, and the value from A to J is increasing. The value of the sample can correspond to one of the intervals, so as to achieve the purpose of grading. The same applies to other state nodes. The distance to the railway has an interval length of 4740 m, as shown in Figure 8, and the state of intervals A and B, about 9 km within range, has the greatest attraction of arable land. With the increase in distance, the influence is more and more weak. The domestic railway's furthest distance influence is also not reduced to 0, which shows that the influence of the railway is wide, but the impact is only large in a certain neighborhood. The distance to the highway is 570 m, the distance to the scenic spot is 2690 m and the distance to the commercial center is 1710 m. The reason why the length of the interval is different is because of the limitation of computing speed and memory. As can be seen from the figure, the influence rules of the three are roughly the same. The optimal impact distance of highway is interval A, about 500 m; the best impact distance of the scenic spot is between A and B, which is about 5 km. The optimal impact distance of the commercial center is between A and B, which is about 3.5 km. In terms of natural endowment, in view of the accessibility of data, elevation, slope and rainfall are selected as state nodes. Among them, rainfall can be seen as a general trend, and regions with greater rainfall have better conditions for agricultural development. Compared with the data integrity of other modules, the anomaly in interval F in the figure may be due to insufficient rainfall data sampling. The probability of elevation's influence on land use state transition presents a "cliff" distribution, that is, more than 67.3% of the cultivated land is distributed below 800 m, which indicates that greater elevation will have a great limitation on the current agricultural development in Chongzhou. The slope is the same. Moreover, 70.1% of the cultivated land is below 5°, and more than 5° will have great restrictions on mechanized farming. The slope is divided into five grades according to the relevant provisions of topographic slope classification technology in the second national land survey: I (0°~2), II (2°~6), III (6~15), IV (15~25) and V (>25), and above 25 degrees will forbid land reclamation. In fact, the cultivated land in Chongzhou is below the degree, so the slope has a great limitation on agricultural development, which is a factor that must be taken into account in the study. The influence of water sources is similar to that of roads, and it has a greater influence on cultivated land within the range of about 1.7 km between A and B and decreases with the increase in distance. In terms of land use evaluation, mechanized tillage does not play a decisive role in land state. Areas with good natural endowment are more likely to be converted to arable land.

As can be seen from Figure 9, 59.5% of the construction land elevation is mainly kept below 800 m, and 57.3% of the construction land slope is kept below 5°. The optimal impact distance of the water area is within the range of about 1.7 km between A and B. In the two state intervals of A and B, within about 9 km, the railway has the greatest attraction to urban construction land. The optimal impact distance of highway is interval A, about 500 m; the best impact distance of the scenic spot is between A and B, which is about 5 km. The optimal impact distance of the commercial center is between A and B, which is about 3.5 km.

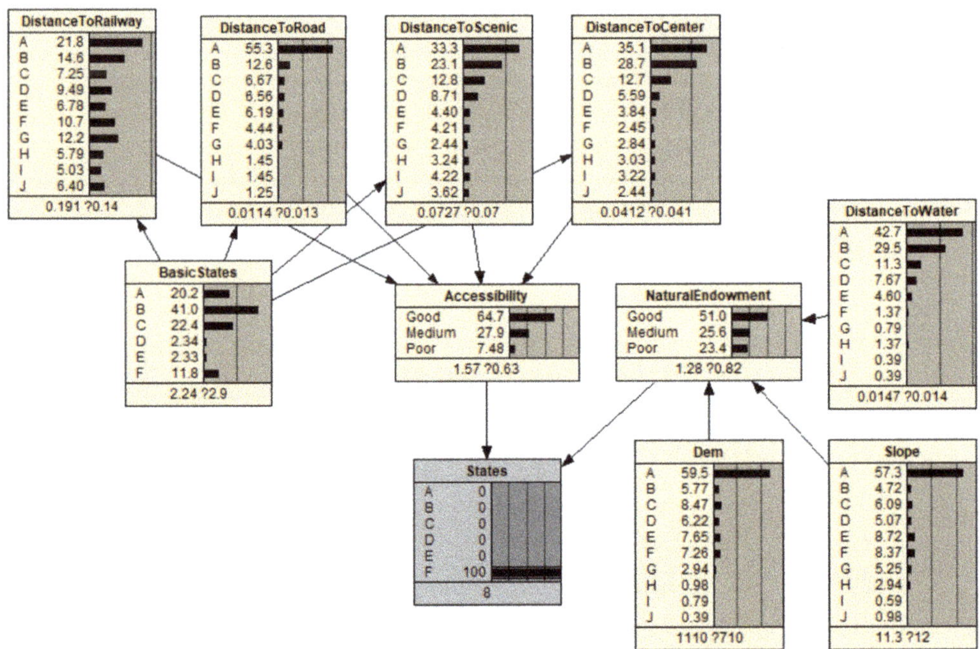

Figure 9. Probability diagram of influencing factors of urban construction land: "DistanceToRailway" means the distance to the railway, "DistanceToRoad" means the distance to the road, "DistanceToScenic" means the distance to the scenic spot, "DistanceToCenter" means the distance to the commercial center, "NaturalEndowment" refers to natural abilities or qualities, "Tractorploughing" refers to mechanized farming or not, "BasicStates" refers to initial land use status and "States" refers to simulated land use status.

4.2. Comparative Analysis of Influence Factors

It is generally believed that all influencing factors, except elevation, slope and rainfall, should present the distribution as shown in Figure 10a if their influences on land use are equidistant attenuations. However, this is not the case in fact. In this study, the influence of various factors on land use under uncertain conditions was obtained through a Bayesian network, and different results were obtained, as shown in Figure 10b. Firstly, the influence of various factors on land use is not inversely proportional to the distance. Secondly, the influence rules of various factors are not consistent. For example, the attenuation rate of railway influence is slow, while that of highway influence is faster. In order to express the influence law of each factor on land use, Gaussian radial basis function is used to express the non-inversely proportional and different influences. According to the Bayesian network constructed in Netica, the mean value and standard deviation of the influence probability of each factor can be obtained to determine the Gaussian radial basis function to express the influence probability of each factor. Figure 11 shows the Gaussian radial basis function curve of the influence factors of cultivated land drawn in MATLAB R2019a. Figure 12 shows the Gaussian radial basis function curve of influencing factors of urban construction land. Figures 11 and 12a from left to right in turn show: road (green), water (magenta), business center (yellow) and attractions (red) and railway (blue). It can be seen that under the same coordinate measurement, the influence of railway is one of the widest, the highway has the narrowest scope of influence but rail impacts few and far between. Although the road influence area is narrow, the road network is complex, and its influence cannot be underestimated. Second, the influence of the water range is relatively narrow, and the publicizing states of drainage systems are mainly the Golden Horse river, Minjiang river is flow and Black Rock stream (into Dujiangyan). The condition (within the territory of the longest) and the size of the tributary of about 180. Figure 11b shows rainfall (magenta, in millimeters), Figure 11c shows slope (red, in degrees) and Figure 11d shows elevation (blue, in meters). Slope (red, in degrees) is shown in Figure 12b and elevation (blue, in meters) is shown in Figure 12c. The analysis is described in Section 4.1.

4.3. Affect Probability Distribution

According to the characteristics of the Bayesian network, the probabilities of each node are independent from each other, so the joint influence probability of each factor can be calculated, which is used to express the relationship between each factor and is conducive to the transformation of independent influence probability to comprehensive influence probability. Figure 13 shows the spatial distribution of the joint influence probability of cultivated land. Figure 14 shows the spatial distribution of probability of joint impact of urban construction land. As you can see, in terms of the probability rule, arable land and urban construction land are in competition, but from the point of spatial distribution, cultivated land is the best space in central and southern areas, and the near eastern area transition probability is low and water systems are mainly distributed in the central, southern and western states. The water supply in the east is not only relatively weak, but it is also located in the direction leading to the center of Chengdu and a cluster of industrial parks. The suitability probability of urban construction land is relatively smooth, except for the low probability of the western protection area (brown-red part), the other parts of the suitability probability have little difference. The low suitability probability of a small piece of cultivated land and urban construction land in the south is caused by water area factors. Compared with other areas, this area is far from water area, so the overall probability will be slightly affected.

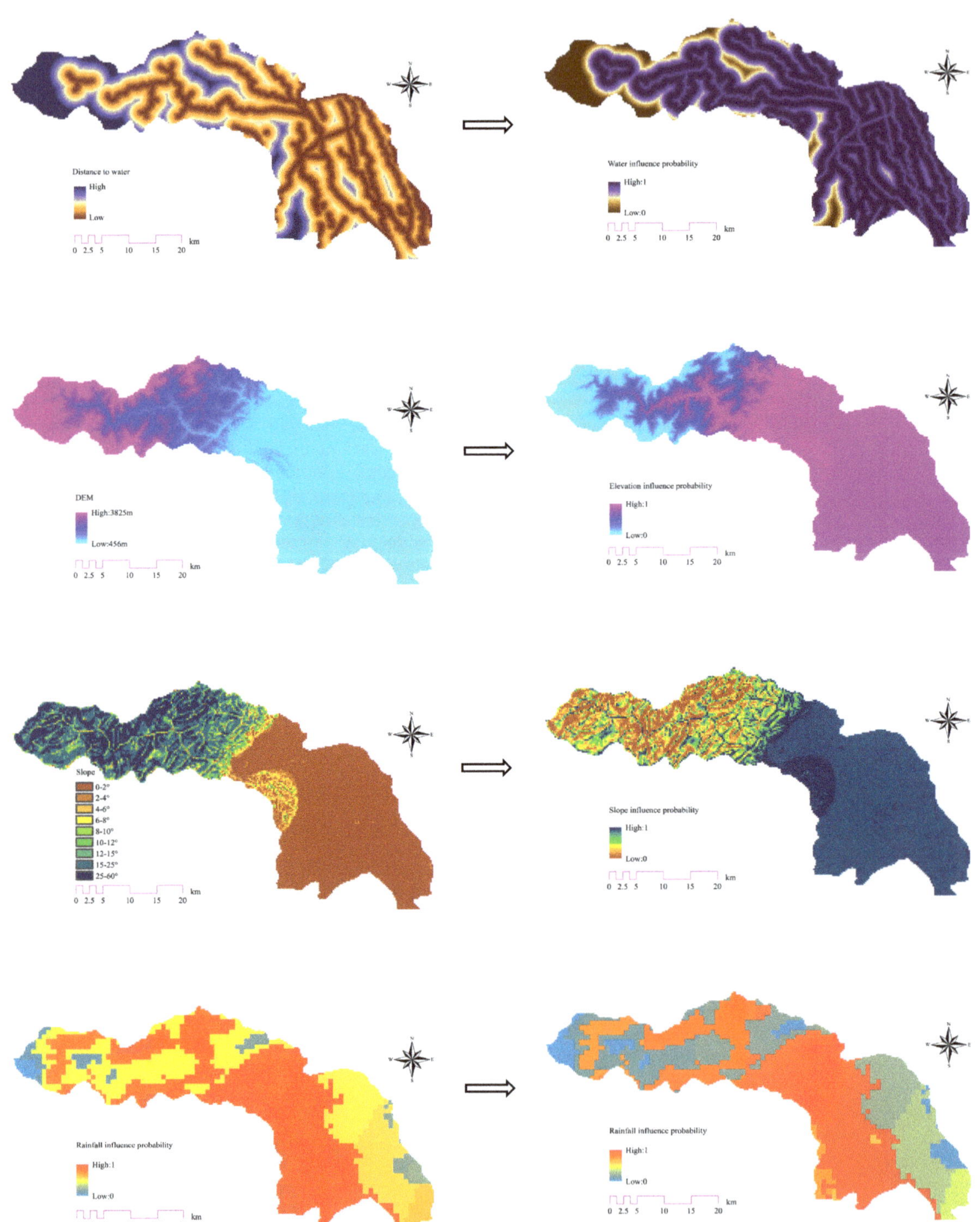

Figure 10. Spatial expression of various influencing factors (**a**) and influence probability distribution (**b**).

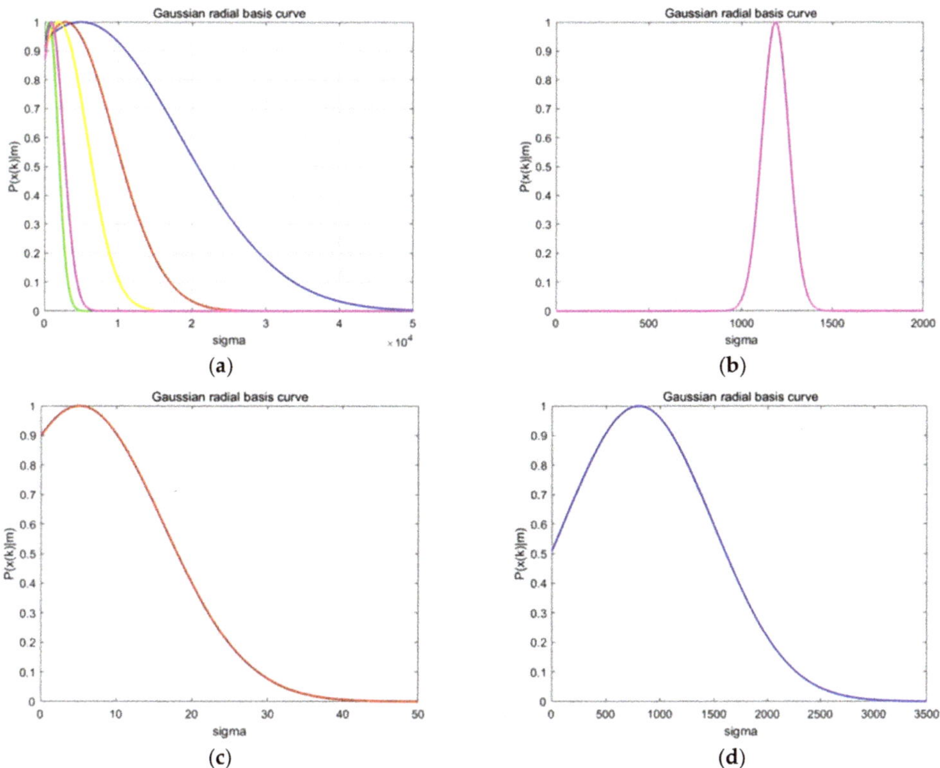

Figure 11. (**a**–**d**) Gaussian radial basis function curves of influencing factors of cultivated land.

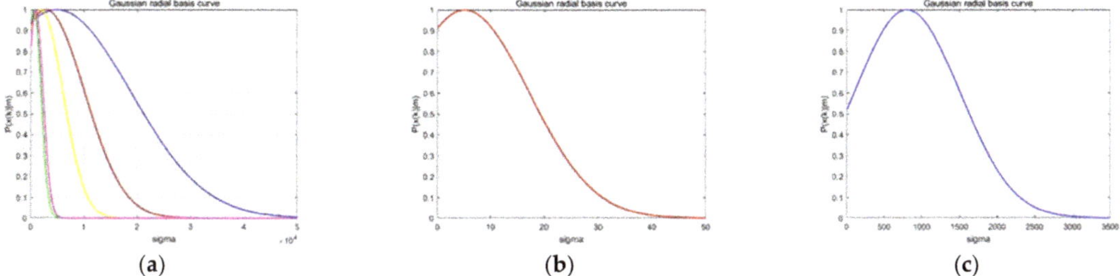

Figure 12. (**a**–**c**) Gaussian radial basis function curves of influencing factors of urban construction land.

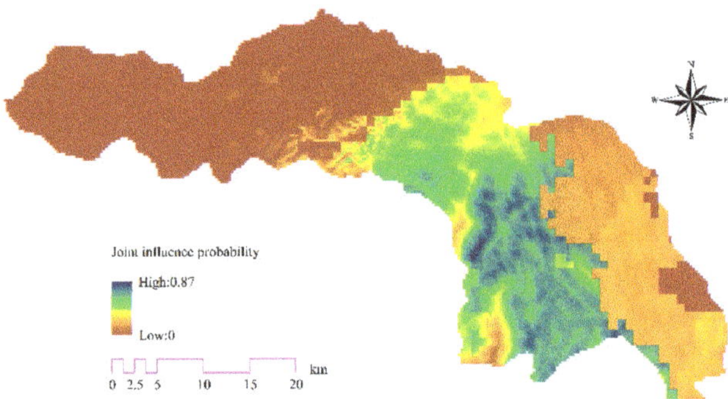

Figure 13. Probability distribution of joint influence of cultivated land.

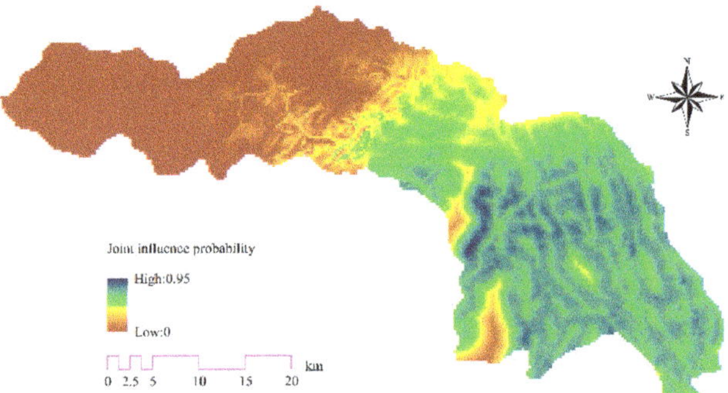

Figure 14. Probability distribution of joint influence of urban construction land.

4.4. Optimization Analysis of Land Use Structure

To solve the fuzzy mathematical programming problem in MATLAB R2019a, it is necessary to turn the fuzzy mathematical programming problem into the general programming problem. The parameters of constraint conditions include three types, including environmental parameters, economic parameters, social parameters, planning parameters and parameters of objective function, as shown in Tables 1 and 2. Finally, the value of λ is 2×10^{-8}; x_1 = 436.87 km^2, that is, the area of cultivated land is 436.87 km^2; x_2 = 69.42 km^2, that is, the area of construction land is 69.42 km^2; $z = 98.38$ CNY/kg, that is, the economic benefit brought by 1 kg carbon emission is CNY 98.38. The original construction land scale of Chongzhou was 68.5 km^2, while the construction land scale is controlled to 70 km^2, and the optimized result is 69.42 km^2. Therefore, it can be seen from the solution results that the construction land scale of Chongzhou in 2017 is within a reasonable range and can be increased appropriately but should not be too much. The cultivated land protection red line is 338 km^2, and the optimized results show that Chongzhou can have 436.87 km^2 of cultivated land. It is suggested that in addition to the reclamation of wasteland, greater efforts can be made in land consolidation and reclamation to make full use of high-quality agricultural resources. The cultivated land protection red line is 338 km^2. According to the calculation, the economic benefit of 1 kg carbon emission in Chongzhou is CNY 98.38, which belongs to the high economic carbon efficiency. In addition, Chongzhou is not a heavy industrial city with key carbon emission departments as its pillar industry. Its agriculture and tertiary industries are developed, and the secondary industry is dominated by green and intelligent industries. The annual carbon sink of Chongzhou is 2.3×10^9 kg, with a strong natural carbon neutralization field, which to a certain extent ensures that the carbon emissions generated in Chongzhou will not cause climate pressure in Wenchuan, Dujiangyan, Dayi and other surrounding areas and can be neutralized by itself.

Table 1. Constraint parameters.

Constraint Type	Constraint Parameters	Unit	Value
Environmental parameters	Soil erosion rate	%	2
	Erodible area	km^2	8
	Cultivated wastewater per unit area	kg/km^2	6.48×10^6
	Municipal wastewater per unit area	kg/km^2	18.91×10^6
	Total capacity of wastewater treatment	kg/year	$[1 \times 10^{10}, 1.2 \times 10^{10}]$
	Solid waste per unit area	kg/km^2	58.39×10^3
	Total capacity of solid waste treatment	kg	$[7 \times 10^7, 7.3 \times 10^7]$
Economic parameters	Unit farmland fixed assets investment	$yuan/km^2$	4.16×10^6
	Unit urban fixed assets investment	$yuan/km^2$	4.01×10^8
	Import investment	yuan	6.84×10^8
	Export investment	yuan	0
	Total investment	yuan	$[2.93 \times 10^{10}, 3 \times 10^{10}]$
Social parameters	Unit grain output	kg/km^2	2.51×10^5
	Total grain demand	kg	$[8.97 \times 10^7, 9 \times 10^7]$
	Unit cultivated labor	$people/km^2$	251
	Unit urban labor force	$people/km^2$	6 312
	Total labor force	people	$[5.42 \times 10^5, 6.65 \times 10^5]$
Planning parameters	Cultivated land protection red line	km^2	338
	Limits on the scale of construction land	km^2	70
	Non-negative restriction	km^2	0

Table 2. Objective function parameters.

Objection Parameters	Unit	Value
Economic benefits per unit of land area	$yuan/km^2$	8.60×10^6
	$yuan/km^2$	4.41×10^8
Carbon emissions from farmland wastewater treatment	kg/km^2	6.80×10^3
Carbon emissions from municipal wastewater treatment	kg/km^2	19.85×10^3
Carbon emissions from solid waste disposal	kg/km^2	62.28×10^3
Carbon emissions from livestock and poultry production	kg	9.59×10^7
Carbon emissions from people ploughing land	kg/km^2	19.85×10^3
Carbon emissions from urban construction land	kg/km^2	498.68×10^3
Carbon emissions from energy use on cultivated land	kg/km^2	1.07×10^3
Carbon emissions from urban energy use	kg/km^2	2.89×10^6

Finally, considering the application of two stage results, based on the parameters obtained from the Bayesian networks and fuzzy mathematical programming, we build the cellular automata (CA) by considering the impact of land use change of three kinds of basic situations, that is, the role of the neighborhood, the areas of conservation should be prevented from developing into urban areas as well as planning restrictions, as mentioned in Section 2.1.3. The spatial simulation of the optimized results is obtained. As shown in Figure 15, the gray area represents the protected areas and water areas that are prohibited from development and utilization, the green area represents the cultivated land, and the red area represents the construction land. By comparing the land use changes before and after simulation, it can be clearly seen that the construction land increment circled in yellow is located in the central and western parts of Chongzhou, as shown in Figure 15a. Correspondingly, the amount of cultivated land in this region will decrease, but the total area of cultivated land has space to increase. From the point of urban agglomeration, the region in the radiation zone between the state and the Chengdu core are more likely to develop into urban construction land, but the zone influenced by publicizing internal stated may also be influenced by external conditions, such as the urban siphon effect. Small cities such as Chongzhou are more susceptible to the siphon effect of large cities such as Chengdu, which is reflected in space.

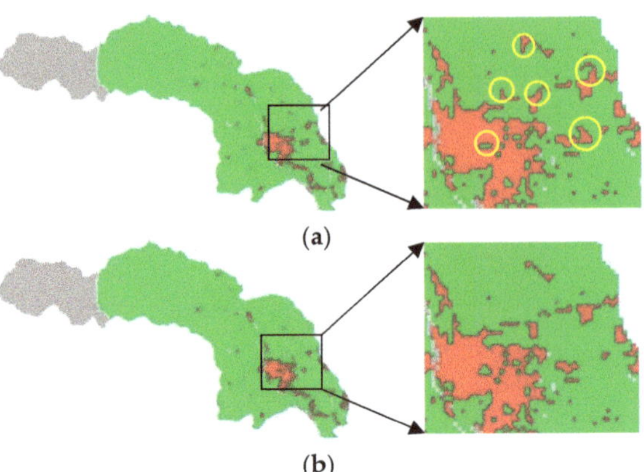

Figure 15. (**a**,**b**) Simulation results of land use spatial optimization of Chongzhou in 2017.

4.5. Behavior Rule Validation

In the 1940s, the Monte Carlo method was proposed. Monte Carlo is an experimental probabilistic method, which is based on a large number of randomized trials to describe the complex uncertainty problems encountered in practice [47]. It is a method that can be systematically and comprehensively explored, which is superior to the parameters of isolation, effectiveness analysis and single-class models in performance, especially in the ability to show the coupling error caused by coupling through the Monte Carlo test. Therefore, this study uses Monte Carlo thought to verify the transformation rules and completes simulation and calculation with ArcGIS and Excel. The Monte Carlo method determines the transition state of land use for each unit based on probability, is a more scientific method to deal with uncertain factors of land use. Considering the size of the study area, this study sampled 100,000 random points in the urban area, which can basically cover the whole area closely. Figure 16 shows the distribution of random points in a part of the area. In addition to the reserve, other non-urban cells into urban construction land cells have certain regularity and transformation rules. In order to verify the rationality of the conversion rules obtained in this study, according to the Monte Carlo idea, when the number of random points is large enough, the situation close to the actual situation can be calculated, that is, when the joint influence probability reaches that value, the cell transformation will occur. The verification results show that if policy restrictions are not taken into account, when the probability of the joint impact formulated in this study exceeds 50%, the state of cell can be changed under the condition of natural development and converted into a cell of urban construction land with an accuracy of 91.99%.

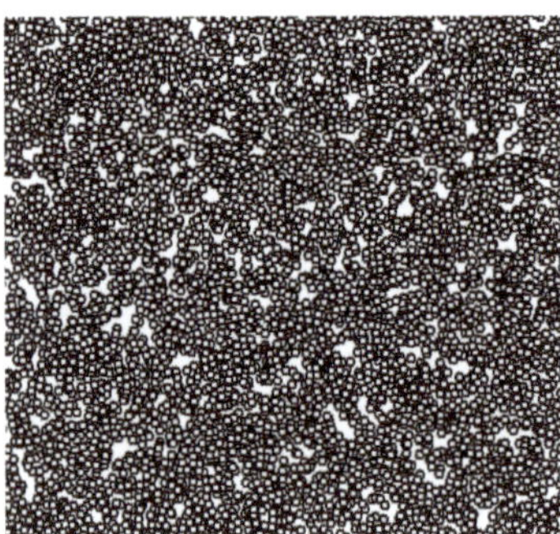

Figure 16. Schematic diagram of Monte Carlo random sampling.

5. Discussion

Based on the current situation and shortage of urban land use structure and spatial arrangement, this study analyzes the influencing mechanism of land use and optimize the land use structure under uncertainties by using a Bayesian network and fuzzy mathematical programming. Based on the results of the two stages of analysis, the cellular automata simulation is completed. The results indicated that, according to the joint influence probability of arable land and urban construction land, the best land space for cultivated land is in the middle and the south, while the conversion probability of the area near the east is low. Except for the western protection area, the joint impact probability of construction land in the other parts is generally similar. From the perspective of structure, the construction land increment of 1.5 km^2 can achieve the control of the construction land scale. After the optimization of fuzzy planning, the optimal construction land scale is 69.42 km^2. Under the condition that the cultivated land red line is guaranteed, there is still 98.87 km^2 of space for the increase in cultivated land. In order to make full use of high-quality agricultural resources conditions, the obtained parameters were applied to the simulation, which appeared that the increase in construction land would occur in the central and western part of Chongzhou, which may be caused by the urban siphon effect. According to Monte Carlo verification, when the conversion probability exceeds 50%, the cultivated land could be turned into urban construction land, with an accuracy of 91.99%. Therefore, this proposed framework is helpful to understand the process of land use and provides a reference for making scientific and reasonable territorial spatial planning and guiding land use practice under uncertainties.

Under conventional constraints, this study uses a small entry point to express one aspect of the uncertainty, namely, the uncertainty of the target conditions for economic carbon efficiency. The data were compared with Van et al.'s (2012) simulated land supply with known drivers of land use, and then the uncertainty was analyzed using the Monte Carlo method. Meyer et al. (2012) used a Bayesian network to help assess the uncertainties brought about by stakeholder knowledge modeling. This study considers the uncertainty in the optimization of land use structure and applied the results obtained by the uncertainty method to the spatial simulation.

As shown in the above results, the two-stage land use analysis framework constructed in this study has a good optimization effect. According to the analysis of spatial influencing factors, in the spatial pattern, cultivated land and urban construction land constitute a competitive relationship, and cultivated land is at a competitive disadvantage. If the construction land is not controlled, the agricultural space is bound to cause pressure. From the perspective of land use structure, Chongzhou can realize the growth of construction land and cultivated land area under this optimization framework, but the spatial distribution may change dynamically. Taking construction land as an example, this study simulated its spatial distribution, and the results shows that the increase could occur in the central and western part of Chongzhou, which may be caused by the radiation force of various conditions inside Chongzhou and the radiation force of the external core area of Chengdu. Although the development of the protected area is restricted, it does not constitute the conditions for the development of the urban area. Based on the analysis results of land use in Chongzhou, it can be seen that the spatial influence factors can objectively reflect the law of land use in Chongzhou, and the driving effect of all kinds of spatial influence factors on land use in Chongzhou is significant. The constructed Bayesian networks can capture the uncertainty of spatial influencing factors, especially in the case that all land use information cannot be captured at present, which highlights its advantages in applicability. In addition, fuzzy mathematical planning can supplement the role of land use planning on land use to a certain extent, quantify the impact including economy, environment and policy, help to understand the development of land use in an all-round way and provide some support for optimizing land use so as to promote the orderly, controllable and diversified development of cities. Based on the results of this study, this paper will provide some thoughts for urban land use from the following aspects: Firstly, based on the concept of sustainable development, more consideration should be given to the environmental benefits and carbon emissions of land use under uncertain and flexible conditions, so as to formulate more promising land plans for cities. Second, based on the influencing factors of land use, planning should correctly grasp the law of land use development, strictly protect high-quality land and guard against the expansion of inefficient use of construction land. The third aspect is to rationally plan functional areas according to different geographical locations of urban space, give full play to regional resource endowments and advantages and formulate more targeted land schemes.

Due to the limitation of current technology and foundations, there is still room for improvement and optimization in this study. The goal of this study is to provide support for the decision-making system, which can reflect the information and elements needed in the decision-making process to a certain extent. The Bayesian network and fuzzy mathematical programming method can capture more uncertainty, which is verified to be correct to in some sense, but the construction of the model is limited to the primary stage and cannot reflect all information and components. The real system is often much more complex than the idea, so it needs to be further explored to clarify more details inside the model, which have been rarely probed so far. The current challenges also include the quantification of external influencing factors, the accessibility of spatial data and the reasonable setting of fuzzy parameters. The uncertainty of land use will continue to exist in the future. Although, considering specific problems, the choice of models and methods may be different, it is still necessary to ensure that models and methods do not have a dominant impact on the research results, and more attention should be paid to practical problems.

6. Conclusions

From the perspective of land supply-side structural reform, this study took cultivated land and urban construction land in Chongzhou in 2017 as the main research objects, carried out a two-stage study on land use optimization and drew the following conclusions. The spatial distribution of joint impact probability of cultivated land and urban construction land shows that the best land space of cultivated land is in the middle and south, while the regional conversion probability near the east is low, which may be caused by the distribution of water sources and rainfall. The joint impact probability of urban construction land is relatively smooth. Except for the probability of the western reserve being low, the joint impact probability of other parts is generally similar. Compared with other areas, this area is far from the water area, so the overall probability is slightly affected. From the perspective of spatial performance, the form of urban construction land is relatively concentrated, while cultivated land is located in the periphery of urban construction land. In the long run, urban construction land has the dynamic characteristics of "catching up" with cultivated land. If the scale of urban construction land is not controlled and high-quality cultivated land resources are not protected, there will be a risk of imbalance in urban spatial structure, and the space of forest land, irrigated grassland and protected areas will be more tense. In addition to the reclamation of wasteland, Chongzhou can also make greater efforts in land consolidation and reclamation to make full use of high-quality agricultural resource conditions. Construction land can be appropriately increased, but it should not exceed the scale control of construction land. The carbon emissions from production, construction and living in Chongzhou can be fully absorbed by the local carbon sink, which can be self-sufficient, and the carbon emissions are well controlled. Therefore, Chongzhou plays an important role as an ecological conservation area in Chengdu. The results show that the increase in construction land can occur in the central and western parts of Chongzhou, which may be caused by the radiation force of various conditions inside Chongzhou and the external radiation force of the core area of Chengdu. The Monte Carlo verification results show that when the conversion probability is more than 50%, the cultivated land can be turned into urban construction land, with an accuracy of 91.99%, passing the test. In addition, the study provides a methodological reference for the formulation of more sustainable and ecologically friendly land schemes, explores the land use impact mechanism and land use impact under uncertain conditions, expands the perspective and method of land use impact mechanism research, further deepens the understanding of uncertainty in land use process and provides certain theoretical significance for land use development research.

Author Contributions: J.Y. and B.Q. conceived and designed the research; A.D. and B.Q. collected, managed, A.D. verified the data; M.Z. and S.L. calculated and analyzed the data and the results; J.Y. and M.Z. wrote the manuscript. All authors have read and agreed to the published version of the manuscript.

Funding: This research was supported by National Natural Science Foundation of China No. 41401631.

Institutional Review Board Statement: Not applicable.

Informed Consent Statement: Not applicable.

Data Availability Statement: Not applicable.

Conflicts of Interest: The authors declare that they have no conflict of interest.

References

1. Verstegen, J.A.; Karssenberg, D.; van der Hilst, F.; Faaij, A. Spatio-temporal uncertainty in Spatial Decision Support Systems: A case study of changing land availability for bioenergy crops in Mozambique. *Comput. Environ. Urban Syst.* **2012**, *36*, 30–42. [CrossRef]
2. Jiang, S.; Meng, J.J.; Zhu, L.K.; Cheng, H.R. Spatial-temporal pattern of land use conflict in China and its multilevel driving mechanisms. *Sci. Total Environ.* **2021**, *801*, 149697. [CrossRef]
3. Liang, X.; Guan, Q.F.; Clarke, K.C.; Liu, S.S.; Wang, B.Y.; Yao, Y. Understanding the drivers of sustainable land expansion using a patch-generating land use simulation (PLUS) model: A case study in Wuhan, China. *Comput. Environ. Urban Syst.* **2021**, *85*, 101569. [CrossRef]
4. Zhou, Y.; Li, X.H.; Liu, Y.S. Land use change and driving factors in rural China during the period 1995–2015. *Land Use Policy* **2020**, *99*, 105048. [CrossRef]
5. Alexander, P.; Prestele, R.; Verburg, P.H.; Arneth, A.; Baranzelli, C.; Batista e Silva, F.; Brown, C.; Butler, A.; Calvin, K.; Dendoncker, N.; et al. Assessing uncertainties in land cover projections. *Glob. Chang. Biol.* **2017**, *23*, 767–781. [CrossRef] [PubMed]
6. Grebitus, C.; Steiner, B.; Veeman, M. The roles of human values and generalized trust on stated preferences when food is labeled with environmental footprints: Insights from Germany. *Food Policy* **2015**, *52*, 84–91. [CrossRef]
7. Gao, P.P.; Li, Y.P.; Gong, J.W.; Huang, G.H. Urban land-use planning under multi-uncertainty and multiobjective considering ecosystem service value and economic benefit—A case study of Guangzhou, China. *Ecol. Complex.* **2021**, *45*, 100886. [CrossRef]
8. Ma, S.H.; Wen, Z.Z. Optimization of land use structure to balance economic benefits and ecosystem services under uncertainties: A case study in Wuhan, China. *J. Clean. Prod.* **2021**, *311*, 127537. [CrossRef]
9. Hersperger, A.M.; Oliveira, E.; Pagliarin, S.; Palka, G.; Verburg, P.; Bolliger, J.; Grădinaru, S. Urban land-use change: The role of strategic spatial planning. *Glob. Environ. Chang.* **2018**, *51*, 32–42. [CrossRef]
10. Harvey, D. *Space as a Keyword*; Blackwell Publishing Ltd.: Hoboken, NJ, USA, 2008.
11. Van der Hilst, F.; Verstegen, J.A.; Karssenberg, D.; Faaij, A.P.C. Spatiotemporal land use modelling to assess land availability for energy crops—Illustrated for Mozambique. *Glob. Chang. Biol. Bioenergy* **2012**, *4*, 859–874. [CrossRef]
12. Meyer, S.R.; Johnson, M.L.; Lilieholm, R.J.; Cronan, C.S. Development of a stakeholder-driven spatial modeling framework for strategic landscape planning using Bayesian networks across two urban-rural gradients in Maine, USA. *Ecol. Model.* **2014**, *291*, 42–57. [CrossRef]
13. Chuai, X.W.; Huang, X.J.; Wang, W.J.; Zhao, R.Q.; Zhang, M.; Wu, C.Y. Land use, total carbon emission's change and low carbon land management in Coastal Jiangsu, China. *J. Clean. Prod.* **2015**, *103*, 77–86. [CrossRef]
14. Han, D.; Qiao, R.L.; Ma, X.M. Optimization of Land-Use Structure Based on the Trade-Off Between Carbon Emission Targets and Economic Development in Shenzhen, China. *Sustainability* **2019**, *11*, 11. [CrossRef]
15. Xue, M.G.; Ma, S.H. Optimized Land-Use Scheme Based on Ecosystem Service Value: Case Study of Taiyuan, China. *J. Urban Plan. Dev.* **2018**, *144*, 04018016. [CrossRef]
16. Feng, Y.; Yang, Q.; Cui, L.; Liu, Y. Simulation and Prediction of Urban Land Use Change with Spatial Autoregressive Model Based Cellular Automata. *Geogr. Geo-Inf. Sci.* **2016**, *32*, 37–44.
17. Pedrielli, G.; Matta, A.; Alfieri, A.; Zhang, M.Y. Design and control of manufacturing systems: A discrete event optimisation methodology. *Int. J. Prod. Res.* **2018**, *56*, 543–564. [CrossRef]
18. Resat, H.G.; Turkay, M. A bi-objective model for design and analysis of sustainable intermodal transportation systems: A case study of Turkey. *Int. J. Prod. Res.* **2019**, *57*, 6146–6161. [CrossRef]
19. Du, G.; Zhang, Y.Y.; Liu, X.J.; Jiao, R.J.; Xia, Y.; Li, Y. A review of leader-follower joint optimization problems and mathematical models for product design and development. *Int. J. Adv. Manuf. Technol.* **2019**, *103*, 3405–3424. [CrossRef]

20. Xu, X.; Wu, Z.; Lin, Q.; Xie, L. Models on the topological optimization of crop mixing planting. *Syst. Sci. Compr. Stud. Agric.* **2003**, *19*, 63–65, 70.
21. Wang, F.; Zhou, B.; Xu, J. Land suitability spatial analysis and optimal exploitation pattern in tidal flat. *Trans. Chin. Soc. Agric. Eng.* **2008**, *24*, 119–123.
22. Zhang, H.-B.; Zhang, X.-H. Land use structural optimization of Lilin based on GMOP-ESV. *Trans. Nonferr. Met. Soc. China* **2011**, *21*, S738–S742. [CrossRef]
23. Sun, J.; Li, Y.P.; Suo, C.; Liu, Y.R. Impacts of irrigation efficiency on agricultural water-land nexus system management under multiple uncertainties—A case study in Amu Darya River basin, Central Asia. *Agric. Water Manag.* **2019**, *216*, 76–88. [CrossRef]
24. Ren, C.; Li, Z.; Zhang, H. Integrated multi-objective stochastic fuzzy programming and AHP method for agricultural water and land optimization allocation under multiple uncertainties. *J. Clean. Prod.* **2019**, *210*, 12–24. [CrossRef]
25. Ou, G.; Tan, S.; Zhou, M.; Lu, S.; Tao, Y.; Zhang, Z.; Zhang, L.; Yan, D.; Guan, X.; Wu, G. An interval chance-constrained fuzzy modeling approach for supporting land-use planning and eco-environment planning at a watershed level. *J. Environ. Manag.* **2017**, *204*, 651–666. [CrossRef] [PubMed]
26. Ma, S.; Xue, M.; Zhou, H. A method for planning regional ecosystem sustainability under multiple uncertainties: A case study for Wuhan, China. *J. Clean. Prod.* **2019**, *210*, 1545–1561. [CrossRef]
27. Bhuiyan, T.H.; Moseley, M.C.; Medal, H.R.; Rashidi, E.; Grala, R.K. A stochastic programming model with endogenous uncertainty for incentivizing fuel reduction treatment under uncertain landowner behavior. *Eur. J. Oper. Res.* **2019**, *277*, 699–718. [CrossRef]
28. Miç, P.; Koyuncu, M.; Hallak, J. Primary Health Care Center (PHCC) Location-Allocation with Multi-Objective Modelling: A Case Study in Idleb, Syria. *Int. J. Environ. Res. Public Health* **2019**, *16*, 811. [CrossRef]
29. Türk, E.; Zwick, P.D. Optimization of land use decisions using binary integer programming: The case of Hillsborough County, Florida, USA. *J. Environ. Manag.* **2019**, *235*, 240–249. [CrossRef]
30. Ouattara, P.D.; Kouassi, E.; Egbendewe, A.Y.G.; Akinkugbe, O. Risk aversion and land allocation between annual and perennial crops in semisubsistence farming: A stochastic optimization approach. *Agric. Econ.* **2019**, *50*, 329–339. [CrossRef]
31. Rodias, E.C.; Lampridi, M.; Sopegno, A.; Berruto, R.; Banias, G.; Bochtis, D.D.; Busato, P. Optimal energy performance on allocating energy crops. *Biosyst. Eng.* **2019**, *181*, 11–27. [CrossRef]
32. Santé-Riveira, I.; Crecente-Maseda, R.; Miranda-Barrós, D. GIS-based planning support system for rural land-use allocation. *Comput. Electron. Agric.* **2008**, *63*, 257–273. [CrossRef]
33. Aljanabi, A.A.; Mays, W.L.; Fox, P. Optimization Model for Agricultural Reclaimed Water Allocation Using Mixed-Integer Nonlinear Programming. *Water* **2018**, *10*, 1291. [CrossRef]
34. Hocaoğlu, M.F. Weapon target assignment optimization for land based multi-air defense systems: A goal programming approach. *Comput. Ind. Eng.* **2019**, *128*, 681–689. [CrossRef]
35. Nie, Y.; Avraamidou, S.; Xiao, X.; Pistikopoulos, E.N.; Li, J.; Zeng, Y.; Song, F.; Yu, J.; Zhu, M. A Food-Energy-Water Nexus approach for land use optimization. *Sci. Total Environ.* **2019**, *659*, 7–19. [CrossRef]
36. Lin, B.Q.; Zhu, J.P. Impact of China's new-type urbanization on energy intensity: A city-level analysis. *Energy Econ.* **2021**, *99*, 11. [CrossRef]
37. Howes, A.L.; Maron, M.; McAlpine, C.A. Bayesian Networks and Adaptive Management of Wildlife Habitat. *Conserv. Biol.* **2010**, *24*, 974–983. [CrossRef]
38. Mitraka, Z.; Doxani, G.; Frate, F.D.; Chrysoulakis, N. Uncertainty Estimation of Local-Scale Land Surface Temperature Products Over Urban Areas Using Monte Carlo Simulations. *IEEE Geosci. Remote Sens. Lett.* **2016**, *13*, 917–921. [CrossRef]
39. Chen, S.H.; Pollino, C.A. Good practice in Bayesian network modelling. *Environ. Model. Softw.* **2012**, *37*, 134–145. [CrossRef]
40. Liu, Y.; Martin, M.; Liu, Y.; He, J. Landuse change model based on cellar automata of decision-making with grey situation. *Remote Sens.* **2004**, *5232*, 575–586.
41. Shouzhang, P. *1-km Monthly Precipitation Dataset for China (1901–2017)*; National Tibetan Plateau Data Center: Beijing, China, 2020.
42. Peng, S.; Ding, Y.; Liu, W.; Li, Z. 1 km monthly temperature and precipitation dataset for China from 1901 to 2017. *Earth Syst. Sci. Data* **2019**, *11*, 1931–1946. [CrossRef]
43. Song, R.; Zhu, J.; Hou, P.; Wang, H. *Getting Every Ton of Emissions Right: An Analysis of Emission Factors for Purchased Electricity in China*; World Resources Institute: Beijing, China, 2013.
44. Zhang, C.; Zhong, L.J.; Fu, X.T.; Wang, J.; Wu, Z.X. Revealing Water Stress by the Thermal Power Industry in China Based on a High Spatial Resolution Water Withdrawal and Consumption Inventory. *Environ. Sci. Technol.* **2016**, *50*, 1642–1652. [CrossRef] [PubMed]
45. Chavez, A.; Ramaswami, A. Articulating a trans-boundary infrastructure supply chain greenhouse gas emission footprint for cities: Mathematical relationships and policy relevance. *Energy Policy* **2013**, *54*, 376–384. [CrossRef]
46. Liu, Z.; Davis, S.J.; Feng, K.; Hubacek, K.; Liang, S.; Anadon, L.D.; Chen, B.; Liu, J.; Yan, J.; Guan, D. Targeted opportunities to address the climate–trade dilemma in China. *Nat. Clim. Chang.* **2016**, *6*, 201–206. [CrossRef]
47. Papadopoulos, C.E.; Yeung, H. Uncertainty estimation and Monte Carlo simulation method. *Flow Meas. Instrum.* **2002**, *12*, 291–298. [CrossRef]

Article

Multi-Scenario Simulations of Land Use and Habitat Quality Based on a PLUS-InVEST Model: A Case Study of Baoding, China

Nan Hu [1], Dong Xu [1], Ning Zou [1], Shuxin Fan [1], Peiyan Wang [2] and Yunyuan Li [1,*]

[1] School of Landscape Architecture, Beijing Forestry University, 35-Qinghua East Road, Beijing 100083, China
[2] School of Art and Design, Beijing Forestry University, 35-Qinghua East Road, Beijing 100083, China
* Correspondence: bllyy@bjfu.edu.cn

Abstract: Habitat quality and ecosystem service value (ESV) are important foundations for sustainable development. Baoding, as the strategic hinterland of Beijing–Tianjin–Hebei, is of great significance to regional ecological conservation and sustainable urban development. Based on land-use data from 2000 to 2020, the land-use scenarios of natural development (ND), water protection (WP), forest rehabilitation (FR), and cultivated land protection (CP) in 2030 were predicted by the PLUS model and adopt the InVEST model and equivalent ESV table to assess ecological sustainability. The results show that: (1) From 2000 to 2020, the construction land in Baoding has increased by 812 km^2, and the cultivated land and forest land decreased by 708 km^2 and 154 km^2. Habitat quality is obviously deteriorating in 4.66% of the city. (2) Under different scenarios, the order of habitat quality is CP > FR > WP > ND. The habitat quality under each scenario is dominated by medium habitat quality. (3) Under different scenarios, the order of ESV is FR > CP> WP > ND. The fluctuation of forest land and cultivated land scale is affecting the ESV. (4) CP and FR will form a land-use pattern that has "high ecological quality and value", which better balances the economic development and ecological protection of Baoding. This research study will provide a reference for the effective allocation of land resources and will guide the formulation of urban land space planning policy in Baoding.

Keywords: land use; habitat quality; ecosystem service value; PLUS-InVEST model; multi-scenario simulations

Citation: Hu, N.; Xu, D.; Zou, N.; Fan, S.; Wang, P.; Li, Y. Multi-Scenario Simulations of Land Use and Habitat Quality Based on a PLUS-InVEST Model: A Case Study of Baoding, China. *Sustainability* **2023**, *15*, 557. https://doi.org/10.3390/su15010557

Academic Editors: Qingsong He, Jiayu Wu, Chen Zeng and Linzi Zheng

Received: 18 November 2022
Revised: 22 December 2022
Accepted: 26 December 2022
Published: 28 December 2022

1. Introduction

To achieve urban–rural integration in the new era, harmonizing regional ecological quality and ecological values is key to promoting regional economic sustainability at the macro level [1,2]. Thus, it is important to accurately assess the future evolution of regional habitat quality trends and calculate the future value of regional ecosystem services. Land-use change represents an important link between human socio-economic activities and natural environment evolution, habitat quality, and ecosystem service values (ESV) [3,4]. Currently, the analysis and prediction of land-use change can be achieved using various models; for example, Logistic-CA [5], CLUES [6,7], and FLUS [8,9], which are based on meta-cellular automata (CA) [10], have been widely used for the spatial optimization of land-use patterns as well as ecological red line delineation. The PLUS model, which is a novel land-use simulation model, exhibits higher simulation accuracy than that of the above-mentioned models and can be better integrated with planning policies to support regional sustainable development studies [11]. For instance, Li et al. coupled the grey multi-objective optimization model (GMOP) and the PLUS model to analyze the spatial and temporal evolution of ESV in the Sichuan–Yunnan region for 2026 under three scenarios: the business-as-usual scenario, the ecological development priority scenario, and the ecological and economic balance scenario [9]. Moreover, Lin et al. obtained the spatial and temporal distributions of carbon stocks in Guangdong Province from 1990 to 2020 by using the InVEST and PLUS models [12], while Lin et al. used the PLUS model for simulating land-use and land-cover change in the Fuxian Lake basin [13]. The above-mentioned studies demonstrated the applicability of the PLUS model for the analysis and prediction of land-use change.

Land-use change tends to affect the regional habitat quality, which is the ability of an ecosystem to provide space for sustainable survival and the development of suitable individuals and populations within a certain spatial and temporal scale [14,15]. Stable habitats are the prerequisite and basis for ecosystem services and functions, and they are essential for maintaining and enhancing biodiversity [16]. Habitat quality assessment methods have evolved from single-indicator analyses to multi-indicator integrated measurements and then to distributed model dynamic assessments. With the increasing concern for urban ecological environments and the quality of life, researchers constructed mathematical models such as ARIES [17], SolVES [18], and InVEST [19,20] to carry out a long time-series dynamic assessment, among which the InVEST model based on the distributed algorithm of 3S technology has been widely used. Furthermore, changes in land-use type can affect the structures, processes, and functions of ecosystems, which in turn significantly affect the ESV [21]. Accordingly, scholars studied spatial and temporal changes in ESV in different study areas [22,23], watersheds [24], and ecological zones [25]. Previous studies used the InVEST model to measure habitat quality in different regions, which effectively assesses the impact of changes in natural and social conditions on ecosystem service quality. Consequently, combining the InVEST and ESV will explore the synchronicity of ecological "quality" and "value" to characterize the relationship between ecological development and ecological economy.

With economic development and changes in urban spatial structure, Baoding, as an important city in the Beijing–Tianjin–Hebei city cluster, is facing ecological and environmental problems such as air pollution, water shortage, degradation, and the shrinkage of wetland and loss of biodiversity [26,27], which have become bottlenecks limiting the development of the region. In 2017, it was listed by the Ministry of Housing and Construction as the third batch of pilot cities for "Double Urban Repairs" [28]. To relieve the pressure on the ecological environment of Baoding and promote the development of ecological industries and the ecological integration of Beijing, Tianjin, and Hebei, this paper checks the ecological development-related policy documents of Baoding between 2000 and 2020 and innovatively extracts four policy focuses as the basis for setting scenarios, while combining the results of the scenarios with the highest ecological quality and value for re-validation to provide a reference for the effective allocation of land resources and the spatial planning of land under ecological protection.

2. Materials and Methods

This paper uses the PLUS model and multi-period high-precision land-use data to identify the land-use change patterns in Baoding from 2000 to 2020 and to simulate the land-use situation in Baoding in 2030 under the guidance of four important urban development policies. On this basis, the InVEST model was used to calculate the habitat quality of Baoding in different periods and situations, and the ecological service value equivalence table was introduced to estimate the value of ecosystem services in different situations in Baoding to identify the land-use pattern of "high ecological quality-high ecological value", and to further validate this land-use pattern. The land-use model was further validated to provide reference for the effective allocation of land resources and land space planning under ecological protection (Figure 1).

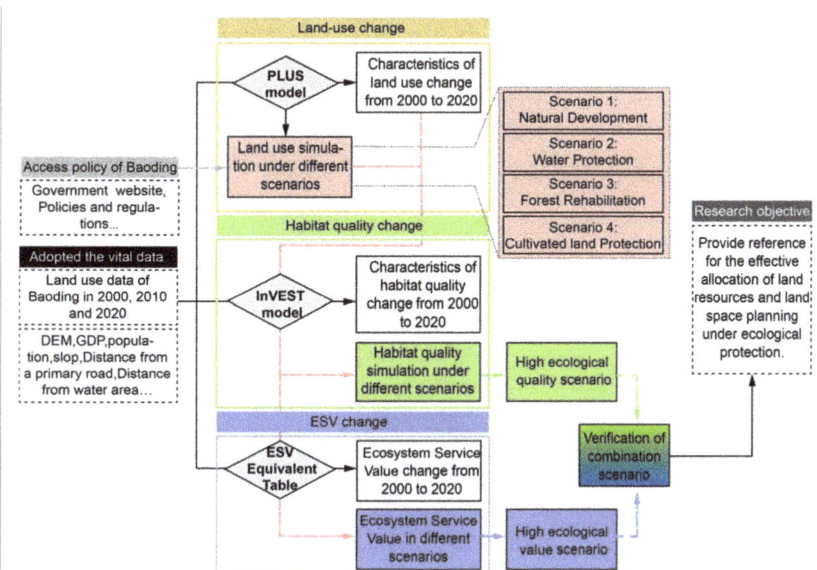

Figure 1. Research Framework.

2.1. Overview of the Study Area

Baoding is in the western part of the central Hebei province and between the eastern foothills of Taihang Mountains and the western part of the Jizhong Plain (113°45′32″–116°19′41″ E and 38°14′29″–39°57′3″ N) (Figure 2). It is adjacent to Beijing and Zhangjiakou in the north, Langfang and Cangzhou in the east, Shijiazhuang and Hengshui in the south, and Shanxi in the west. The topography of the city is high in the northwest and low in the southeast, while the landscape consists of mountains and plains and an uneven regional distribution of river resources. Furthermore, 5 municipal districts, 4 county-level cities, and 15 counties are under its jurisdiction, thereby exhibiting a total area of 22,190 km^2, a resident population of 11,546,000 in 2020, a GDP of CNY 335.33 billion (accounting for 9.38% of the total GDP of Hebei Province) and one of the best economic growth rates in the province [29].

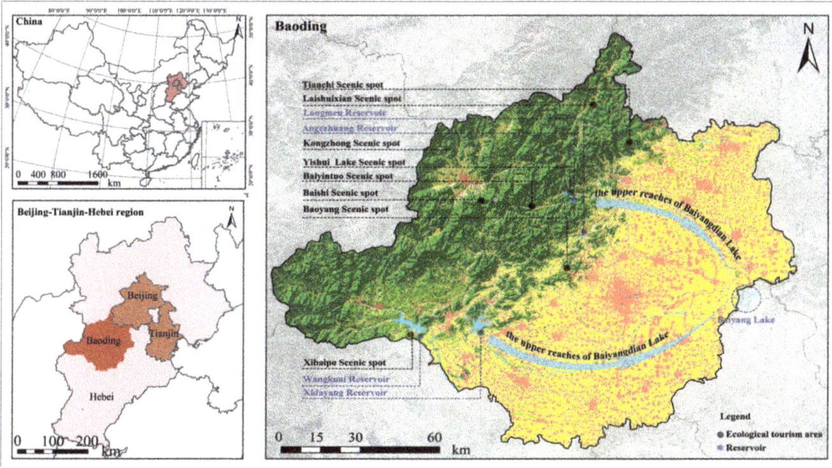

Figure 2. Location of the study area.

2.2. Data Source

The main data sources used in this study include the following: (1) Baoding grain production data and acreage data from the Baoding Statistical Yearbook and grain price data from the National Compilation of Cost and Benefit Information on Agricultural Products [30]; (2) land-use data from the Globeland30 global surface use database (http://www.globallandcover.com, accessed on 10 September 2022), in which the images for the land cover classification of development and update of GlobeLand30 are mainly 30 m multispectral images, including Landsat TM5 ETM+ [31]; (3) elevation data from the geospatial data cloud platform of the Computer Network Information Center of the Chinese Academy of Sciences (http://www.gscloud.cn, accessed on 10 September 2022), with a spatial resolution of 30 × 30 m; (4) temperature and precipitation data from the National Science and Technology Infrastructure Platform at the National Earth System Science Data Centre—Loess Plateau Sub-Center (http://loess.geodata.cn, accessed on 12 September 2022) [32]; (5) GDP and population data from the Resource and Environment Science Data Registration and Publication System; (6) roads, water, and railway data are cited from the National Geographic Information Resources Catalogue Service (www.webmap.cn). To unify the spatial accuracy of the data, the above data were processed by using the cropping and resampling tools of ArcGis 10.5 to convert them into 30 m × 30 m raster files.

2.3. Research Methodology

2.3.1. PLUS Model

Based on the metacellular automata model, the PLUS model integrates a land expansion strategy analysis module and a simulation model of future land-use changes via multi-class random patch seeding [11]. First, it analyzes the spatial characteristics of various land-use expansion forces and elucidates the driving factors between the two phases of land-use data. Second, it employs the random forest algorithm to sample and calculate the land-use expansion forces one by one to obtain the development probability of each land-use type. Finally, based on the roulette wheel strategy, it combines random patch generation, the transition transfer matrix, and the decreasing threshold mechanism to achieve optimization and to determine the final land-use pattern.

- Selection of land-use change factors

Land-use change factors are affected by multiple factors. On the basis of considering the accuracy and reality of the model, the factors of the natural environment, social economy, and traffic accessibility are comprehensively considered. According to the principles of availability, quantification, and the consistency of driving factors, 10 driving factors were selected (Table 1).

Table 1. Land-use change factors.

Influencing Factors	Driving Factors	Data Source
Natural environmental	Elevation Slope	Computer Network Information Center of the Chinese Academy of Sciences (http://www.gscloud.cn, accessed on 10 September 2022)
	Average annual precipitation Average annual temperature	National Science and Technology Infrastructure Platform-National Earth System Science Data Centre–Loess Plateau Sub-centre (http://loess.geodata.cn, accessed on 12 September 2022)
	Distance to water	National Geographic Information Resources Catalogue Service (www.webmap.cn, accessed on 8 September 2022)
Social economy	GDP Population	Resource and Environment Science Data Registration and Publication System (https://www.resdc.cn, accessed on 9 September 2022)
Traffic accessibility	Distance to railway Distance to high speed Distance to the road	National Geographic Information Resources Catalogue Service (www.webmap.cn, accessed on 8 September 2022)

- Setting neighborhood effects

The neighborhood effects can reflect the expansion capacity of different land types [33]. When applying the PLUS model, it is necessary to determine the ease of interconversion between different land-use types by using neighborhood weights as follows [11]:

$$\Omega_{i,k}^{t} = \frac{\mathrm{con}\left(c_i^{t-1} = k\right)}{n \times n - 1} \times w_k \tag{1}$$

where $\mathrm{con}\left(c_i^{t-1} = k\right)$ represents the total number of grid cells occupied by land-use type k at the last iteration within the $n \times n$ window (n is generally taken as 3) [34]; and w_k is the weight among the different land-use types between $[0, 1]$. The neighborhood weights are set based on existing research results and the actual conditions of Baoding (Table 2).

Table 2. Neighborhood weights parameters.

Type of Land-Use	Neighborhood Weights	Type of Land-Use	Neighborhood Weights
Cultivated land	1.00	Forestland	0.69
Grassland	0.11	Shrubland	0.00
Wetland	0.00	Water	0.01
Construction land	0.24		

- Verification of model accuracy

The PLUS model is often tested using point-by-point comparisons or random validations. Random validations are generally suitable for large-scale simulations, whereas point-by-point comparisons are suitable for relatively small areas [35]. In this study, based on the actual land-use data in the Baoding, this study uses a Kappa coefficient and overall accuracy to test the simulation accuracy of the model. The overall accuracy (OA) refers to the ratio of the number of correctly classified class cells to the total number of classes. The Kappa coefficient is often used to compare the similarity between two images [13]. The expression of the kappa coefficient is as follows:

$$\mathrm{kappa} = \frac{P_0 - P_c}{P_p - P_c} \tag{2}$$

where P_0 is the observed agreement rate between the reference map and the simulation results; P_c is the expected proportion of correct simulations in the random case; and P_p is the proportion of correct simulations in the ideal classification case and is generally taken as 1.

The results show that the Kappa coefficient of the predicted images in 2020 is 0.869 and the overall accuracy is 0.925, indicating that the PLUS simulation results are spatially consistent and exhibit good applicability. Thus, this approach can be used for future land-use simulation models based on Baoding.

2.3.2. Scenario Setting

Baoding is the ecological environment support area of Beijing–Tianjin–Hebei. According to the characteristics of land-use change in Baoding and the guidance of urban development policies, this study set up 4 scenarios, natural development, water protection, forest rehabilitation, and cultivated land protection, to predict the land-use change in Baoding in 2030. Simulation parameters include the land-use demand, conversion constraint, transition matrix, and neighborhood weights.

1. Natural development: This scenario simulates habitat quality and EVS development until 2030 when the land is not affected by any policies and fully complies with the current situational changes from 2000 to 2020, so no restricted area and land-use change restrictions are set in this scenario. Based on the high-precision land-use data

released by the government in 2000, 2010, and 2020, the Markov chain of the PLUS model was used to identify the changing rules for various types of land in the past 20 years, and the identification results were used as the basis for land-use changes in this scenario. It is also the control group.

2. Water protection: Baoding's rich wetland resources are important for the ecological management of northern China and the Beijing–Tianjin–Hebei ecosystem [36]. The scenario simulates habitat quality and EVS development until 2030 under water protection policies. The scenario is based on the State Council's "Wetland Protection Plan for Hebei Province [37]" and the "Baoding Ecological and Environmental Protection Plan [38]", which propose increasing the wetland and water protection rate in Baoding by 9% by 2030 and to enhance the stability of wetland and water ecosystems, as well as the core strategy of converting degraded cultivated land into wetland and water. Firstly, this condition was input into the model to adjust the land demand predicted by a Markov chain. Then, the existing wetland and water were used as restricted areas and the change of wetland and water was restricted. Next, neighborhood weight parameters were set according to Table 2. Lastly, the proportion and spatial distribution of land-use under this scenario were obtained.
3. Forest rehabilitation: Forestland is the root of the survival and development of forests and wildlife and has an important position in maintaining ecological security. Forestland resources in Baoding play an important role in the ecological space of Beijing–Tianjin–Bao and the water conservation of the Yanshan–Taihang Mountains. The scenario simulates habitat quality and ESV development to 2030 under forestland conservation policies. The scenario is based on the government's documents, including the "Baoding Forest and Water Ecosystem Construction Plan [39]", which proposes an optimal proportion of forested land in Baoding of 35% by 2030, and the State Council document "New Round of Returning Cultivated Land to Forest [40]" and "14th Five-Year Plan for Forestry and Grassland Protection and Development in Hebei Province", which specifically propose the conversion of severely sandy land to forestland. The core strategy involves converting cultivated land into forest land. Firstly, this condition was input into the model to adjust the land demand predicted by the Markov chain. Then, the existing forestlands in the nature reserve were used as restricted areas and the change of forestland was restricted. Next, neighborhood weight parameters were set according to Table 2. Lastly, the proportion and spatial distribution of land-use under this scenario were obtained.
4. Cultivated land protection: Cultivated land is an important foundation for consolidating and improving food production capacity and ensuring national food security [36], and Baoding's cultivated land resources are important for the construction of the Beijing–Tianjin Agricultural Circle, the Yanshan–Taihang Mountains, and the pre-mountain agricultural area. The scenario simulates habitat quality and ESV development up until 2030 under cultivated land protection policies [41]. The scenario is based on the State Council's documents 24 and 44 of 2020 and the Baoding government's "Spring Thunder Action" for cultivated land protection from 2021, which strictly controls the extent of existing cultivated land and focuses on improving its quality according to the optimal protection model. Firstly, this condition was input into the model to adjust the land demand predicted by the Markov chain. Then, the existing permanent basic cultivated land was used as restricted areas and the change of cultivated land was restricted. Next, neighborhood weight parameters were set according to Table 2. Lastly, the proportion and spatial distribution of land-use under this scenario were obtained.

2.3.3. InVEST Model

The habitat quality module of the InVEST model assesses habitat quality, while reflecting the genetic variation and species reproduction potential; this is achieved by considering the sensitivity of each land cover type relative to threat factors as well as the intensity of external threats [42]. Consequently, the statuses of different resources and conditions in the environment are determined for the survival and development of individuals or populations. The relative impacts of each threat and the distance between the habitat grid and the threat need to be calculated before habitat quality can be calculated using the InVEST model. The formula is as follows:

$$i_{rxy} = 1 - \left(\frac{d_{xy}}{d_{r\ max}}\right) \text{ if linear} \quad (3)$$

$$i_{rxy} = exp\left(-\left(\frac{2.99}{d_{r\ max}}\right)d_{xy}\right) \text{ if exponential} \quad (4)$$

where d_{xy} is the linear distance between grid cells x and y, and $d_{r\ max}$ is the maximum effective distance of threat r's reach across space.

Habitat degradation is calculated as follows:

$$\mathrm{D_{xj}} = \sum_{r=1}^{R}\sum_{y=1}^{Y_r}\left(\frac{w_r}{\sum_{r=1}^{R} w_r}\right) r_y i_{rxy} \beta_x S_{jr} \quad (5)$$

where y indexes all grid cells on r's raster map, and Y_r indicates the set of grid cells on r's raster map.

The final measure of biodiversity and ecosystem function is based on a habitat quality index, which is calculated as follows:

$$\mathrm{Q_{xj}} = \mathrm{H_j}\left(1 - \frac{\mathrm{D_{xj}^z}}{\mathrm{D_{xj}^z} + \mathrm{k^z}}\right) \quad (6)$$

where Q_{xj} is the habitat quality index for land-use type j in grid x, and H_j is the habitat suitability for land-use type j, D_{xj} is the habitat degradation for land-use type j in grid x, k is the half-saturation constant, and z is the default parameter of the model (usually taken as 2.5).

Since the sensitivity of different land-use types varies toward habitat quality threats, only the impact of human activities on the habitat is considered; the more sensitive the land-use unit, the lower its ability to resist disturbances and the more severe the degradation. Habitat suitability and its sensitivity to different threat factors were determined based on relevant information regarding Baoding and previous studies in the neighboring areas [43] (Table 3). Four types of areas with high population activity were selected as stressors, namely: cultivated land, construction land, roads, and railways (Table 4).

Table 3. Sensitivity parameters of different land types to habitat threat factors.

Type of Land-Use	Habitat Suitability	Threats Factors			
		Cultivated Land	Construction Land	Railways	Roads
Cultivated land	0.5	0.3	0.9	0.7	0.7
Forestland	1	0.8	0.9	0.8	0.7
Grassland	0.4	0.3	0.3	0.55	0.55
Shrubland	1	0.5	0.7	0.35	0.35
Wetland	0.7	0.7	0.8	0.4	0.4
Water	0.9	0.8	0.85	0.7	0.65
Construction land	0	0	0	0	0

Table 4. Threat factor parameters.

Threats Factors	Maximum Duress Distance/km	Weight
Cultivated land	3	0.5
Construction land	5	0.8
Railways	5	1
Roads	3	0.6

2.3.4. Calculating the ESV

The ESV in Baoding was determined by referring to the model proposed by Costanza et al. [21]. Then, it was combined with the ESV equivalent factor proposed by Xie et al. [22,23]. The standard equivalent is the economic value of the natural food produced by one hectare of cropland in a year at the national average yield level. It is generally considered to be equivalent to one-seventh of the economic value of food produced on a hectare of cropland at market prices [44,45]. Table 5 shows the ESVs of grain yield per unit area of the ecosystem in Baoding (1567.76 CNY/hm^2/a).

$$ESV = \sum_{f=1}^{m}\sum_{i=1}^{n}(C_f \times E_{fi} \times A_i) \tag{7}$$

where ESV is the ecosystem service value in CNY/yr, i is the land-use type and n is the number of land-use types (n = 7 in this study), f is the ecosystem service type and A_i is the area of land-use type i in hm^2. E_{fi} is the equivalent value of ecosystem service f for land-use type i, m is the number of ecosystem service types (m = 11 in this study), and C_f is the economic value of 1 unit of ecosystem services in CNY hm^{-2}yr^{-1}. This equation can also be represented as follows:

$$E = \frac{1}{7}\sum_{i=1}^{n}\frac{m_i p_i q_i}{TotalArea} \tag{8}$$

where E is the economic value of 1 unit of ecosystem services, i is the grain crop type, m_i is the average price of grain crop i in Baoding, p_i is the yield of grain crop i in kg/hm^2, q_i is the planting area of grain crop i in hm^2, and TotalArea is the total area planted in grain crops (hm^2).

Table 5. ESVs per unit area of the ecosystem (CNY/hm^2).

Types of Ecosystem Services		Cultivated Land	Forest Land	Grass Land	Water
Supply Services	Food production	1332.60	395.86	365.81	1026.89
	Raw material production	627.11	909.31	538.27	572.24
	Water supply	31.36	470.33	297.88	8528.65
Mediation Services	Gas regulation	1050.40	2990.52	1891.77	2092.97
	Climate regulation	564.40	8948.03	5001.18	4617.08
	Purifying the environment	156.78	2622.09	1651.38	7172.54
	Hydrological regulation	423.30	5855.61	3663.35	99,137.77
Support Services	Soil conservation	1614.80	3641.14	2304.62	2539.78
	Maintaining nutrient circulation	188.13	278.28	177.68	195.97
	Biodiversity	203.81	3315.83	2704.40	8168.07
Cultural Services	Aesthetic Landscape	94.07	1454.10	877.95	5189.31

3. Results

3.1. Analysis of Land-Use Change

3.1.1. Characteristics of Land-Use Change from 2000 to 2020

During 2000–2020, the land-use of Baoding was dominated by cultivated land, forestland, and construction land, with the sum of the areas of these land types accounting for over 90%, cultivated land and forestland accounted for 49% and 34% of the total area on average per year, respectively (Table 6). Over the past 20 years, forestland in the middle- and low-lying mountainous areas of Baoding was gradually converted to cultivated land and construction land (Figure 3). Meanwhile, the grassland area in the hilly region, which acts as a border between the mountainous area and the plain area, slightly increased. Overall, the expansion of construction land in the plain area is prominent, and the main urban area shows a clear trend of expansion toward the north.

Table 6. Proportion of land-use types in Baoding from 2000 to 2020 (%).

Year	Type of Land-Use						
	Cultivated Land	Forest Land	Grass Land	Shrub Land	Wet Land	Water	Construction Land
2000	49.96	33.85	5.37	0.01	0.03	0.46	10.32
2010	50.18	33.93	5.18	0.01	0.03	0.44	10.24
2020	46.77	33.16	5.56	0.02	0.03	0.49	13.97

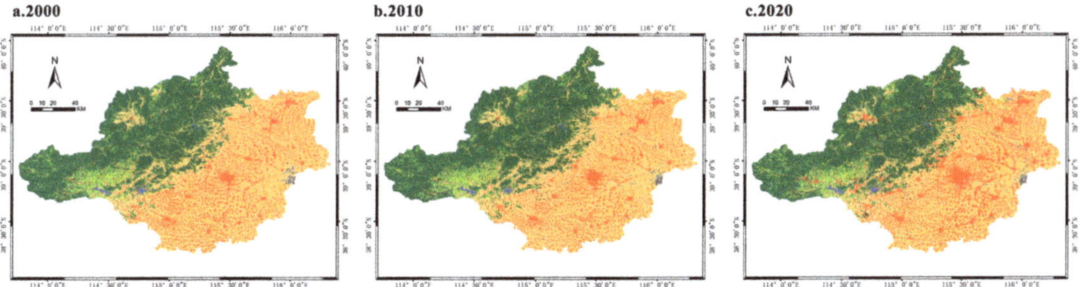

Figure 3. Spatial distribution of land-use from 2000 to 2020.

Cultivated land and forestland account for the highest share of the total area of Baoding. However, the area of cultivated land and forestland slightly increased during 2000–2020, which was followed by a significant reduction; that is, cultivated land decreased from 11,090 km^2 in 2000 to 10,380 km^2 in 2020, while forestland decreased from 7510 km^2 in 2000 to 7360 km^2 in 2020. Accordingly, there is a dire need to conserve cultivated land and forestland. In 20 years, the area of construction land grew by 812 km^2, which represents the highest increase in the area of any land type. Meanwhile, the area of wetland has decreased by 0.53 km^2, and the area of water areas has increased by 5 km^2; this contrasting trend shows that some wetland gradually transformed into water lands, and that the ecological protection of the wetland is necessary. Grassland and shrubland slowly increased over these 20 years by 41 km^2 and 4 km^2, respectively. Overall, the land-use patterns in Baoding show a significant increase in construction land, a significant decrease in cultivated land and forestland, and a slight shift in other land types.

3.1.2. Land-Use Simulation under Different Scenarios

Land-use simulations for 2030 were carried out based on the neighborhood weights in Table 2 and the transfer cost matrix under different scenarios. Compared with the 2020 land-use data, the change in the land-use areas in Baoding in 2030 varied under different scenarios (Table 7); the specific scenario settings and corresponding results have been mentioned below.

Table 7. Proportion (%) and decadal change (km^2) of land-use types in Baoding under different scenarios in 2030.

Type of Land-Use	Natural Development		Water Protection		Forest Rehabilitation		Cultivated Land Protection	
	Proportion	Decadal Change	Proportion	Decadal Change	Proportion	Decadal Change	Proportion	Decadal Change
Cultivated land	44.33	−540.54	44.33	−540.54	42.99	−837.91	46.77	0.00
Forestland	32.58	−129.71	32.62	−120.78	35.00	407.82	33.87	156.46
Grassland	5.69	28.64	5.69	28.64	5.69	28.64	4.82	−164.25
Shrubland	0.02	−0.03	0.02	−0.12	0.02	−1.21	0.02	−0.12
Wetland	0.01	−3.60	0.03	1.66	0.02	−2.47	0.03	0.00
Water	0.51	5.69	0.55	13.32	0.52	7.82	0.49	0.11
Construction land	16.86	639.55	16.76	617.83	15.76	397.31	14.01	7.80

Natural development scenario: The increase in construction land area is higher than the other three scenarios, and cultivated land and forestland are still the main contributors to the increase in construction land area. Under this scenario, the forestland area in Baoding decreased by 130 km^2, and the cultivated land area decreased by 541 km^2 (becoming the largest land type to be transformed). Meanwhile, the area of construction land increased by 636 km^2, whereas the area of other land types only changed slightly. Based on the spatial distribution of land-use changes (Figure 4), there is a tendency for cultivated land to transform into forestland in the northwestern part of the mountainous area and at the edge of the nature reserve in the southwestern part of Baoding. Nevertheless, construction land in the northern part of the hilly area, the northeastern part of the plain area, and the central city will continue to expand outward along the urban edges, whereas construction land in the southern part of the plain area is likely to transform into cultivated land.

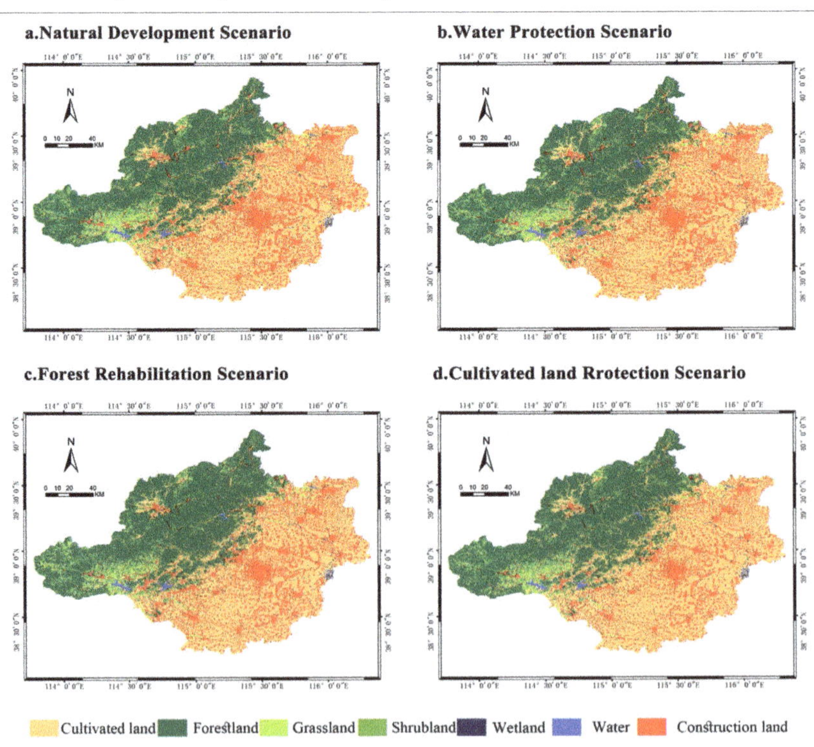

Figure 4. Spatial distribution of land-use change under different scenarios in 2030.

Water protection scenario: The land-use change trend in this scenario is similar to ND, the reduction in forestland and cultivated land has not been effectively curbed, but the transfer of ecological land is lower and the range of change is smaller. Meanwhile, the water body and wetland will, respectively, increase by 13 km^2 and 2 km^2, because a series of ecological restoration and comprehensive management measures have been carried out in the upstream rivers of Baiyangdian Lake, the water area along the rivers and lakes in the mountainous and hilly areas has increased.

Forest rehabilitation: The area of cultivated land in Baoding was further compressed compared with the ND, with a reduction of 838 km^2, which was the largest reduction in the four scenarios. The expansion of construction land was still obvious, but its expansion rate was effectively controlled from 3% in the ND to 2% because the development of construction land will also be restricted to a certain extent when returning cultivated land to forest. Meanwhile, cultivated land in mountainous and hilly areas (mainly concentrated in the southwest with complex terrain and developed water systems) transferred to woodland.

Cultivated land protection: Under the CP, land use was maintained correspondingly with the land-use situation in 2020. The area of cultivated land remained unchanged, the area of forestland increased by 156 km^2, the area of grassland decreased by 164 km^2, and the area of construction land only increased by 8 km^2. With respect to the areas of land-use change concentrated along the southwest waters of mountainous areas, cultivated land protection manifested as the transfer of grassland to forestland at the edge of the water body and the transfer of grassland to construction land at the junction of hilly area and plain area, and the expansion of construction land in plain areas no longer has the phenomenon of large-scale diffusion due to the strict restriction of cultivated land protection systems and policy guidance with respect to national stock development.

3.2. Analysis of Habitat Quality Change

3.2.1. Characteristics of Habitat Quality Change from 2000 to 2020

Based on previous studies on habitat quality and the actual situation in Baoding [46], the natural breakpoint method was used to classify habitat quality into five categories: very low (0.0–0.1), low (0.1–0.4), medium (0.40–0.68), high (0.68–0.90), and very high (0.90–1.0). Table 8 shows that the habitat quality in Baoding is relatively stable between 2000 and 2010, exhibiting an average change of 67 km^2 or <1% for all land types. However, during 2010–2020, the changes in habitat quality increased sharply, with an average annual change of 413 km^2 being observed for all land types. In particular, the low habitat quality area increased by 827 km^2, the medium habitat quality area decreased by -874 km^2, and the habitat quality in other areas changed relatively slightly (1% on average). Therefore, a continuous transformation trend of medium habitat quality areas to low habitat quality areas was observed.

Table 8. Proportion of each habitat quality category in Baoding from 2000 to 2020 (%).

Year	Habitat Quality Rating				
	Very Low	Low	Medium	High	Very High
2000	10.32	5.36	49.97	0.31	34.03
2010	10.25	4.68	50.71	0.33	34.04
2020	13.97	5.59	46.77	0.34	33.32

To further analyze the changes in habitat quality, the spatial distribution of habitat quality in Baoding was calculated (Figure 5). The land-use type is mainly woodland and the ecological environment faces decreased human disturbance, thereby exhibiting strong stability. Areas with low and medium habitat quality are mainly concentrated in mountainous counties and towns because these contain large construction, agricultural, and forestlands; since they exhibit a strong heterogeneity, habitat degradation is higher. Habitat quality in hilly areas is mainly influenced by the land-use types controlled by natural factors within the area; for example, water and grasslands dominate the land-use types along the irrigation areas and reservoirs in the southwestern part of Baoding, with patches of woodland being fragmented and interspersed with grasslands, thereby resulting in a low overall habitat quality. In plain areas, due to the significant increase in urban land-use and human activity, habitat quality is degrading and has been mainly categorized as low and medium.

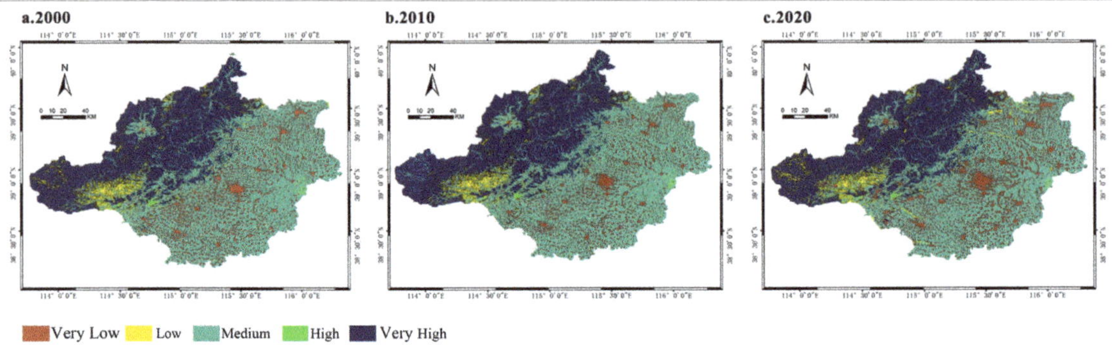

Figure 5. Spatial distribution of habitat quality from 2000 to 2020.

The ESV in Baoding from 2000 to 2020 showed a decreasing trend year by year, which was CNY 340, 339 and 332 billion. Forestland has the highest ESV, with an average value of CNY 230 billion (accounting for 68% of the total value); this indicates that although forestland is not the largest land type in Baoding, it plays a more important role in ecological protection and development. Cultivated land, which is the largest land type in terms of area, has an average ESV of CNY 68 million (accounting for 20% of the total value). Because of economic, social, and regional development, cultivated land is more prone to encroachment. Furthermore, it plays the role of restraining the uncontrolled expansion of construction land, which has a significant positive impact on ecological development. Variations in the ecological value of water, and grassland are low, indicating that changes in these regions have less influence on the ESV in Baoding. Therefore, when formulating relevant policies, government departments can consider improving or protecting forestland and cultivated land to improve the ecological quality of Baoding's regional environment to maintain its ecosystem.

3.2.2. Habitat Quality Simulation under Different Scenarios

Under the natural development scenario, the area of medium and high habitat quality changed to low habitat quality (Table 9). The area of low habitat quality expanded by 636 km^2, the area of medium habitat quality decreased by 539 km^2, and the area of high habitat quality decreased by 124 km^2. The natural development focuses on meeting the needs of future population growth and urban construction, while the infrastructure of the old city has difficulty bearing the load and cannot meet the conditions for the large-scale development of new projects, which results in the annexation of cultivated land and forestland by construction land to maintain the expansion trend, resulting in a significant decline in habitat quality. The ESV of the scenario is calculated to be CNY 33 billion (Table 10).

Table 9. Proportion (%) and decadal change (km^2) of habitat quality in Baoding under different scenarios in 2030.

Type of Land-Use	Natural Development		Water Protection		Forest Rehabilitation		Cultivated Land Protection	
	Proportion	Decadal Change	Proportion	Decadal Change	Proportion	Decadal Change	Proportion	Decadal Change
Very low	16.84	636.02	16.77	619.55	15.76	397.37	13.35	−137.85
Low	5.68	20.12	5.68	20.56	5.67	19.10	4.81	−173.46
Medium	44.34	−539.33	44.34	−539.76	43.01	−835.66	47.44	147.63
High	0.38	7.06	0.38	8.04	0.35	0.16	0.34	0.10
Very high	32.76	−123.87	32.83	−108.39	35.21	419.03	34.06	163.58

Table 10. ESV in Baoding under different scenarios in 2030 (CNY billion).

Type of Land-Use	2020	2030 Scenario Simulation			
		Natural Development	Water Protection	Forest Rehabilitation	Cultivated Land Protection
Cultivated land	6.52	6.18	6.18	6.00	6.52
Forestland and shrubland	22.74	22.34	22.37	24.00	23.22
Grassland	2.40	2.46	2.46	2.46	2.08
Water and wetland	1.59	1.62	1.80	1.66	1.59
Total	33.25	32.60	32.80	34.11	33.42

Under the water protection scenario, the changing trend with respect to habitat quality was similar to natural development (Figure 6), but the decrease in habitat quality was lower. The low habitat quality area increased by 620 km^2, the high habitat quality area decreased by 108 km^2, and the higher habitat quality increased by 8 km^2. Meanwhile, the habitat improvement area is concentrated on the coast of Yishui Lake in the central part of Baoding where social capital is introduced to protect and comprehensively develop the lake, and a national water conservancy scenic spot will be established so that the habitat quality along the coast of Yishui Lake will continue to improve. The total value of ESV in this scenario is CNY 33 billion, an increase of CNY 207 million compared with the natural development.

Under the forest rehabilitation scenario, the area of low habitat quality area increased by 397 km^2, and the area of high habitat quality area increased by 419 km^2. The project of returning cultivated land to forest in Baoding involves 17 counties and 210,000 farmers. The entire area has a wide coverage and the mountainous area has also been protected in a large area. Therefore, the area of habitat quality improvement has increased significantly, and it is mostly concentrated on key areas such as both sides of traffic arteries, around towns, around lakes and reservoirs, and eco-tourism areas, initially forming the first ecological barrier of sandstorm invasions to the south. The ESV reached CNY 34 billion, an increase of CNY 859 million over 2020.

Under the cultivated land protection scenario, the low and very low habitat quality areas respectively decreased by 138 km^2 and 173 km^2, and the medium and high habitat quality areas increased by 147 km^2 and 164 km^2. Cultivated land protection is related to national food security. Baoding has carried out strict non-agricultural and non-grain supervision on permanent basic cultivated land. Thus, the amount of cultivated land is effectively guaranteed in this scenario, slowing down the expansion rate of construction land. However, the cultivated land use in the western mountainous and hilly areas of Baoding is greatly affected by natural conditions, and its own ecological environment is relatively fragile, which makes the middle habitat quality level area increase compared with other scenarios. The ESV of the cultivated land protection scenario was CNY 33 billion, and it increased by CNY 164 million compared with 2020.

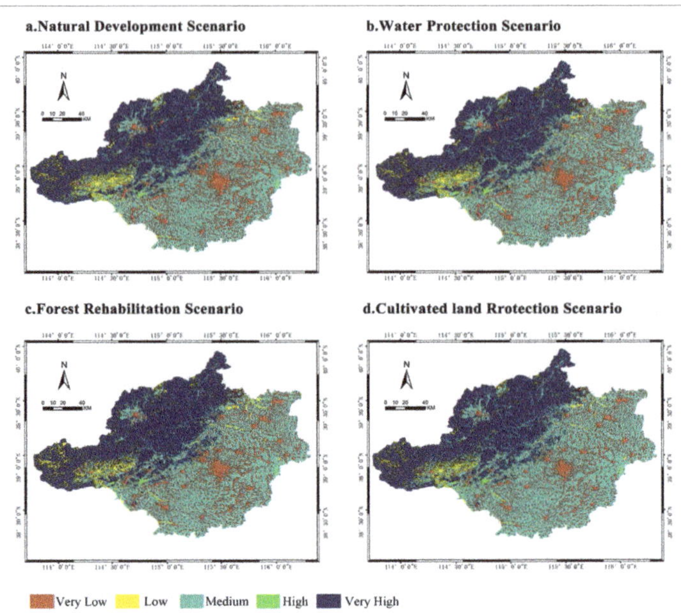

Figure 6. Spatial distribution of habitat quality change under different scenarios in 2030.

3.3. Verification of Combination Scenario

The results of the four scenarios show that the habitat quality of the cultivated land protection scenario is the best and the ecosystem service value of the forest rehabilitation scenario is the highest, so combining the cultivated land protection and forest rehabilitation scenarios can balance urban economic development and ecological protection. In order to verify this conclusion, a combination scenario was established, which combined the cultivated land protection and forest rehabilitation scenarios as follows: the forestland cover reached 35%; the probability of transferring cultivated land to forestland appropriately increased; the existing forestland was used as the restricted conversion area; the probability of transferring cultivated land to grassland, water, and construction land was reduced; the permanent basic agricultural land was used as the restricted conversion area. The results are shown in Figure 7.

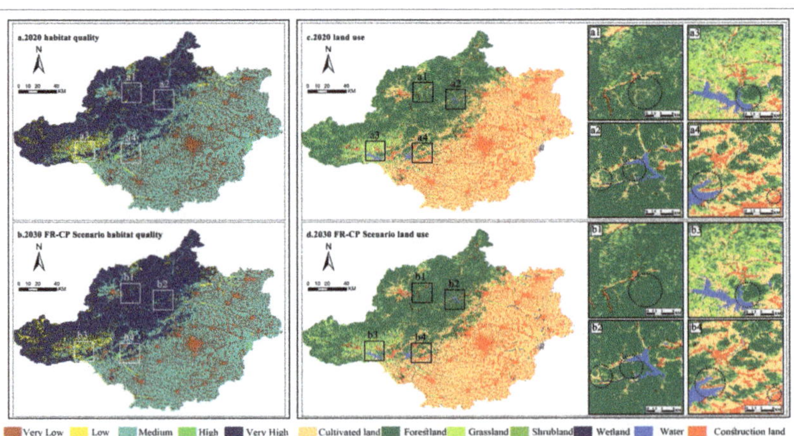

Figure 7. Spatial distribution of habitat quality and land-use in 2020 and 2030 under combination scenario.

The combination scenario increased the area of forestland and water by 408 km^2 and 10 km^2, decreased the area of cultivated land by 246 km^2, and increased the area of construction land by 6 km^2. The low and medium habitat quality areas decreased by 153 km^2 and 193 km^2, respectively, and the high habitat quality area increased by 432 km^2, and this is the scenario with the highest decrease in low habitat quality and increase in high habitat quality area. The value of ecosystem services for this scenario is CNY 34.2 billion, an increase of CNY 886 million from 2020, which is higher than the other four scenarios. This is a land-use model that meets the goal of "high ecological quality-high ecological value".

In order to explore the specific changes in land use, four scenic areas with a significant shift in habitat quality levels from low to high were selected. For example, a1 in Figure 7 is the Baishi Mountain Scenic Area. The area has been actively reforested in recent years. The forest plantation of Chinese herbs is used as an effective method for adjusting the structure of plantation and increasing farmers' income. Therefore, b1 shows that the cultivated land in this scenic area is transformed into forestland. As a2 is the scenic area of Longmen Mountain, the area has been gradually greening and building up the surrounding forestable barren hills in recent years. Therefore, b2 shows that the forestland in this scenic area increased. For example, a3 is the Xibaipo scenic area (national 5A level scenic area). In recent years, driven by the scenic spot, the surrounding villages and towns actively carried out the construction of a beautiful countryside and strengthened the protection and construction of the forest around the village. Therefore, b3 shows the change from grassland to forestland at the edge of water bodies in this scenic area. For example, a4 is Xidayang Reservoir. The function of this reservoir has changed from flood control and water supply to water conservation reserve. Therefore, b4 shows the increase of forestland on the north side of the reservoir. Since the cultivated land in this area is a restricted conversion area, b4 shows the conversion from grassland to forestland.

4. Discussion

The research data of this paper are from the National Bureau of Statistics. It includes statistical monitoring data with a 10-year cycle, which is consistent with the national development process and the implementation cycle of relevant policies. Therefore, the research also takes a 10-year cycle to observe the dynamic changes of land use, habitat quality, and ESV. However, in future, if the real-time ecological assessment with shorter cycles can be supplemented, it will be more effective for supporting assessments. We hope to continue to explore this content in future work.

This paper provides model-driven research results, which are different from the very complex actual ecological and habitat changes and issues. However, via a comparative study, it can be observed that the changing trend with respect to urban land use, habitat quality, and ESV expressed by the model is consistent with the actual situation. In future, we hope that more accurate simulation results can be obtained by improving data accuracy and model simulation accuracy in order to guide urban development more specifically.

5. Conclusions and Recommendations

In this paper, the PLUS-InVEST model and ESV were used to simulate and analyze the evolution of land use and habitat quality in Baoding. The spatio-temporal change and driving mechanism have the following characteristics:

1. Cultivated land and forestland are the two types of land with the largest proportion in Baoding. The combination scenario verifies that the overall habitat quality and ESV of the city are optimal under the priority protection of cultivated land and forestland. In the future, cultivated land protection and forest rehabilitation scenarios should continue to be used as the main policies to guide urban development, strictly delineate the scope of cultivated land protection, and improve the forestland's coverage to the optimum level.
2. Habitat quality in Baoding contrasts significantly between the forestland in the north-west and the cultivated and built-up land in the southeast. The existing ecological

tourism area and reservoir within the forestland are important ecological barriers for protecting the city from infringement and should be protected with emphasis. The land-use situation in the cultivated land and construction land area is complex, and the efficiency of land resource use should be improved by strengthening rural land improvement and urban infrastructure construction.

Author Contributions: Conceptualization, N.H. and D.X.; methodology, N.H.; software, N.Z.; validation, D.X., N.H. and Y.L.; formal analysis, D.X.; investigation, N.H. and D.X.; resources, N.H.; data curation, D.X.; writing—original draft preparation, N.H.; writing—review and editing, S.F. and P.W.; visualization, N.Z.; supervision, Y.L.; project administration, N.H. All authors have read and agreed to the published version of the manuscript.

Funding: This research was funded by the Beijing Municipal Science and Technology Planning Project (No. D171100007117003).

Institutional Review Board Statement: Not applicable.

Informed Consent Statement: Not applicable.

Data Availability Statement: The data will be available upon request to the corresponding author.

Conflicts of Interest: The authors declare no conflict of interest.

References

1. Pastor, A.V.; Tzoraki, O.; Bruno, D.; Kaletová, T.; Mendoza-Lera, C.; Alamanos, A.; Brummer, M.; Datry, T.; De Girolamo, A.M.; Jakubínský, J.; et al. Rethinking ecosystem service indicators for their application to intermittent rivers. *Ecol. Indic.* **2022**, *137*, 108693. [CrossRef]
2. Awasthi, P.; Bargali, K.; Bargali, S.S.; Jhariya, M.K.J.L.D. Structure and functioning of Coriaria nepalensis dominated shrublands in degraded hills of Kumaun Himalaya. I. Dry matter dynamics. *Land Degrad. Dev.* **2022**, *33*, 1474–1494. [CrossRef]
3. Moreno-Conn, L.M.; Rodriguez-Hernandez, N.S.; Arguello, J.O.; Baquero, J.E.; Bernal-Riobo, J.H.; Arango, M. Land use change and its effect on ecosystem services in an Oxisol of the eastern High Plains of meta department in Colombia. *Front. Environ. Sci.* **2022**, 2113. [CrossRef]
4. Feng, S.; Sun, R.H.; Chen, L.D. Spatio-temporal characteristics of habitat quality based on land-use changes in Guangdong Province. *Acta Ecol. Sin.* **2022**, *42*, 6997–7010. [CrossRef]
5. Dongjie, G.; Zulun, Z.; Jing, T. Dynamic simulation of land use change based on logistic-CA-Markov and WLC-CA-Markov models: A case study in three gorges reservoir area of Chongqing, China. *Environ. Sci. Pollut. Res.* **2019**, *26*, 20669–20688.
6. Abdurrahim, A.; Remzi, E. Future land use/land cover scenarios considering natural hazards using Dyna-CLUE in Uzungöl Nature Conservation Area (Trabzon-NE Türkiye). *Nat. Hazards* **2022**, *114*. [CrossRef]
7. Shahidul, I.; Yuechen, L.; Mingguo, M.; Anxu, C.; Zhongxi, G. Simulation and Prediction of the Spatial Dynamics of Land Use Changes Modelling Through CLUE-S in the Southeastern Region of Bangladesh. *J. Indian Soc. Remote Sens.* **2021**, *49*, 2755–2777.
8. Haiyan, H.; BingBing, Z.; Fengsong, P.; Guohua, H.; Zhongbo, S.; Yijian, Z.; Han, Z.; Yukun, G.; Meng, L.; Xia, L. Future Land Use/Land Cover Change Has Nontrivial and Potentially Dominant Impact on Global Gross Primary Productivity. *Earth's Future* **2022**, *10*, e2021EF002628.
9. Yingxue, L.; Zhaoshun, L.; Shujie, L.; Xiang, L. Multi-Scenario Simulation Analysis of Land Use and Carbon Storage Changes in Changchun City Based on FLUS and InVEST Model. *Land* **2022**, *11*, 647.
10. Chaturvedi, V.; de Vries, W.T. Machine Learning Algorithms for Urban Land Use Planning: A Review. *Urban Sci.* **2021**, *5*, 68. [CrossRef]
11. Liang, X.; Guan, Q.; Clarke, K.C.; Liu, S.; Wang, B.; Yao, Y. Understanding the drivers of sustainable land expansion using a patch-generating land use simulation (PLUS) model: A case study in Wuhan, China. *Comput. Environ. Urban Syst.* **2021**, *85*, 101569. [CrossRef]
12. Tong, L.; Muzhuang, Y.; Dafang, W.; Feng, L.; Jinhai, Y.; Yingjia, W. Spatial correlation and prediction of land use carbon storage based on the InVEST-PLUS model- A case study in Guangdong Province. *China Environ. Sci.* **2022**, *42*, 4827–4839. [CrossRef]
13. Zhiqiang, L.; Shuangyun, P. Comparison of multimodel simulations of land use and land cover change considering integrated constraints—A case study of the Fuxian Lake basin. *Ecol. Indic.* **2022**, *142*, 109254. [CrossRef]
14. Jihwan, K.; Heejoon, C.; Wonhyeop, S.; Jiweon, Y.; Youngkeun, S. Complex spatiotemporal changes in land-use and ecosystem services in the Jeju Island UNESCO heritage and biosphere site (Republic of Korea). *Environ. Conserv.* **2022**, *49*, 272–279.
15. Katherine, H.P.; Patricio, P.; Mauricio, F. Sixty years of land-use and land-cover change dynamics in a global biodiversity hotspot under threat from global change. *J. Land Use Sci.* **2021**, *16*, 467–478.
16. Eggleton, P.; Vanbergen, A.J.; Jones, D.T.; Lambert, M.C.; Rockett, C.; Hammond, P.M.; Beccaloni, J.; Marriott, D.; Ross, E.; Giusti, A. Assemblages of Soil Macrofauna across a Scottish Land-Use Intensification Gradient: Influences of Habitat Quality, Heterogeneity and Area. *J. Appl. Ecol.* **2005**, *42*, 1153–1164. [CrossRef]

17. Masteel; Ling, P.; Zenghui, M. Ecosystem Service Evaluation Model and Empirical Research. *Anhui Agric. Bull.* **2021**, *27*, 141–143. [CrossRef]
18. Hongjuan, Z.; Yan, G.; Yawei, H.; Yanyan, L.; Yue, Z.; Kang, L. Response of a SolVES model value transfer method to different spatial scales. *Acta Ecol. Sin.* **2019**, *39*, 9233–9245. [CrossRef]
19. Salata, S.; Garnero, G.; Barbieri, C.A.; Giaimo, C. The Integration of Ecosystem Services in Planning: An Evaluation of the Nutrient Retention Model Using InVEST Software. *Land* **2017**, *6*, 48. [CrossRef]
20. Rakesh, K.; Chandrakant, G.; Ankush, R.; Asheesh, S.; Chandrasekhar, M.; Rajesh, B. Quantification of heat mitigation by urban green spaces using InVEST model—A scenario analysis of Nagpur City, India. *Arab. J. Geosci.* **2021**, *14*, 82. [CrossRef]
21. Costanza, R.; d'Arge, R.; de Groot, R.; Farber, S.; Grasso, M.; Hannon, B.; Limburg, K.; Naeem, S.; O'Neill, R.V.; Paruelo, J.; et al. The value of the world's ecosystem services and natural capital. *Ecol. Econ.* **1998**, *25*, 3–15. [CrossRef]
22. Gaodi, X.; Caixia, Z.; Leiming, Z.; Wenhui, C.; Shimei, L. Improvement of the Evaluation Method for Ecosystem Service Value Based on Per Unit Area. *J. Nat. Resour.* **2015**, *30*, 1243–1254. [CrossRef]
23. Gaodi, X.; Chunxia, L.; Method, C.P.; Du, Z.; Shuangcheng, L. Value evaluation of ecological assets in qinghai-tibet plateau. *J. Nat. Resour.* **2003**, *18*, 189–196.
24. Alves, F.; Roebeling, P.; Pinto, P.; Batista, P. Valuing ecosystem service losses from coastal erosion using a benefits transfer approach: A case study for the Central Portuguese coast. *J. Coast. Res.* **2009**, *56*, 1169–1173.
25. Ning, W.; Guang, Y.; Xueying, H.; Guangpu, J.; Feng, L.; Tao, L.; Jianing. Land Use Change and Ecosystem Service Value in Inner Mongolia from 1990 to 2018. *J. Soil Water Conserv.* **2020**, *34*, 244–250. [CrossRef]
26. Ren, Y.; Ruitong, L. Rural Comprehensive Land Consolidation and Territorial Ecological Restoration:Cohesion and Integration. *Mod. City Res.* **2021**, *3*, 23–32. [CrossRef]
27. Jie, J. Spatio-temporal Differentiation of Ecosystem Service Value in Dianchi Basin and Its Influencing Factors Based on land Use. *J. Soil Water Conserv.* **2022**, *29*, 344–351. [CrossRef]
28. Zhaoxu, L. Study on the Conservation Planning of Historic and Cultural Cities under the Concept of "Double Urban Repairs". Master's Thesis, Hebei Agricultural University, Baoding, China, 2019.
29. Baoding Statistics Bureau. *Baoding Statistical Yearbook*; Baoding Statistics Bureau: Baoding, China, 2020; p. 59. [CrossRef]
30. National Development and Reform Commission. *Compilation of National Agricultural Product Cost Income Data*; China Statistics Press: Beijing, China, 2020; p. 625.
31. Liu, T.; Chen, X. Application of deep learning in globeland30-2010 product refinement. *Int. Arch. Photogramm. Remote Sens. Spat. Inf. Sci.* **2018**, *42*. [CrossRef]
32. Xinliang, X. Kilometre grid dataset of China's population spatial distribution. *Resour. Environ. Sci. Data Regist. Publ. Syst.* **2017**. [CrossRef]
33. Liu, X.; Liang, X.; Li, X.; Xu, X.; Ou, J.; Chen, Y.; Li, S.; Wang, S.; Pei, F. A future land use simulation model (FLUS) for simulating multiple land use scenarios by coupling human and natural effects. *Landsc. Urban Plan.* **2017**, *168*, 94–116. [CrossRef]
34. Feng, H.; Yan, Z.; Yu, G.; Panpan, Z.; Shuai, L.; Changchun, Z. Spatial and temporal changes in land use and habitat quality in the Weihe River Basinbased on the PLUS and InVEST models and predictions. *Geogr. Arid. Area* **2022**, *45*, 1125–1136.
35. Rui, Z.; Xuemin, L. A Study on the Spatial and Temporal Pattern Evolution of "Three Generations" in Beijing Tianjin Hebei Metropolitan Circle and Its Driving Forces. *Ecol. Econ.* **2021**, *37*, 201–208.
36. Shandong, N.; Xiao, L.; Guozheng, G. What Is the Operation Logic of Cultivated Land Protection Policies in China? A Grounded Theory Analysis. *Sustainability* **2022**, *14*, 8887.
37. The People's Government of Hebei Province. Regulations of Hebei Province on Wetland Protection. *Hebei For.* **2016**, 6–9.
38. Zhengyan, L.; Xianqiang, M. Research on the Key Directions and Tasks of Ecological Environment Protection during the Fourteenth Five Year Plan Period. *China Environ. Manag.* **2019**, *11*, 40–45. [CrossRef]
39. Tianni, Z.; Tongwen, J. Baoding Re energizes Standard Xiong An Chuangshen. *Hebei For.* **2018**, 26–27.
40. Qi, W. In 20 years, 500 million mu of farmland has been returned to forests and grasslands nationwide. *Land Green.* **2019**, 5–6.
41. Xiaoyan, Y.; Yinjun, C. Analysis on the influencing factors of farmers' conversion to cultivated land and their "non grain" planting behavior and scale—Based on the survey data of farmers in Zhejiang and Hebei provinces. *Rural. Obs. China* **2010**, 2–10+21.
42. Luijten, J.C. A systematic method for generating land use patterns using stochastic rules and basic landscape characteristics: Results for a Colombian hillside watershed. *Agric. Ecosyst. Environ.* **2003**, *95*, 427–441. [CrossRef]
43. Yifan, G.; Jianghui, D.; Aomeng, L. Spatio-temporal difference of coupling coordinative degree of productive ecological carrying capacity and ecological function in Baoding, Hebei Province. *Acta Ecol. Sin.* **2020**, *40*, 7175–7186.
44. Xuesong, Z.; Wei, R.; Hongjie, P. Urban land use change simulation and spatial responses of ecosystem service value under multiple scenarios: A case study of Wuhan, China. *Ecol. Indic.* **2022**, *144*, 109526.
45. Jingheng, W.; Yecui, H.; Rong, S.; Wei, W. Research on the Optimal Allocation of Ecological Land from the Perspective of Human Needs—Taking Hechi City, Guangxi as an Example. *Int. J. Environ. Res. Public Health* **2022**, *19*, 12418.
46. Mengdi, Z.; Fen, Z.; Xiong, L. Evaluation of Habitat Quality Based on InVEST Model: A Case Study of Tongzhou District of Bejjing, China. *Landsc. Archit.* **2020**, *27*, 95–99. [CrossRef]

 *sustainability*

Article

Does Agroforestry Correlate with the Sustainability of Agricultural Landscapes? Evidence from China's Nationally Important Agricultural Heritage Systems

Menghan Zhang [1] and Jingyi Liu [2,*]

1 School of Landscape Architecture, Beijing Forestry University, Beijing 100083, China; zmh1993@bjfu.edu.cn
2 College of Forestry and Landscape Architecture, South China Agricultural University, Guangzhou 510642, China
* Correspondence: liujingyi@scau.edu.cn; Tel.: +86-151-2009-1302

Abstract: Compared with industrial monoculture, agroforestry has been perceived as a more sustainable approach to landscape management that provides various landscape-specific benefits. However, little is known about agroforestry's influence on the comprehensive sustainability of agricultural landscapes. This study focused on the importance of agroforestry and its influence on landscape sustainability, using 118 China National Important Agricultural Heritage Systems (China-NIAHS) as cases. In each China-NIAHS, we evaluated the importance of agroforestry and the landscape's comprehensive sustainability and explored their correlation. The findings indicate that agroforestry is important in most China-NIAHS. Agroforestry's importance is strongly correlated with most sustainability indicators, including biodiversity, income diversity, resource utilization, hydrogeological preservation, and water regulation. Based on the findings, we discuss the role of agroforestry in promoting sustainability and provide suggestions for sustainable management and policymaking for agricultural landscapes on a national scale.

Keywords: agricultural landscape; agricultural heritage; agroforestry; landscape sustainability

Citation: Zhang, M.; Liu, J. Does Agroforestry Correlate with the Sustainability of Agricultural Landscapes? Evidence from China's Nationally Important Agricultural Heritage Systems. *Sustainability* **2022**, *14*, 7239. https://doi.org/10.3390/su14127239

Academic Editors: Qingsong He, Jiayu Wu, Chen Zeng and Linzi Zheng

Received: 19 April 2022
Accepted: 10 June 2022
Published: 13 June 2022

1. Introduction

More than 10% of the earth's land surface is covered by agricultural landscapes [1]. These landscapes are shaped by agricultural practices that have been adapted to specific environments for centuries [2]. Agricultural heritage landscapes are prominent examples of the results of these traditional human–nature interactions that continue to be used today [3]. They are appreciated worldwide as being rich in interrelated natural and cultural values [4,5]. To promote the understanding and conservation of agricultural heritage landscapes, in 2002, the Food and Agriculture Organization of the United Nations (FAO) launched the Globally Important Agricultural Heritage Systems (GIAHS). In response to the GIAHS, since 2012, China's Ministry of Agriculture and Rural Affairs has engaged in a long-term project to survey and protect agricultural heritage landscapes throughout China, proposing a list of China's National Important Agricultural Heritage Systems (China-NIAHS). So far, 118 sites have been identified in the China-NIAHS (Figure 1). They are referred to as the "national treasures," or "pearls of traditional wisdom" [6].

However, the current affection for modern agricultural approaches has been threatening the sustainability of many agricultural heritage landscapes [7–9]. Given that only nine plant species account for almost two-thirds of the total crop yield of agriculture [10], industrial monoculture has been a prevailing alternative to traditional agriculture. Despite its recognized economic benefits, the spread of industrial monoculture has been criticized for its negative social and environmental impacts [11]. For example, industrial monoculture is associated with soil depletion [12], inefficiency in capturing nutrients and water [13], exotic species invasion [14], biodiversity loss [15], susceptibility to pests and diseases [16], and deterioration of cultures and local livelihoods [13]. Furthermore, these issues appear to have been aggravated as a result of climate change and resource scarcity.

Figure 1. Examples of China-NIAHS that are characterized by agroforestry: (**a**) Hani Rice Terraces System; (**b**) Xiajin Ancient Mulberry Grove System; (**c**) Diebu Zhagana Agriculture-Forestry-Animal Husbandry Composite System; (**d**) Xin'an Traditional Cherry Terraces System; (**e**) Yangbi Walnut-crop Mixed System; (**f**) Yangshan Peach-Poultry Agricultural System (https://www.ciae.com.cn/list/zh/agricultural_heritage/video.html, (accessed on 22 February 2022)).

The multiple problems associated with modern agriculture prompt an investigation of the sustainability of agricultural heritage landscapes [17]. Sustainability emphasizes a long-lasting reconciliation of human development and environmental protection [18]. To understand and assess sustainability, the ecological, economic, and social benefits people obtain from landscapes have been identified [18,19]. The United Nations Sustainable Development Goals (SDG) outline a global agenda for sustainability [20]. While sustainability is relevant at multiple scales, there has been agreement that sustainability research and application to guide sustainable agricultural and rural development would be more operational at a landscape scale [21,22]. In this regard, instead of just calculating the values of assessment indicators, many studies focus on identifying the determinants shaping the sustainability of agricultural heritage landscapes [23], including those relating to labor involvement (including available laborers, intergenerational inheritance, off-farm activities, laborer's level of management experience, etc.) [24], production diversity (including biodiversity, structure of mixed crop-livestock systems, etc.) [25], landscape management (including forest conservation, irrigation systems, etc.) [26], and overall environment (including regional natural and cultural contexts, global climate change and droughts, the dynamics of public policies and local agencies, etc.) [27].

Although previous research has investigated a variety of factors that may influence landscape sustainability, the sustainability of agricultural heritage landscapes is usually assessed by the impact of different landscape management approaches [28,29]. A prominent landscape management approach characterizing agricultural heritage landscapes is agroforestry. Agroforestry, in which woody perennials are deliberately grown on the same land management units as crops and/or animals in spatial or temporal arrangements [30], is assumed to be important for the following two reasons: first, because GIAHS and China-NIAHS are named after crops, trees, and livestock, agroforestry appears to be the mainstay of the agricultural heritage landscapes. Second, agroforestry, involved in diverse landscape managements approaches, is an integral part of a sustainable working system that reconciles the values of various components (land, plant, animal, labor, irrigation, etc.) [31–33]. However, there is a lack of systematic research clearly explaining the relationship between agroforestry and sustainability compared with other determinants.

Derived from indigenous ecological practices within the context of traditional agriculture, agroforestry offers various benefits by combining the most desirable attributes of crops, trees, and livestock [34]. For example, agroforestry offers environmental benefits (including carbon sequestration, biodiversity conservation, soil enrichment, and water quality [35]), enhances local livelihoods [36], and provides social value [37]. In addition, it represents a valid means of supporting climate change mitigation while maintaining traditional systems and cultural services [38], as it entails a deep understanding of the human role in landscapes, considering indigenous knowledge and landscape management, aesthetic values, inherited cultural patterns, and people's spiritual identities [39]. Although agroforestry has been increasingly studied in recent years, most previous studies focused on individual cases, particularly on production and ecological functions. For example, He et al. (2020) studied the agrobiodiversity potential of the Shexian Dryland Stone Terraces [40]. Bai et al. (2016) investigated the water regulation of the Hani Rice Terraces [41]. Prior research has seldom, however, considered agroforestry as a sustainable approach to landscape management which embraces the complexity of socioecological systems [42]. Nor have agroforestry's effects on sustainability at a national or regional scale been examined using a larger sample size. Meanwhile, the economic and social aspects of agroforestry were frequently ignored in assessing its sustainability.

Considering agroforestry as a landscape management approach, this paper examined agroforestry on the Chinese national scale using 118 of the China-NIAHS as cases (most do not have a clear boundary, with scales varying from approximately 0.6 to 25,000 km^2). We explored agroforestry's importance and impact on the sustainability of agricultural heritage landscapes. Specifically, we addressed two research questions: (1) How important is agroforestry in China-NIAHS? (2) How does agroforestry influence the landscape sustainability of China-NIAHS?

To address the research questions, we collected and interpreted information about China-NIAHS from multiple sources. For each site, we measured the importance of agroforestry and evaluated landscape sustainability using multidimensional indicators. The impacts of agroforestry's importance on landscape sustainability were then examined using correlation analysis. Finally, we discussed the mechanisms of the impacts and the implications for the sustainable development of agricultural heritage landscapes at the national scale.

2. Materials and Methods

Four research phases were undertaken to investigate the importance of agroforestry and it impacts on the landscape sustainability of China-NIAHS (Figure 2):

1. Collection of detailed information about China-NIAHS;
2. Evaluation of the importance of agroforestry;
3. Establishment of the assessment indicator system to measure the landscape sustainability of China-NIAHS;

4. Correlation analysis between the importance of agroforestry and the landscape sustainability of China-NIAHS.

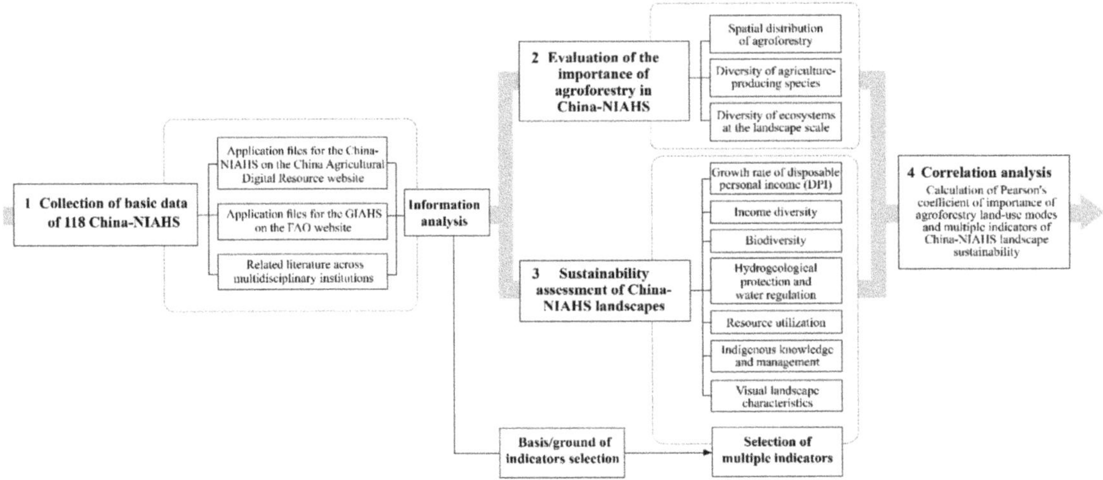

Figure 2. Schematic diagram of the research process.

2.1. Data Sources

In the first phase, a review of all the application files for the China-NIAHS (available at the China Agricultural Digital Resource website) and GIAHS (available at the FAO website) was conducted to obtain detailed information about the importance of agroforestry and its impact on landscape sustainability [43,44]. As of March 2022, there were 118 sites on the list of China-NIAHS. Besides these two authoritative platforms, we studied the literature in this field across multidisciplinary institutions (including scientific institutions, colleges, and community groups) [45]. We comprehensively analyzed the descriptive text information, research results, and photographic and video materials based on the database from those diverse sources. Considering such a wide range of sources avoids oversimplification and subjective bias, allowing for a more thorough assessment [46].

2.2. Evaluation of the Importance of Agroforestry in China-NIAHS

The second phase was to determine whether agroforestry is important in the 118 China-NIAHS. As importance is a value judgment that may be assessed by detailed quantitative and qualitative data, this phase primarily focused on the manual interpretation that converts the descriptive data (including numeric/text descriptions and visual information) into the indicative score. Some measures were taken to increase the objectivity of this process. First, the interpretation was guided by three explicit standards that were closely related to the importance of agroforestry based on the existing literature [47,48]: (1) spatial distribution of agroforestry (at edges, scattered, or aggregated), (2) diversity of agriculture-producing species, and (3) diversity of ecosystems at the landscape scale (farms, orchards, fish ponds, pastures, natural forests, etc.). The specific criteria for scores ranging from 1 to 5 are summarized in Table 1. Second, to reduce bias, the scoring process for each of the 118 China-NIAHS was based on the multiple information sources mentioned above as cross-references.

Table 1. Standard and score table of the importance of agroforestry in the China-NIAHS.

The Criterion of the Importance of Agroforestry Land-Use Modes			Given Score		Importance Score
Agroforestry Does not Exist in the China-NIAHS			0		0
Agroforestry exists in the China-NIAHS	Spatial distribution of agroforestry	Only a small amount distributed at the edges	1	X1	The average of X1, X2, X3
		Scattered with low density	3		
		Aggregated with high density	5		
	Diversity of agriculture-producing species	Little diversity/varieties in single agriculture (≤2)	1	X2	
		More diversity/varieties in single agriculture (>2)	2		
		Involved in any two: agriculture, forestry, and husbandry, with little diversity/varieties (≤4)	3		
		Involved in any two: agriculture, forestry, and husbandry, with more diversity/varieties (>4)	4		
		Involved in agriculture, forestry, and husbandry with more diversity/varieties	5		
	Diversity of ecosystems at the landscape scale	Involved in a small number of ecosystems (<3)	1	X3	
		Involved in a medium number of ecosystems (3 or 4)	3		
		Involved in a large number of ecosystems (≥5)	5		

2.3. *Landscape Sustainability Assessment of China-NIAHS*

2.3.1. Indicator Selection

The third phase established an indicator system after a well-rounded study of diverse elements that reveal landscape sustainability. The FAO chose five official criteria for identifying the GIAHS: (1) food and livelihood security; (2) agrobiodiversity; (3) local and traditional knowledge systems; (4) cultural value systems and social organizations; and (5) landscape and seascape features. A survey by Santoro et al. (2020) measured the contribution of agroforestry to sustainable developments of 59 GIAHS using five indicators: (1) timber, fuelwood, food, and by-products; (2) biodiversity; (3) landscape; (4) hydrogeological protection and water regulation; and (5) structural/management characteristics [49]. Zhao et al. (2020) adopted five indicators: (1) economic contribution; (2) social equity; (3) environmental protection; (4) ecological resources; and (5) disaster resilience [50].

Based on the principles and research mentioned above, we selected seven indicators: (1) growth rate of disposable personal income (DPI); (2) income diversity; (3) biodiversity; (4) hydrogeological protection and water regulation; (5) resource utilization; (6) visual landscape characteristics; and (7) indigenous knowledge and management (Figure 3). The selection of indicators followed several principles:

(1) **Comprehensiveness**. Sustainability assessment needs to consider multiple factors that reflect the needs of humans [32,42]. The Sustainable Development Goals (SDG) are part of a program designed by the United Nations to steer global development policies and funds, from 2015 to 2030, to achieve long-term social, economic, and environmental goals [51]. As illustrated in Figure 3, the seven selected indicators are well-matched with SDG indicators and cover sustainability's ecological, economic, and social dimensions, making the indicator system more comprehensive.

(2) **Validity**. The designation of China-NIAHS is the result of an examination procedure based on qualitative criteria. The seven selected indicators are well-matched with official criteria for identifying the China-NIAHS. Therefore, the system's validation is ensured by the engagement of stakeholders in this field.

(3) **Operability**. The growth rate of the DPI was taken from each county's 2020 National Economic and Social Development Bulletin. Information on other indicators was found in the text and graphical descriptions of all application files, because these files

conform to the official criteria for recognizing China-NIAHS and GIAHS. To ensure the reliability of the evaluation results, we also referred to relevant research literature.

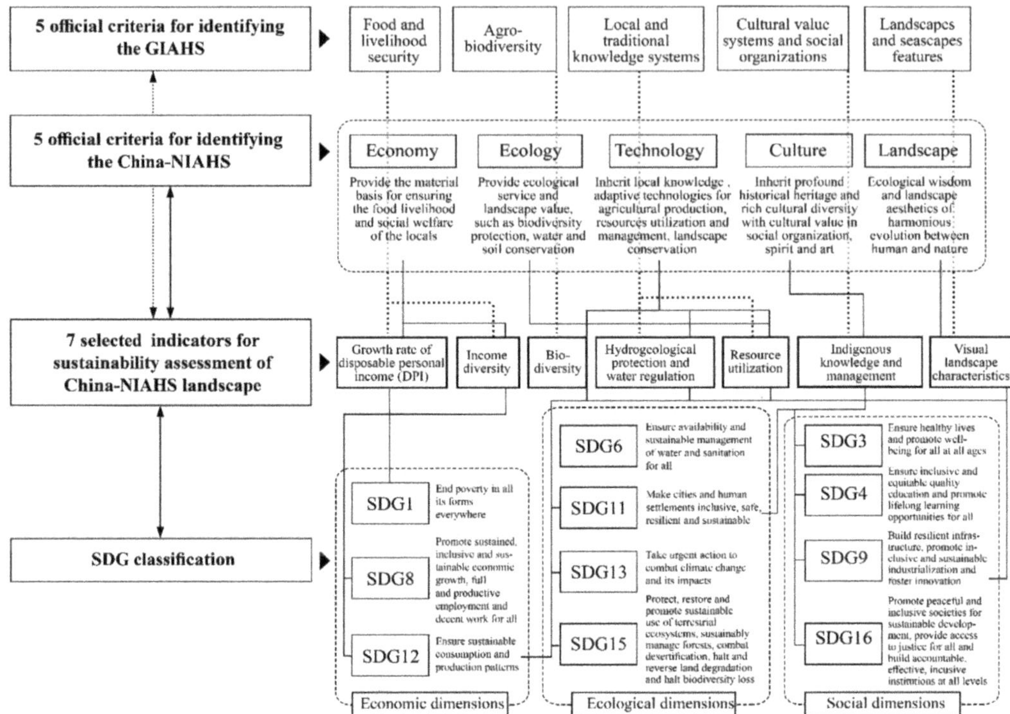

Figure 3. Matching diagram of landscape sustainability assessment indicators and official criteria for identifying the China-NIAHS, GIAHS, and Sustainable Development Goals.

2.3.2. Indicator Weight Assignment

Our assessment system for landscape sustainability uses qualitative indicators to evaluate China-NIAHS. Thus, it relies on the participation of field specialists and repeated multidisciplinary surveys. This study integrated the Delphi method (DM) with an analytic hierarchical process (AHP) to calculate the indicators' weights [52,53]. The qualitative questionnaires were distributed individually to ten specialists using the DM. The relative importance of each indicator was determined through comparison using AHP, based on qualitative questionnaires conducted by field specialists. Then, we gathered the specialists' opinions and checked for consistency (<0.10) [54]. After three rounds of anonymous consultation and feedback adjustment, a highly consistent opinion was reached (consistency test result = 0.0176). Table 2 shows the finalized sustainability assessment indicator system and weights.

Table 2. Indicator system of landscape sustainability assessment of the China-NIAHS.

Goal Level	Orientation Level	Weight	Indicator Level	Content	Weight
Landscape Sustainability of China-NIAHS	Economic sustainability	0.2099	the growth rate of DPI	The growth rate of disposable personal income that 2020 National Economic and Social Development Bulletin collected from each county (city, district).	0.0350
			income diversity	Income sources that are provided by a variety of products, including but not limited to food, timber, fuelwood, fruits, herbs, fertilizers, and feed.	0.1749
	Environmental sustainability	0.5499	biodiversity	Biodiversity closely related to agricultural production that supports agroecosystems, including indigenous knowledge and adaptive technologies, in agricultural production and utilization of biological resources.	0.1321
			hydrogeological protection and water regulation	The ability to regulate water supply, improve water quality, fix soil, and protect topsoil, including indigenous knowledge and adaptive techniques in water and soil resource management and landscape conservation.	0.3024
			resource utilization	The degree of accessing and using environmental resources such as soil, water, light, and heat in a vertical or horizontal space, and their biological interaction.	0.1154
	Social sustainability	0.2402	visual landscape characteristics	The significance and protection of landscape features that reflect the ecological wisdom and landscape aesthetics of harmonious evolution between humans and nature.	0.0400
			indigenous knowledge and management	The scale and influence of belief and worship, cultural taboos, traditional customs, festival activities, and the management system based on indigenous knowledge inherited from previous generations.	0.2002

2.3.3. Indicator Score Evaluation

To better understand the comprehensive landscape sustainability in each nominated area, this study assessed the degree of each indicator using primary data from the China-NIAHS, assigning a score from 0 to 3.

The indicator of the growth rate of DPI was linearly transformed to standard values, with the formula for data adjustment in the range of 0–3 being:

If $x > \sim 10\%$, $x' = 3$; If $x < 10\%$,

$$x' = \frac{x - lower(x)}{upper(x) - lower(x)} * 3$$

where, x is the original data, and x' is the recalibrated standard value with a threshold between 0–3. Negative values are equivalent to 0.

Except for the growth rate in DPI, we manually judged the six indicators according to application files from an authoritative platform (China Agricultural Digital Resource website) with multi-source information. This is because it was impossible to individually measure such a large number of China-NIAHS, each operating at different scales and without clear boundaries, with a comprehensive assessment. In addition to operability, several measures were taken to ensure the appropriateness of the procedure:

(1) **Collection of reliable data.** The application files we used were produced by different stakeholders (including agricultural and cooperative departments, agricultural organizations, research institutions and local governments) and examined by Chinese experts qualified in the field. Thus, the information derived from China-NIAHS application files not only reflects diverse values, but is also reliable, with less bias.

(2) **Cross-reference of multi-source information.** We verified the data by seeking multi-source information (including heritage reports, literature, websites, etc.) as a cross-reference. Such a wide range of sources can also avoid oversimplifying the complexity of agricultural heritage, allowing for a more comprehensive assessment of sustainability.

(3) **Clear and consistent scoring standard.** Rather than being a goal or condition, sustainability is viewed as a process for sustainability-enhancing decision making [55]. In response, we converted the descriptive data based on detailed China-NIAHS information into indicative data. As brief examples show in Table 3, the relative degrees were scored from 0 to 3 with a clear standard, consistent across all sites.

2.4. Correlation between Agroforestry and the Landscape Sustainability of China-NIAHS

In the fourth phase, Pearson's correlation coefficient (P) was used to analyze the degree of correlation between the scores showing the importance levels of agroforestry and the scores of multiple indicators of China-NIAHS sustainability. Pearson's correlation coefficient is one of the most common parametric tests for understanding bivariate inferential relationships [56]. Its value, which corresponds to the statistical significance of the coefficient, varies from -1 to 1, with a value closer to 1 suggesting a stronger bivariate correlation. The calculation formula is as follows:

$$P = \frac{cov(X,Y)}{\sigma_X \sigma_Y} = \frac{E((X-\mu_X)(Y-\mu_Y))}{\sigma_X \sigma_Y} = \frac{E(XY) - E(X)E(Y)}{\sqrt{E(X^2) - E^2(X)}\sqrt{E(Y^2) - E^2(Y)}}$$

where, $cov\ (X, Y)$ is the covariance (an indicator reflecting the degree of correlation between two random variables), and $\sigma_X \sigma_Y$ is the standard deviation.

Table 3. Examples of score standard for the sustainability assessment indicator of China-NIAHS.

Name of China-NIAHS	Interpretation of Indicator	Income Diversity I	Bio-Diversity II	Hydrogeological Protection and Water Regulation III	Resource Utilization IV	Visual Landscape Characteristics V	Indigenous Knowledge and Management VI
Huzhou Mulberry-Dyke and Fishpond System, Zhejiang	Ponds planted with mulberry trees on dykes providing leaves for silkworm rearing, with silkworm feces feeding fish. Every winter, the rich mud at the bottom floats up to the dykes as mulberry fertilizer, improving dyke soil (I, II, IV, V); making full use of water resources in low land with both high yield and adaptability to both drought and flood (I, III); indigenous knowledge survives, with sericulture folk activities and festivals flourishing, and tourism, silk, and freshwater aquaculture developed (I, VI).	3	3	3	3	3	3
Xinghua Duotian Agrosystem, Jiangsu	Compound production of the forest, crop, pond, and fish (I, II, IV, V); take both water storage management and flood control into account (III); unique and impressive landscape (V); indigenous knowledge survives and the tourism income is abundant (I, VI).	2	2	3	2	3	3
Zhangqiu Onion Cultivation System, Shandong	The main production mode of scallion and wheat rotation for more than two years (I, II, IV); traditional deep furrow Yongpei technology (VI).	1	1	0	1	0	1
Hami Cantaloupe Cultivation System, Xinjiang	The cultivation system of cantaloupe (I, II, IV); soil and water conservation (III); develops ecological agriculture system with tourism based on regional characteristics (VI).	0	0	1	0	1	2

Note: The multi-indicator score of landscape sustainability: 0 = ignored, 1 = slight, 2 = obvious, and 3 = very obvious.

3. Results

3.1. Importance of Agroforestry to the China-NIAHSs

Agroforestry can be found at a total of 109 China-NIAHS (91.5%), indicating that the practice is widespread throughout China. A total of 10 sites received a score of 0 (8.5%), 15 received a score of 1~<2 (12.7%), 24 received a score of 2~<3 (20.3%), 26 received a score of 3~<4 (22.0%), and 43 received a score of 4~<5 (23.7%). In general, the importance of agroforestry is higher in southern China than in northern China (particularly in arid/semi-arid northern China and in the northeast plain). Although the importance of agroforestry in China-NIAHS varies, the results indicate that agroforestry plays a fundamental role in more than half of the China-NIAHS (scores of 3~<5) (Figure 4).

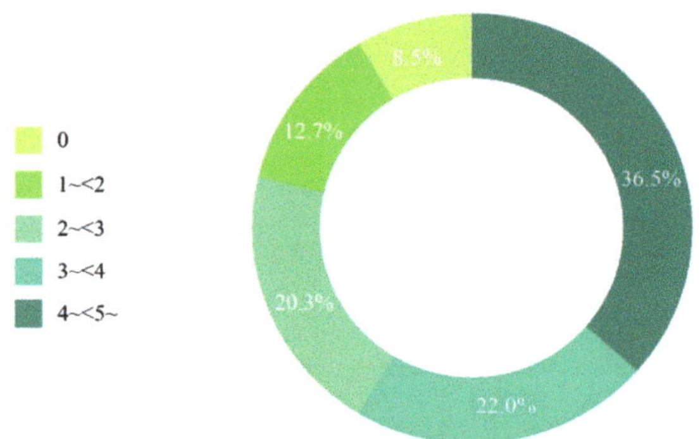

Figure 4. Proportion of different scores showing the importance of agroforestry in China-NIAHS.

3.2. Correlation between the Importance of Agroforestry and the Landscape Sustainability of China-NIAHS

Figure 5 maps the correlation between the importance of agroforestry and landscape sustainability in 118 China-NIAHS. It shows that China-NIAHS that accord greater importance to agroforestry (green solid circles with larger diameters) usually have higher landscape sustainability scores (open circles with larger diameters).

Considering the wide spatial variety in natural and cultural conditions among the 118 China-NIAHS, we analyzed the importance of agroforestry and landscape sustainability within a specific context. According to the distribution of the results in the eight agricultural zones (compiled by the National Agricultural Regionalization Committee), the landscape sustainability of China-NIAHS varies consistently with the importance of agroforestry (Figure 6). It suggests that the correlation applies to different regions of China.

Specifically, as agroforestry gains importance, most indicators and their weighted sums rise, indicating a positive correlation (Figure 7). Among these indicators, agroforestry is deemed crucial for income diversity, biodiversity, resource utilization, hydrogeological protection, and water regulation, while achieving slightly lower values for visual landscape characteristics and indigenous knowledge and management. However, it is difficult to identify the correlation between the importance of agroforestry and the growth rate of DPI.

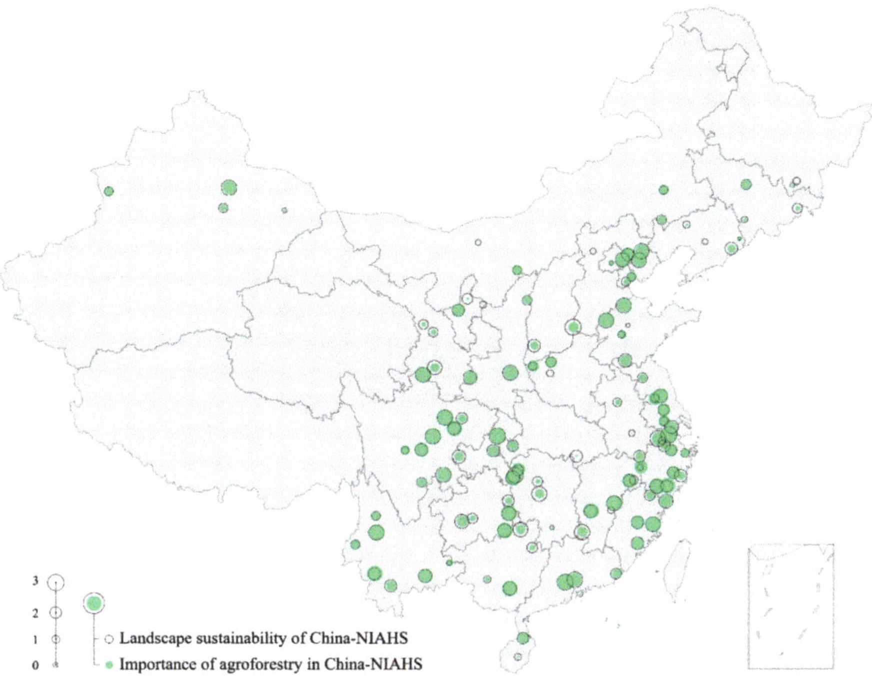

Figure 5. Map of the evaluation results and the relationship between the importance of agroforestry and China-NIAHS landscape sustainability.

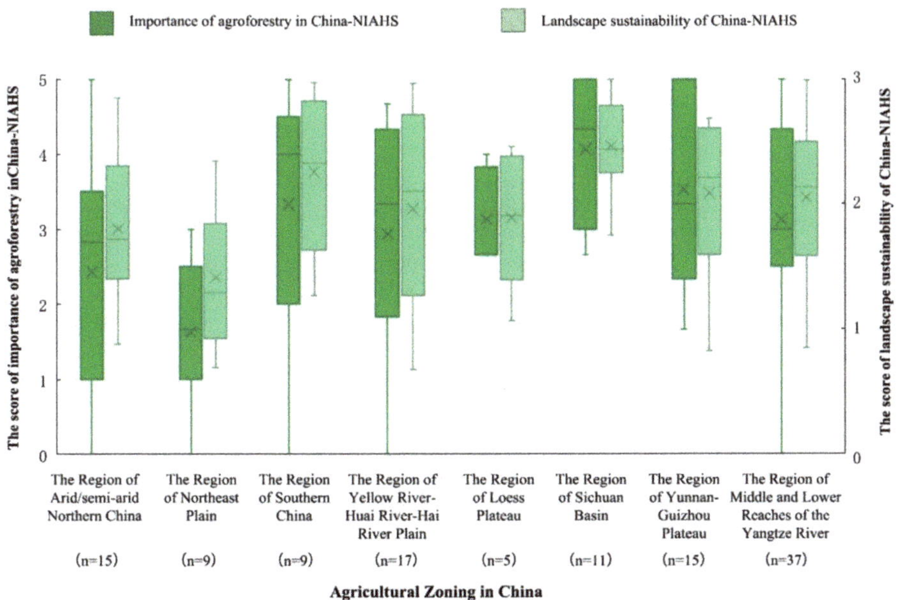

Figure 6. Distributions of the scores regarding the importance of agroforestry and landscape sustainability of China-NIAHS in different agricultural zones.

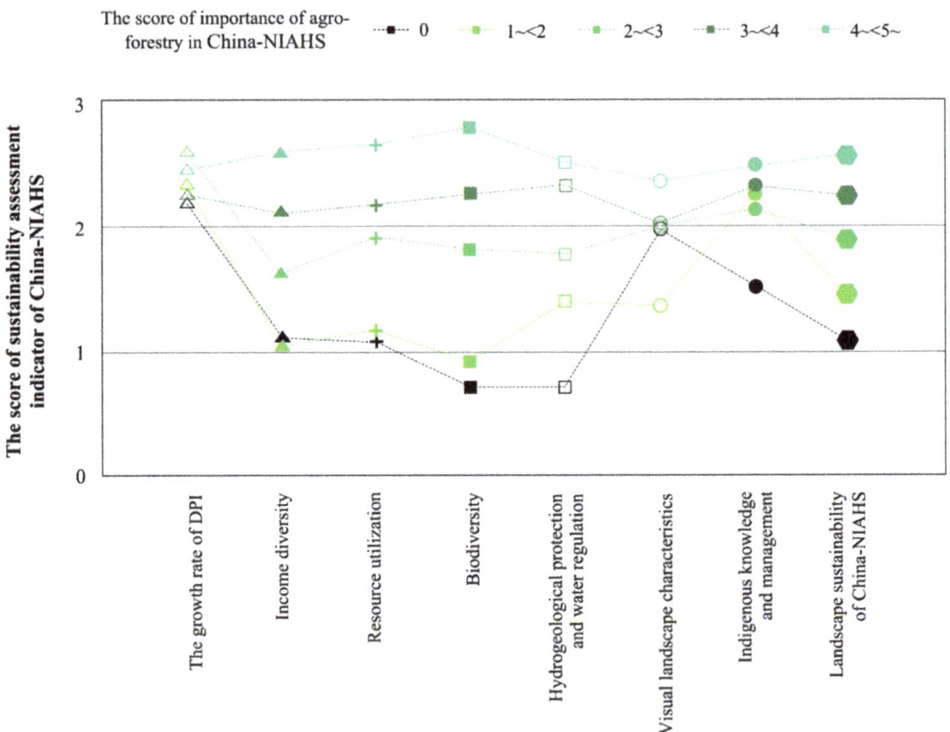

Figure 7. Diagram of the relationship between the importance of agroforestry and multiple indicators of China-NIAHS's landscape sustainability.

Figure 8 shows the Pearson's correlation coefficient between any two indicators. There is a statistically significant correlation between the importance of agroforestry and the weighted sum of China-NIAHS landscape sustainability ($p = 0.76$), despite a wide range of values for any two indicators ($0.02 \leq p \leq 0.94$). Four out of seven indicators suggested a strong correlation between the importance of agroforestry and the landscape sustainability of China-NIAHS due to their high coefficient values ($p \geq 0.50$). These indicators are biodiversity ($p = 0.78$), resource utilization ($p = 0.74$), hydrogeological preservation and water regulation ($p = 0.67$), and income diversity ($p = 0.52$). These findings suggest that agroforestry provides substantial ecological and economic value.

The importance of agroforestry is seen to have a medium correlation with indigenous knowledge and management (0.31) and visual landscape characteristics (0.34), as the value falls between 0.30 and 0.50. According to the findings, the social benefit that agroforestry could provide appears to be less significant than the ecological and economic benefits. However, social indicators may be more challenging to measure than ecological and economic indicators because they depend on complex, long-term processes that are difficult to identify.

Neither the importance of agroforestry nor any of the configuration indicators analyzed have a significant correlation with the growth rate of DPI ($0.02 \leq p \leq 0.10$). Compared with the strong correlation between agroforestry and income diversity ($p = 0.52$), this slight correlation ($p = 0.05$) indicates a gap between income diversity and the growth rate of DPI. This should be addressed with more ecological subsidies or mechanism adjustments, which will require joint efforts by the stakeholders in multiple fields.

Furthermore, there was a strong correlation ($p \geq 0.75$) between any combination of any two of the following—income diversity, biodiversity, and resource utilization. This correlation revealed the co-benefit of ecological and economic indicators.

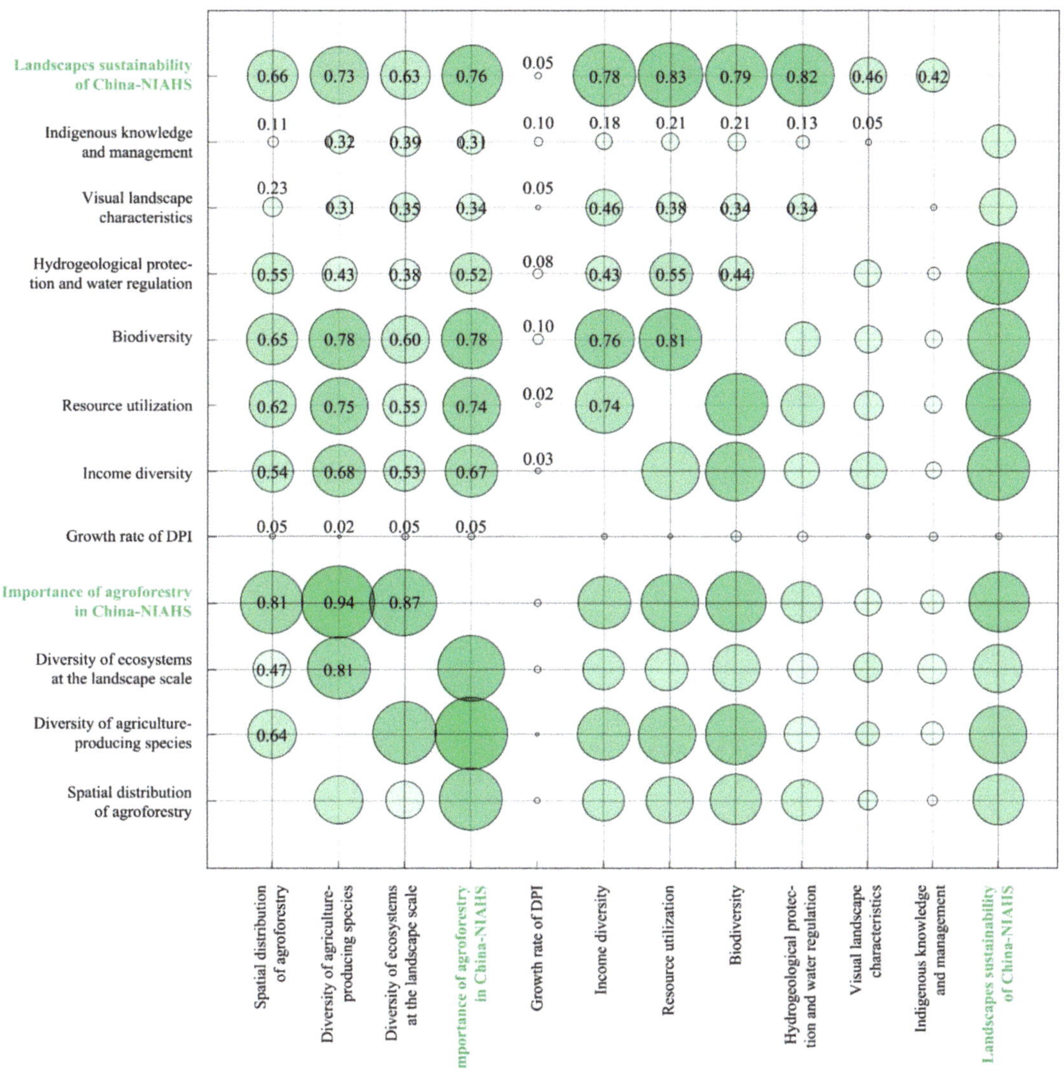

Figure 8. Pearson's correlation coefficient plots of the importance of agroforestry and multiple indicators of landscape sustainability. Note: Pearson's correlation coefficients are commonly classified into very high degree/strong correlation ($|p| \geq 0.75$), high degree/strong correlation ($0.50 \leq |p| < 0.75$), moderate degree/medium correlation ($0.30 \leq |p| < 0.50$), low degree/small correlation ($|p| < 0.30$), and no correlation ($p = 0$).

4. Discussion

4.1. The Specific Roles of Agroforestry in the Landscape Sustainability of China-NIAHS

The results revealed a strong correlation between the importance of agroforestry and the landscape sustainability of China-NIAHS. The analysis of the China-NIAHS dossiers and many studies highlighted the fundamental roles that agroforestry plays in influencing landscape sustainability, which can be discussed and categorized into the following three different types.

First, agroforestry is crucial for hydrogeological protection and water regulation ($p = 0.52$). This is consistent with previous research conducted on the GIAHS in Asia [49]. This function of agroforestry can be seen in the China-NIAHS that are characterized by steep cultivated slopes with terraced landscapes, such as the Hani Rice Terraces System (Yunnan) and the Pidu Forest Farming Culture System (Sichuan). At these sites, forests are mainly located at the upper fringes overlooking the terraces, where they can regulate the impact of precipitation and prolong the water supply. They retain water and slowly release it to the terraces below, avoiding excessive water flow. At the same time, the forest roots help to stabilize soil and prevent landslides, as in the Xiajin Ancient Mulberry Grove System (Shandong) and the Jiaxian Traditional Chinese Date Gardens (Shaanxi). In addition, some agroforestry land use has an indirect role in water resource regulation. For example, in the Wannian Traditional Rice Culture System (Jiangxi), mountain springs flow from the surrounding mountains and forests bringing nutrients (especially tree litter and soil minerals) that nurture high-quality rice crops.

Second, agroforestry optimizes resource utilization in China-NIAHS ($p = 0.74$). According to the ecological niche theory, the distinct niches of different species in agroforestry actively complement each other in space and interaction [57]. The aboveground canopy and underground root system, in particular, may fully utilize light, water, and nutrients by generating a spatially vertical stratification [58,59]. Compared to common field farms, the Qianxi Chestnut Compound Cultivation System (Hebei) and the Pinggu Juglans Hopeiensis Production System (Beijing) enhance light energy consumption rates by 13% to 31% and water utilization rates by 10.3% to 15.2%, respectively [58,60]. In addition, the optimal resources utilization by agroforestry is also linked to the microclimate created by forests. This microclimate is conducive to the growth of cultivated crops in the undercanopy. In the Kuaijishan Ancient Chinese Torreya Community (Zhejiang) and the Pu'er Ancient Tea Garden and Tea Culture System (Yunnan), woody plants provide moderate shading and regulate temperature, light, wind speed, and humidity [58,61].

Last but not least, agroforestry plays a productive role in local communities. On the one hand, it can promote yield through the above-mentioned efficient use of environmental resources [59]. For example, the increase in yield due to the forest network of northern China is generally 25–50% [62]. On the other hand, the wide range of products it provides (such as timber, firewood, fruit, food, herbs, fodder, etc.) contributes significantly to income diversity ($p = 0.67$). In the Diebu Zhagana Agriculture-Forestry-Animal Husbandry Composite System (Gansu), local forests (composed mainly of conifer species, including the protected *Taxus chinensis*) provide grazing for pigs, timber for house construction, firewood for heating, as well as fertilizers, such as bedding and wood ash. Additionally, medicinal herbs and around 88 species of medicinal and edible fungi are found in local forests [63]. Significantly, this income diversity contributes to residents' livelihoods. For example, in the Dazhai area of the Longji Terraced Fields in Longsheng (Guangxi), the income provided by terraced crops and forest by-products provides residents with a 97.7% higher income than non-residents [64].

4.2. Implications and Suggestions for China-NIAHS-Related Landscape Management and Policymaking

This study not only provides a detailed picture of China-NIAHS, but also gives insight into management strategies and policymaking. On the one hand, the findings assist in identifying China-NIAHS that lack sustainability. Most of these are located in North and Central China, where effective conservation plans should be implemented. To better protect and manage these at-risk China-NIAHS, the policy and economic pathways for rural development should vary according to the differences in economic, ecological, and social sustainability.

On the other hand, landscape sustainability has a strong correlation with the importance of agroforestry, and much more so with agriculture-producing species, as we discovered in our study. This finding supports previous studies indicating that agricultural diversity can improve landscape sustainability [65]. Considering agroforestry's multiple fundamental roles in promoting landscape sustainability, this realization should help promote agroforestry and its role as an integral part of a worldwide multifunctional working landscape. In response, agroforestry can be intensified or expanded to include more mixed species. Due to the interactions among elements, it may supply a great variety of benefits. As a result, agroforestry is more reliable than grant policies and subsidies for a single element when adopted in practice.

Agroforestry was shown to have a lower correlation with indigenous knowledge and management ($p = 0.31$) than with ecological and economic variables. This is conceivable because China-NIAHS are huge, complex systems with differing temporal dynamics. Thus, it is difficult to obtain information regarding the roles of indigenous knowledge and management, which are frequently inadequately described.

Even so, the analysis of the China-NIAHS dossiers and many studies has highlighted the widespread presence of indigenous knowledge and management, regardless of the importance assigned to agroforestry [66,67]. While some practices are slowly disappearing due to radical social and economic changes, most still survive and play a fundamental role in maximizing the sustainable use of resources [68,69], which is regulated by prohibitions on land-use tied to indigenous belief systems. For example, given that forests conserve soil and water, indigenous villages have consistently developed a belief system centered on sacred forest worship as part of their long-term cultivation of crops and habitation of settlements. Thus, logging and hunting are not permitted in forest areas managed by local communities under village regulations. These belief-based prohibitions on land-use are essential to terraced China-NIAHS, such as the Congjiang Rice-Fish-Duck System, the Diebu Zhagana Agriculture-Forestry-Animal Husbandry Composite System, and the Hani Rice Terrace System. This kind of bottom-up approach tailored to local conditions outperformed top-down rural planning regarding sustainability [70,71]. Consequently, we suggest that indigenous knowledge and management, as well as related belief systems collected from local stakeholders, could be integrated into planning and policymaking involving China-NIAHS [72].

4.3. Limitations and Outlook

This study built an approach to evaluating sustainability that aids in developing a trans-regional or national collaboration platform for interdisciplinary China-NIAHS research. However, limitations in terms of the data and objectivity of this study remain.

First, there is a lack of access to certain necessary data. On the one hand, we excluded some indicators for which data is not available, making the assessment system less comprehensive. On the other hand, the data for some indicators might not correspond to the actual situation. For example, the data on indigenous knowledge and management largely depends on the accessibility of information provided by the government and other organizations.

Second, interpretation of the Pearson's correlation coefficient largely depends on the context and purpose of the research. For example, there is likely to be a complex, non-linear relationship between the importance of agroforestry and the growth rate of DPI. Pearson's correlation coefficient could not entirely reflect this.

Third, even though this study was based on a thorough investigation of multi-source data, the indicator values were determined mainly by manual qualitative assessment, which is somewhat subjective. Therefore, future research requires a combination of subjective and objective methods to assign values in the evaluation system. As a prerequisite, more quantitative, site-specific spatial data that explore land-use modes and sustainability at multiple times and geographical scales should be obtained.

As this study aimed to provide general knowledge of agroforestry's effect on landscape sustainability, we did not conduct field surveys on each of the 118 China-NIAHS. Such a large number of sites challenged our attempt to initiate field-derived investigations and in-depth interviews in a traditional manner. Thus, place-specific case studies are needed in the future to gain more specific knowledge concerning the sustainability-enhancing mechanism by which agroforestry shapes the agricultural heritage landscapes.

There are only 118 sites on the China-NIAHS list. Given China's long history of agriculture, more agricultural heritage sites are waiting to be explored. Figure 4 shows that south China is home to the majority of the current China-NIAHS. However, the agricultural diversity of north China, especially northwest China, seems to have been ignored. Combining agricultural diversity and biological abundance datasets will allow more agricultural heritage sites to be designated in the future.

Additionally, since China-NIAHS represent unique regional conjunctions of landscape conservation, green economy, tourism, and cultural interchange [73,74], different contexts should be considered on a national and regional scale. Although we found a correlation between the importance of agroforestry and the landscape sustainability of China-NIAHS applied to different agricultural zoning, future research is needed to investigate spatial variability at a finer scales in response to comprehensive contexts (biological abundance, habitat condition, water system, socioeconomic situation, labor migration, distribution of traditional villages, etc.).This could help with targeted agricultural sustainability and rural planning strategies.

The current enthusiasm for a single agricultural ecosystem, without considering social sustainability, entails a certain risk [75]. This is particularly problematic if applied to agricultural heritage landscapes, given their long-lasting, dynamic interactions between humans and nature, inherited cultural patterns, and people's identities and values. Working as a technique lens for social sustainability, indigenous knowledge and management may enrich sustainability research, as they deepen the understanding of the role of humans in ecosystems. There have been calls for closer communication and cooperation between sustainability research and indigenous knowledge and management, implying a promising field for future research. As the study material shows, there is a lack of a precise and uniform approach to dealing with indigenous knowledge and management. Therefore, the first step is to standardize the description of indigenous knowledge and management of agroforestry based on a detailed scientific investigation in field-derived case studies. Thus, these nonmaterial landscape values can be measured explicitly in a qualitative, quantitative, or spatial way. This will motivate us to work toward a prosperous future, with a productive landscape and the coexistence of humans and nature.

Author Contributions: Conceptualization, M.Z. and J.L.; methodology, M.Z. and J.L.; software, M.Z. and J.L.; validation, M.Z. and J.L.; formal analysis, M.Z.; investigation, M.Z.; resources, M.Z.; data curation, M.Z.; writing—original draft preparation, M.Z.; writing—review and editing, M.Z. and J.L.; visualization, M.Z.; supervision, J.L.; project administration, J.L.; funding acquisition, J.L. All authors have read and agreed to the published version of the manuscript.

Funding: This research was funded by the National Natural Science Foundation of China, grant number 52108051.

Institutional Review Board Statement: Not applicable.

Informed Consent Statement: Not applicable.

Data Availability Statement: The data presented in this study are available on request from the corresponding author.

Conflicts of Interest: The funders had no role in the design of the study; in the collection, analyses, or interpretation of data; in the writing of the manuscript, or in the decision to publish the results.

References

1. Gkoltsiou, A.; Athanasiadou, E.; Paraskevopoulou, A.T. Agricultural Heritage Landscapes of Greece: Three Case Studies and Strategic Steps towards Their Acknowledgement, Conservation and Management. *Sustainability* **2021**, *13*, 5955. [CrossRef]
2. Kizos, T.; Vlahos, G. The evolution of the agricultural landscape. In *Book Reclaiming the Greek Landscape*; Papayannis, T., Howard, P., Eds.; Med-INA: Athens, Greece, 2012; Volume 97, pp. 133–143.
3. Bürgi, M.; Verburg, P.H.; Kuemmerle, T.; Plieninger, T. Analyzing dynamics and values of cultural landscapes. *Landsc. Ecol.* **2017**, *32*, 2077–2081. [CrossRef]
4. Antrop, M. The concept of traditional landscapes as a base for landscape evaluation and planning: The example of Flanders Region. *Landsc. Urban Plan.* **1997**, *38*, 105–117. [CrossRef]
5. Jiao, Y.; Liang, L.; Okuro, T.; Takeuchi, K. Ecosystem services and biodiversity of traditional agricultural landscapes: A case study of the Hani terraces in Southwest China. In *Biocultural Landscapes*; Hong, S.-K., Bogaert, J., Min, Q., Eds.; Springer: Dordrecht, The Netherlands, 2014; pp. 81–88.
6. Liu, P. To establish a protection system for China's famous villages of historic and cultural interest. *J. Peking Univ.* **1998**, *35*, 81–88.
7. Agnoletti, M. The degradation of traditional landscape in a mountain area of Tuscany during the 19th and 20th centuries: Implications for biodiversity and sustainable management. *For. Ecol. Manag.* **2007**, *249*, 5–17. [CrossRef]
8. Van Eetvelde, V.; Antrop, M. Analyzing structural and functional changes of traditional landscapes—Two examples from Southern France. *Landsc. Urban Plan.* **2004**, *67*, 79–95. [CrossRef]
9. Yu, H.; Verburg, P.H.; Liu, L.; Eitelberg, D.A. Spatial analysis of cultural heritage landscapes in Rural China: Land use change and its risks for conservation. *Environ. Manag.* **2016**, *57*, 1304–1318. [CrossRef]
10. Gotsch, N.; Rieder, P. Biodiversity, Biotechnology, and Institutions Among Crops. *J. Sustain. Agric.* **1995**, *5*, 5–40. [CrossRef]
11. Liu, C.L.C.; Kuchma, O.; Krutovsky, K.V. Mixed-species versus monocultures in plantation forestry: Development, benefits, ecosystem services and perspectives for the future. *Glob. Ecol. Conserv.* **2018**, *15*, e00419. [CrossRef]
12. Baltodano, J. Monoculture forestry: A critique from an ecological perspective. In Proceedings of the 6th COP of the FCCC, The Hague, The Netherlands, 13–25 November 2000; Friends of the Earth International: Amsterdam, The Netherlands, 2000; pp. 2–10.
13. Erskine, P.D.; Lamb, D.; Bristow, M. Tree species diversity and ecosystem function: Can tropical multi-species plantations generate greater productivity? *For. Ecol. Manag.* **2006**, *233*, 205–210. [CrossRef]
14. Bremer, L.L.; Farley, K.A. Does plantation forestry restore biodiversity or create green deserts? A synthesis of the effects of land-use transitions on plant species richness. *Biodivers. Conserv.* **2010**, *19*, 3893–3915. [CrossRef]
15. Brockerhoff, E.G.; Jactel, H.; Parrotta, J.A.; Ferraz, S.F. Role of eucalypt and other planted forests in biodiversity conservation and the provision of biodiversity-related ecosystem services. *For. Ecol. Manag.* **2013**, *301*, 43–50. [CrossRef]
16. Nowak, J.T.; Berisford, C.W. Effects of intensive forest management practices on insect infestation levels and loblolly pine growth. *Environ. Entomol.* **2000**, *93*, 336–341. [CrossRef]
17. Li, M.; Zhang, Y.; Xu, M.; He, L.; Liu, L.; Tang, Q. China Eco-Wisdom: A Review of Sustainability of Agricultural Heritage Systems on Aquatic-Ecological Conservation. *Sustainability* **2019**, *12*, 60. [CrossRef]
18. Clark, W.C.; Dickson, N.M. Sustainability science: The emerging research program. *Proc Nat. Acad. Sci. USA* **2003**, *100*, 8059–8061. [CrossRef]
19. Wiens, J.A. Is landscape sustainability a useful concept in a changing world? *Landsc. Ecol.* **2013**, *28*, 1047–1052. [CrossRef]
20. Colglazier, W. Sustainable development agenda: 2030. *Science* **2015**, *349*, 1048–1050. [CrossRef]
21. Plieninger, T.; Muñoz-Rojas, J.; Buck, L.E.; Scherr, S.J. Agroforestry for Sustainable Landscape Management. *Sustain. Sci.* **2020**, *15*, 1255–1266. [CrossRef]
22. Liu, J.; Zhang, M.; Nikita, N. Agent-based design research to explore the effectiveness of bottom-up organizational design in shaping sustainable vernacular landscapes: A case in Hailar, China. *Landsc. Urban Plan.* **2021**, *205*, 103961. [CrossRef]
23. Bond, A.; Morrison-Saunders, A.; Pope, J. Sustainability assessment: The state of the art. *Impact Assess. Proj. Apprais.* **2012**, *30*, 53–62. [CrossRef]
24. Fadul-Pacheco, L.; Wattiaux, M.A.; Espinoza-Ortega, A.; Sánchez-Vera, E.; Arriaga-Jordán, C.M. Evaluation of sustainability of smallholder dairy production systems in the highlands of Mexico during the rainy season. *Agroecol. Sustain. Food Syst.* **2013**, *37*, 882–901. [CrossRef]
25. Ryschawy, J.; Choisis, N.; Choisis, J.P.; Joannon, A.; Gibon, A. Mixed crop-livestock systems: An economic and environmental-friendly way of farming? *Animal* **2012**, *6*, 1722–1730. [CrossRef]
26. Nakajima, E.S.; Ortega, E. Exploring the sustainable horticulture productions systems using the energy assessment to restore the regional sustainability. *J. Clean. Prod.* **2015**, *96*, 531–538. [CrossRef]
27. Yang, L.; Huang, B.; Mao, M.; Yao, L.; Niedermann, S.; Hu, W.; Chen, Y. Sustainability assessment of greenhouse vegetable farming practices from environmental, economic, and socio-institutional perspectives in China. *Environ. Sci. Pollut. Res.* **2016**, *23*, 17287–17297. [CrossRef]
28. Baccar, M.; Bouaziz, A.; Dugué, P.; Gafsi, M.; Le Gal, P.Y. The determining factors of farm sustainability in a context of growing agricultural intensification. *Agroecol. Sustain. Food Syst.* **2019**, *43*, 386–408. [CrossRef]
29. Hu, F.; Gan, Y.; Cui, H.; Zhao, C.; Feng, F.; Yin, W.; Chai, Q. Intercropping maize and wheat with conservation agriculture principles improves water harvesting and reduces carbon emissions in dry areas. *Eur. J. Agron.* **2016**, *74*, 9–17. [CrossRef]
30. Agroforestry. Available online: https://www.fao.org/forestry/agroforestry/80338/en (accessed on 15 October 2021).

31. Fuller, A.M.; Min, Q.; Jiao, W.; Bai, Y. Globally Important Agricultural Heritage Systems (GIAHS) of China: The Challenge of Complexity in Research. *Ecosyst. Health Sustain.* **2015**, *1*, 1–10. [CrossRef]
32. Mastrangelo, M.E.; Weyland, F.; Villarino, S.H.; Barral, M.P.; Nahuelhual, L.; Laterra, P. Concepts and Methods for Landscape Multifunctionality and a Unifying Framework Based on Ecosystem Services. *Landsc. Ecol.* **2014**, *29*, 345–358. [CrossRef]
33. Zhang, Y.; Li, X.; Min, Q. How to Balance the Relationship between Conservation of Important Agricultural Heritage Systems (IAHS) and Socio-Economic Development? A Theoretical Framework of Sustainable Industrial Integration Development. *J. Clean. Prod.* **2018**, *204*, 553–563. [CrossRef]
34. Kremen, C.; Merenlender, A.M. Landscapes That Work for Biodiversity and People. *Science* **2018**, *362*, 6412. [CrossRef]
35. Jose, S. Agroforestry for Ecosystem Services and Environmental Benefits: An Overview. *Agrofor. Syst.* **2009**, *1*, 1–10.
36. Singh, M.; Sridhar, K.B.; Kumar, D.; Dwivedi, R.P.; Dev, I.; Tewari, R.K.; Chaturvedi, O.P. Agroforestry for doubling farmers' income: A proven technology for trans-gangetic plains zone of India. *Indian Farming* **2018**, *68*, 33–34.
37. Van Noordwijk, M. Agroforestry-Based Ecosystem Services: Reconciling Values of Humans and Nature in Sustainable Development. *Land* **2021**, *10*, 699. [CrossRef]
38. Daniel, T.C.; Muhar, A.; Arnberger, A.; Aznar, O.; Boyd, J.W.; Chan, K.M.; Costanza, R.; Elmqvist, T.; Flint, C.G.; Gobster, P.H.; et al. Contributions of cultural services to the ecosystem services agenda. *Proc. Natl. Acad. Sci. USA* **2012**, *109*, 8812–8819. [CrossRef]
39. Schaich, H.; Bieling, C.; Plieninger, T. Linking ecosystem services with cultural landscape research. *Gaia Ecol. Perspect. Sci. Soc.* **2010**, *19*, 269–277. [CrossRef]
40. He, X.L.; Wang, H.F.; Liu, G.X.; Wang, Y.X.; Chen, Y.M.; Jia, H.T.; Wang, L.Y. Protection and utilization of agricultural species diversity and genetic diversity in Shexian Dryland Terrace System. *Chin. J. Eco-Agric.* **2020**, *28*, 1453–1464. (In Chinese)
41. Bai, Y.; Min, Q.; Li, J. Water Conservation Function of Forest Soil in Honghe Hani Rice Terrace System. *J. Soil Water Conserv.* **2016**, *23*, 166–170. (In Chinese)
42. Hölting, L.; Komossa, F.; Filyushkina, A.; Gastinger, M.M.; Verburg, P.H.; Beckmann, M.; Volk, M.; Cord, A.F. Including Stakeholders' Perspectives on Ecosystem Services in Multifunctionality Assessments. *Ecosyst. People* **2020**, *16*, 354–368. [CrossRef]
43. GIAHS Globally Important Agricultural Heritage Systems: Asia and the Pacific. Available online: https://www.fao.org/giahs/giahsaroundtheworld/designated-sites/asia-and-the-pacific/en (accessed on 3 October 2021).
44. China National Important Agricultural Cultural Heritage Online Exhibition. Available online: https://www.ciae.com.cn/list/zh/agricultural_heritage/video.html (accessed on 3 October 2021).
45. Min, Q. Problems and Suggestions in the Application of Important Agricultural Cultural Heritage in China. *J. Herit. Preserv.* **2019**, *4*, 8–11. (In Chinese)
46. Santoro, A.; Venturi, M.; Ben Maachia, S.; Benyahia, F.; Corrieri, F.; Piras, F.; Agnoletti, M. Agroforestry Heritage Systems as Agrobiodiversity Hotspots. The Case of the Mountain Oases of Tunisia. *Sustainability* **2020**, *12*, 4054. [CrossRef]
47. Tilman, D.; Lehman, C.L.; Thomson, K.T. Plant diversity and ecosystem productivity: Theoretical considerations. *Proc. Natl. Acad. Sci. USA* **1997**, *94*, 1857–1861. [CrossRef]
48. Sistla, S.A.; Roddy, A.B.; Williams, N.E.; Kramer, D.B.; Stevens, K.; Allison, S.D. Agroforestry practices promote biodiversity and natural resource diversity in Atlantic Nicaragua. *PLoS ONE* **2016**, *11*, e0162529. [CrossRef]
49. Santoro, A.; Venturi, M.; Bertani, R.; Agnoletti, M. Review of the Role of Forests and Agroforestry Systems in the FAO Globally Important Agricultural Heritage Systems (GIAHS) Programme. *Forests* **2020**, *11*, 860. [CrossRef]
50. Zhao, W.; Wang, S.; Zhang, Y.; Xu, M.; Qie, H.; Lu, Y.; Lu, H.; An, R. A UN SDGs-Based Sustainability Evaluation Framework for Globally Important Agricultural Heritage Systems (GIAHS): A Case Study on the Kuaijishan Ancient Chinese Torreya Community. *Sustainability* **2021**, *13*, 9957. [CrossRef]
51. Barnett, C.; Parnell, S. Ideas, implementation and indicators: Epistemologies of the post-2015 urban agenda. *Environ. Urban.* **2016**, *28*, 87–98. [CrossRef]
52. Saaty, T.L. How to make a decision: The analytic hierarchy process. *Eur. J. Oper. Res.* **1990**, *48*, 9–26. [CrossRef]
53. Liu, X. Delphi Technique in the Assessment of Interdisciplinary Research. *J. Southwest Jiaotong Univ.* **2007**, *2*, 21–25. (In Chinese)
54. Wu, J. Index system construction and evaluation of tourism sustainable development: A case study of Qingdao. *Explor. Econ. Issues* **2013**, *10*, 64–70. (In Chinese)
55. Pollesch, N.; Dale, V.H. Applications of aggregation theory to sustainability assessment. *Ecol. Econ.* **2015**, *114*, 117–127. [CrossRef]
56. Xie, M. The relation of covariance, correlation coefficient and correlation. *Math. Stat. Manag.* **2004**, *3*, 33–36. (In Chinese)
57. Liu, X.Y.; Zeng, D.H. Research advances in interspecific interactions in agroforestry system. *Chin. J. Ecol.* **2007**, *9*, 1464–1470. (In Chinese)
58. Meng, P.; Fan, W.; Song, Z.; Zhang, J.; Xin, X. A Study on the Utilization Efficiency of Water and Energy Resources of Agroforestry System. *For. Res.* **1999**, *3*, 37–42. (In Chinese)
59. Jose, S.; Gillespie, A.R.; Pallardy, S.G. Interspecific interactions in temperate agroforestry. *Agrofor. Syst.* **2004**, *61*, 237–255.
60. Lu, G.; Ma, X.; Zhou, H.; Xu, Z.; Song, Z.; Meng, P.; Liu, Y.; Gong, H. Study on the Farmland Evapotranspiration and the Water Use Efficiency of Crop of the Compound System of Agroforestry. *ACTA Agric. Univ. Pekin.* **1992**, *18*, 26–30. (In Chinese)
61. Hagiwara, H. The Spatial Distribution of Cities, Landscape Change and Traditional Agriculture in the Tokushima Region. *CIRAS Discuss. Pap.* **2019**, *90*, 49–57.
62. Xie, J.; Yu, R.; Hu, Y. An overview research of agroforestry ecosystem. *J. Beijing For. Univ.* **1988**, *1*, 104–108. (In Chinese)

63. Yang, L.; Liu, M.C.; Lun, F.; Min, Q.; Li, W. The impacts of farmers' livelihood capitals on planting decisions: A case study of Zhagana Agriculture—Forestry-Animal Husbandry Composite System. *Land Use Policy* **2019**, *86*, 208–217. [CrossRef]
64. Zhang, Y.; He, L.; Li, X.; Zhang, C.; Chen, Q.; Li, J.; Zhang, A. Why are the Longji Terraces in Southwest China maintained well? A conservation mechanism for agricultural landscapes based on agricultural multi-functions developed by multi-stakeholders. *Land Use Policy* **2019**, *85*, 42–51. [CrossRef]
65. Jackson, L.E.; Pulleman, M.M.; Brussaard, L.; Bawa, K.S.; Brown, G.G.; Cardoso, I.M.; de Ruiter, P.C.; García-Barrios, L.; Hollander, A.D.; Lavelle, P.; et al. Social-ecological and Regional Adaptation of Agrobiodiversity Management across A Global Set of Research Regions. *Glob. Environ. Change* **2012**, *22*, 623–639. [CrossRef]
66. Scott, A. Beyond the conventional: Meeting the Challenges of Landscape Governance within the European Landscape Convention? *J. Environ. Manag.* **2011**, *92*, 2754–2762. [CrossRef]
67. Schulz, B.; Becker, B.; Götsch, E. Indigenous knowledge in a'modern' sustainable agroforestry system—A case study from Brazil. *Agrofor. Syst.* **1994**, *25*, 59–69. [CrossRef]
68. Youn, Y.C. Use of forest resources, traditional forest-related knowledge and livelihood of forest dependent communities: Cases in South Korea. *For. Ecol. Manag.* **2009**, *257*, 2027–2034. [CrossRef]
69. Ramakrishnan, P.S. Traditional forest knowledge and sustainable forestry: A north-east India perspective. *For. Ecol. Manag.* **2007**, *249*, 91–99. [CrossRef]
70. Rovai, M.; Andreoli, M.; Gorelli, S.; Jussila, H. A DSS model for the governance of sustainable rural landscape: A first application to the cultural landscape of Orcia Valley (Tuscany, Italy). *Land Use Policy* **2016**, *56*, 217–237. [CrossRef]
71. Van Berkel, D.B.; Verburg, P.H. Combining exploratory scenarios and participatory backcasting: Using an agent-based model in participatory policy design for a multi-functional landscape. *Landsc. Ecol.* **2012**, *27*, 641–658. [CrossRef] [PubMed]
72. Wong, S.; Dai, Y.; Tang, B.; Liu, J. A New Model of Village Urbanization? Coordinative Governance of State-Village Relations in Guangzhou City, China. *Land Use Policy* **2011**, *109*, 10550. [CrossRef]
73. Chen, B.; Qiu, Z.; Usio, N.; Nakamura, K. Tourism's impacts on rural livelihood in the sustainability of an aging community in Japan. *Sustainability* **2018**, *10*, 2896. [CrossRef]
74. Kajihara, H.; Zhang, S.; You, W.; Min, Q. Concerns and opportunities around cultural heritage in East Asian Globally Important Agricultural Heritage Systems (GIAHS). *Sustainability* **2018**, *10*, 1235. [CrossRef]
75. Redford, K.H.; Adams, W.M. Payment for ecosystem services and the challenge of saving nature. *Conserv. Biol.* **2009**, *23*, 785–787.

Article

Multi-Source Data-Based Evaluation of Suitability of Land for Elderly Care and Layout Optimization: A Case Study of Changsha, China

Jun Yang, Zhifei Lou *, Xinglong Tang and Ying Sun

School of Resources Environment, Hunan Agricultural University, Changsha 430128, China
* Correspondence: loulou@stu.hunau.edu.cn; Tel.: +86-176-1515-1862

Abstract: This paper constructs an evaluation index system for the suitability of community home and institutional elderly care land development, respectively, from different elderly care modes with the data of urban POI, OSM road network, and expert questionnaires in Changsha urban area in 2021, in order to alleviate the pressure of insufficient land for elderly care brought on by the increasingly serious aging problem. The suitability evaluation index system is based on the intersection of Thiessen polygons with the current elderly care facilities as the center point as the supplementary land for the elderly and explores the optimization path of the land for the elderly in combination with the existing residential land in Changsha. The results show the following: ① The spatial variation of land suitability for both community home and institutional elderly facilities is significant, exhibiting a pattern of "high in the middle and low in the surroundings, with high-value areas clustered in the center of the city, decreasing in suitability toward the periphery, and occasional scattered clusters in the suburbs." Among them, Furong District has the highest proportion of suitable areas for the elderly; ② Utilizing Changsha's Tianxin and Yuhua districts as case studies, the optimal path of land use for the elderly are investigated to provide a foundation for land use planning for the elderly in Changsha.

Keywords: elderly land; POI; suitability evaluation; layout optimization; stock land

Citation: Yang, J.; Lou, Z.; Tang, X.; Sun, Y. Multi-Source Data-Based Evaluation of Suitability of Land for Elderly Care and Layout Optimization: A Case Study of Changsha, China. *Sustainability* **2023**, *15*, 2034. https://doi.org/10.3390/su15032034

Academic Editors: Qingsong He, Jiayu Wu, Chen Zeng and Linzi Zheng

Received: 5 December 2022
Revised: 16 January 2023
Accepted: 18 January 2023
Published: 20 January 2023

1. Introduction

Land for the elderly, as we all know, refers to the land inhabited by homes and facilities that provide life care, rehabilitation care, and trusteeship services for the elderly [1]. To be sure, the clear message is that the first half of the 21st century will be the most rapidly developing and problematic period for the aging of China, which may have been due to the population born at the peak in the mid-20th century successively entering old age and the decline of the population fertility rate in recent years [2–4]. Of all the countermeasures that have been put forward in response to this phenomenon, the far-reaching one is the "Tenth Five-Year Plan for the Development of China's Aging Industry" [5], which, of notice, clearly incorporates the aging industry into the plan of economic and social development and sustainable development strategy. In February 2022, the State Council issued the "14th Five-Year Plan" National Aging Care Development and Elderly Service System Plan [6], proposing not only to increase investment in various types of elderly care facilities but also to plan and deploy land scientifically for new elderly care service facilities to support and renovate old-age facilities with stock land.

Indeed, developed countries have entered the aging society earlier [7]. Mounting studies on topics, such as elderly care models [8–10] and land planning for elderly facilities [11], have provided extensive references for the research work of scholars in China. Since China entered the aging society in the 1980s, the research enthusiasm for population aging, pensions, and other related issues has been growing, and the initial research fields mostly involved demography [12], sociology [13], economics [14], and so forth. Only in the 1990s did research on pensions begin based on the perspective of geography [15], covering the spatial and temporal evolution of aging [16], the spatial distribution of elderly facilities [17], and the location of elderly facilities [18]. In recent years, my country's fertility rate has continued to decline and the life expectancy of the elderly has continued to increase. The elderly population has expanded rapidly and the demand for elderly care has continued to increase. This has greatly stimulated the demand for elderly care service facilities by the elderly population in my country. The final analysis is the demand for land for elderly care facilities. Research on the demand for elderly care is mainly reflected in the study of factors influencing the demand for elderly care [19] and the study of matching the supply and demand for elderly care [20]. Research on land use for senior care facilities has focused on two aspects: On the one hand, it focuses on land use policies for eldercare. Scholars have gone through a process of research on land use policies for eldercare facilities, from exploring how to establish norms for eldercare facilities' land use, to analyzing the shortcomings of the existing system [21–23], and then proposing how to effectively guarantee the regulatory system for land use for elderly facilities [24]. Second, the optimization of land use planning and layout of elderly facilities [25]. Scholars have discussed the differences in elderly care models [26], the accessibility of elderly care facilities [27], the balanced layout of elderly care facilities [28], and the architectural design of elderly care facilities [29], respectively.

The current models of care for the elderly in China are traditional family care, community-based home care, and institutional care. The last two pertain to social elderly care. In social care for the elderly, the majority of academic research assesses primarily the single care model in community home care facilities or institutional care facilities. The research results can reflect the conditions of facilities at different levels in a more specific way, but the research results are not comprehensive [30]. From the perspective of elderly care needs, the index evaluation system of the suitability of land use for elderly care for two types of elderly care modes is not perfect [31], as there is a lack of research on the layout optimization strategy proposed for the two types of elderly care modes. From the perspective of the data basis of research, most of them are based on social and economic statistics and questionnaire data. In recent years, they have gradually started to make use of big data, POI data and other multi-source data for Spatio-temporal analysis [20,31]. In light of this, using ArcGIS 10.2 software and data from multiple sources, this paper evaluated the suitability of urban construction land development for elderly care in Changsha City, analyzed its spatial distribution characteristics in-depth, and proposed layout optimization suggestions for Changsha Bureau of Natural Resources. To actualize the creation of a caring city for the elderly and to meet their requirements for a better living, in order to actively respond to the emergence of an aging society, is of immense importance. The specific research ideas of this paper are as follows (Figure 1).

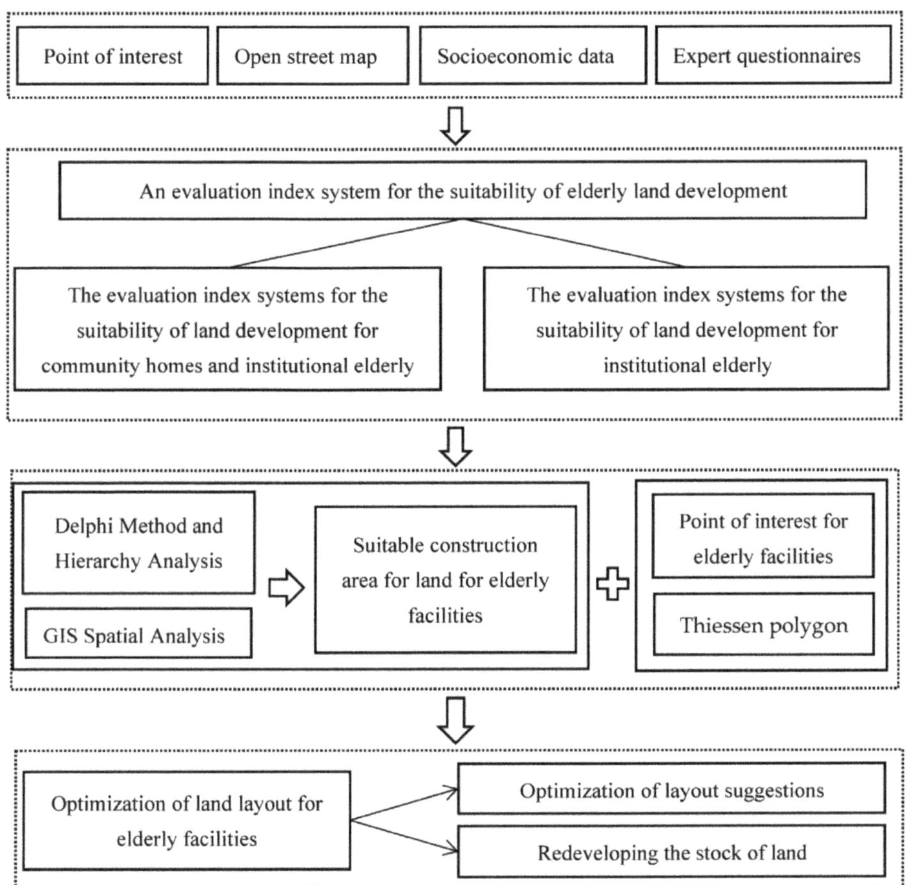

Figure 1. Research ideas.

2. Materials and Methods

2.1. Study Area

Changsha is an essential city for the economic development of central China. Changsha's GDP reached 121 million yuan in 2020, while the city's population reached 7.4729 million by the end of the year. However, the degree of aging is quite severe. It is usually accepted that a region has entered an aging society when the proportion of the elderly population aged 65 and older exceeds 7% of the overall population [32]. According to the findings of Changsha's seventh census [33], in 2020, the elderly population aged 65 and older made up 11.11% of the city's total population, and the city has entered an aging society. In the early stages of the project, the team conducted interviews with the Pension Section of the Civil Affairs Bureau of the Furong District Government and learnt that the demand for elderly care in Changsha's metropolitan region is high but the supply of land for the elderly is inadequate. Therefore, this study takes the urban area of Changsha City as the research area (Figure 2), including Furong District, Yuhua District, Tianxin District, Kaifu District, Yuelu District, Wangcheng District, and Changsha County. This region is the pioneer of Changsha's urban development and also the region with the most rapid economic growth. The urban area is a gradual slope built of terraces of varying heights. The land is more elevated in the south than in the north. The Xiangjiang River travels from south to north through the urban region in the center of China. The urban area's transportation network and infrastructure are relatively complete, and the elderly population density is relatively high (Figure 3). It is the favored location for creating and developing land for senior care facilities.

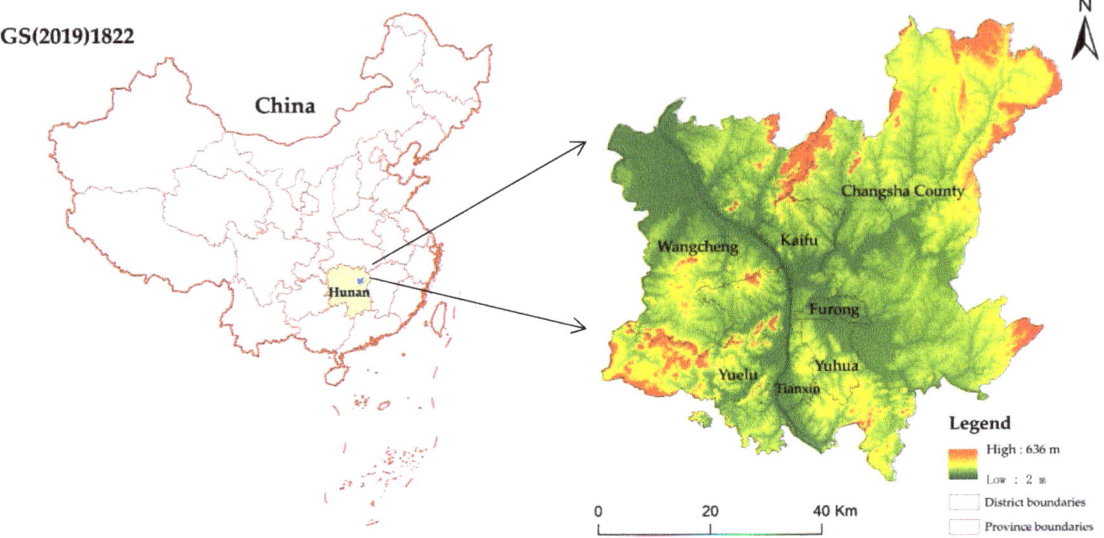

Figure 2. The location of seven districts of Changsha, China, Schematic of elevation.

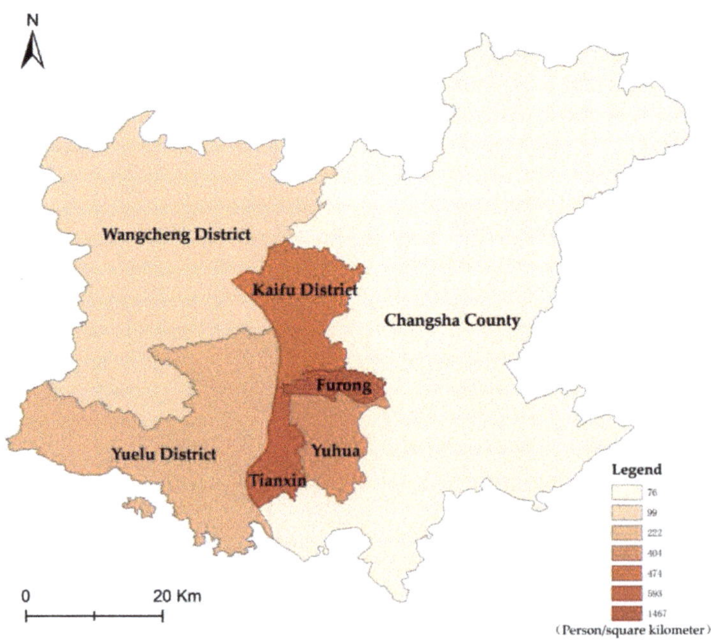

Figure 3. Population density map of Changsha district-level elderly over 65 years old (including 65 years old).

2.2. Data Sources

POI usually called points of interest, it refers to the point data in the Internet electronic map, which basically contains four attributes: name, address, coordinates, and category. Therefore, POI data have the natural advantage of big data mining based on spatial location. Additionally, POI data have strong expression ability and timeliness. In our study, POI data were obtained by applying for a Key on the AutoNavi Open Platform, and using programming code to obtain the POI data of the city in the API (Application Programming Interface) interface of the AutoNavi map. The acquired POI categories are parks, factories, residential communities, bus stations, subway stations, medical and health facilities and elderly service facilities, and the data include basic attribute information such as latitude and longitude coordinates, location information, and facility names. The output coordinates are WGS84 geographic coordinates Tie. The invalid data were eliminated and the final POI data were obtained as shown in Table 1. The urban road network data are obtained from Changsha OSM (Open Street Map) data and visualized by converting the OSM data file format with the Changsha city administrative division vector range as the boundary. We obtain the railroad, traffic roads, and life roads in Changsha, among which traffic roads include main roads, highways, and expressways, etc.; life roads mainly include secondary and tertiary roads in the city. Other data include Changsha City 2022 first quarter stock land data from the Changsha Natural Resources and Planning Bureau website; population data such as the number of the elderly population in each district and county in the study area from the 7th Census; other relevant socio-economic data sources such as the 2020 Changsha City Statistical Yearbook [34].

Table 1. POI data volume.

Data Source	Classification	Quantity
POI data	Park	218
	Factory	1555
	Residence	4262
	Public transport stations	4244
	Subway Stations	135
	Medical and Health Facilities	601
	Senior Care Facilities	200

2.3. Indicator System

An Evaluation Index System for the Suitability of Elderly Land Development

The current study proposed a model for evaluating the suitability of land development for community home and institutional elderly care facilities based on the land requirements of various elderly care models. The majority of the sites for community-based elderly care facilities are located in communities or residential areas, and since certain restrictive factors were avoided in the land use regulations of residential areas, it was unnecessary to separately consider indicators such as noise pollution from railroads and air pollution from factories in the selection of indicators. Adopting the combination of the Delphi method with AHP. This method combines subjectivity and objectivity, making the weights trustworthy. The Delphi method is a significant evaluation based on specialists' comprehensive comprehension of the evaluation item and their own extensive expertise. The analytic hierarchy process is to make pairwise comparisons between indicators to determine the degree of importance so that the evaluation can be endowed with logic. To make expert scoring findings more reasonable and objective, 27 experts in land resource management, land planning, and urban planning-related domains were invited to participate in two rounds of scoring to determine the weight value. The hierarchical analysis approach was executed using yaahp software, and the weights of the detected indicators were compatible with the major trends of the weights of the expert scoring method, as well as the final comprehensive weight values for each indicator (as shown in Tables 2 and 3).

Table 2. Results of evaluation index weights for suitability of land development for institutional elderly facilities.

Target Layer	Guideline Layer	Indicator Layer	Expert Scoring Method Weights	AHP Weights	Final Weights
land development for institutional elderly facilities	Environmental Factor B1	Park Distance C1	0.09	0.12	0.11
		Water system lake distance C2	0.07	0.05	0.06
		Factory distance from C3	0.06	0.03	0.05
		Rail distance C4	0.06	0.02	0.04
		Residential land use distance C5	0.09	0.09	0.09
	Supporting facilities factor B2	Bus stop distance C6	0.07	0.1	0.08
		Subway Station Distance C7	0.06	0.08	0.07
		Distance of traffic road C8	0.06	0.04	0.05
		Living road network density C9	0.07	0.06	0.07
		Hospital distance C10	0.11	0.28	0.20
	Socio-economic factors B3	Population density of the elderly aged 65 and over C11	0.09	0.02	0.05
		GDP per capita C12	0.07	0.04	0.06
		Disposable income per capita C13	0.10	0.06	0.08

Table 3. Results of evaluation index weights for the suitability of land development of community home care facilities.

Target Layer	Guideline Layer	Indicator Layer	Expert Scoring Method Weights	AHP Weights	Final Weights
land development for community home care facilities	Environmental Factor B1	Water system lake distance C1	0.11	0.07	0.09
		Park Distance C2	0.16	0.13	0.15
	Supporting facilities factor B2	Bus stop distance C3	0.08	0.13	0.11
		Subway Station Distance C4	0.07	0.07	0.07
		Traffic road distance C5	0.05	0.03	0.04
		Density of living road network C6	0.08	0.08	0.08
		Hospital distance from C7	0.13	0.29	0.21
	Socio-economic factors B3	Population density of the elderly aged 65 and above C8	0.12	0.07	0.09
		GDP per capita C9	0.08	0.03	0.05
		Disposable income per capita C10	0.10	0.11	0.10

2.4. Methodology

2.4.1. Buffer Analysis

A buffer is an extent that can affect or serve a geospatial target. A buffer is a neighborhood size S of a spatial entity O_i determined by a neighborhood radius R, defined as follows.

$$S_i = \{x | d(x, O_i) \leq R\} \tag{1}$$

In the above equation, S_i denotes the set of points whose distance from O_i is less than or equal to R, i.e., the buffer of O_i; d is the minimum Euclidean distance from any point x within S_i to O_i, which is used to represent the radiation range of the spatial entity to the neighborhood.

In this study, it was assumed that the distance of the land for elderly facilities from parks, water lakes, residential land, bus stations, subway stations, and hospitals was smaller, and the assigned score was higher and decreasing; factories and railroads were polluting, and the further the distance was, the higher the score was, while the main roads of the city were moderate indicators, and the distance was too close to have certain noise pollution, but the distance is too far to reduce the traffic convenience, and the assigned score is larger and then smaller. The buffer distance is set with reference to the Urban Residential

Area Planning and Design Standards, and the delineation of the community living circle is mainly based on the behavioral activities of the residents. The average walking speed of the elderly group is 3–5 km/h, and the walking distance of the general 15-min living circle is about 1250 m. Combined with the relevant studies of previous authors [31,35], the specific distance range and the corresponding scores are shown in Table 4. Buffer zone analysis plays an important role in the planning and layout of pension facilities [25]; therefore, this paper uses buffer zone analysis to evaluate the factors of the land use index system for elderly facilities to pave the way for obtaining suitability evaluation results.

Table 4. Buffer distance setting and assignment results.

Indicators	Buffer Setting Distance (m)	Corresponding Score
Park Distance	500; 1000; 1500; 2000; 2500	5; 4; 3; 2; 1
Water system lake distance	300; 500; 800; 1000; 1200	5; 4; 3; 2; 1
Factory Distance	500; 800; >800	−2; −1; 0
Rail distance	500; 800; >800	−2; −1; 0
Residential land distance	500; 1000; 1500; 2000; 3000	5; 4; 3; 2; 1
Bus stop distance	200; 400; 600; 800; 1000	5; 4; 3; 2; 1
Subway Station Distance	200; 400; 600; 800; 1000	5; 4; 3; 2; 1
Distance from traffic-oriented roads	100; 200; 300; 4000; 500	−1; 4; 3; 2; 1
Hospital Distance	500; 1000; 1500; 2000; 3000	5; 4; 3; 2; 1

2.4.2. Thiessen Polygon

The key point of the Thiessen polygon (Voronoi) algorithm is to construct a triangular network for discrete data points in a reasonable way. Irregular triangular nets (ITNs) conforming to Delauuay's criterion are delineated among all points [36].

Let there be a set of discrete points (X_i, Y_j) on the plane region B (i = 1, 2, 3, ... , k; j = 1, 2, 3, ... , k; k is the number of discrete points) if region B is divided into k mutually adjacent polygons by a set of straight line segments such that.

(1) Each polygon contains and contains only one discrete point;
(2) If any point (x^1, y^1) on region B lies within the polygon containing the discrete point (x_i, y_j), then the following:

$$\sqrt{(x^1 - x_i)^2 + (y^1 - y_j)^2} < \sqrt{(x^1 - x_j)^2 + (y^1 - y_i)^2} \tag{2}$$

holds when $i \neq j$

(3) If the point (x^1, y^1) lies on the common side of two polygons containing the discrete point (x_i, y_j), then the following:

$$\sqrt{(x^1 - x_i)^2 + (y^1 - y_j)^2} = \sqrt{(x^1 - x_j)^2 + (y^1 - y_i)^2} \tag{3}$$

holds.

The resulting polygon is called a Thiessen polygon. The triangle formed by connecting the discrete points within every of two adjacent polygons with a straight line is called a Thiessen triangle.

Thiessen polygon can be used to visualize the service area of senior care facility sites and show the size and service area of each senior care facility site and other characteristics, providing a reference for facility layout optimization. This paper constructs Thiessen polygons with the current status of the elderly care facilities as the center point, and the size of each Thiessen polygon shows the coverage of the elderly care facilities. The different sizes of polygons indicate the unbalanced coverage of elderly care facilities. Therefore, when optimizing the layout, the Thiessen polygons of the current facilities are superimposed on the suitable area, and the candidate points are arranged as much as possible at the intersection of the Thiessen polygons in the suitable area or overlapping sides, to provide a reference for optimizing the layout of elderly care land [37].

2.4.3. Suitability Evaluation

The suitability evaluation of this study is to carry out corresponding raster transformation on the single factor evaluation scores obtained from buffer analysis and road network density analysis, and use the weighted sum tool in ArcGIS platform to carry out weighted summation of data scores and weights, thus obtaining suitability evaluation value [38], which is calculated as follows:

$$S_{i=}\sum_{i=1}^{n}(W_iC_i) \quad (4)$$

S_i is the score of evaluating the suitability of land use for elderly care facilities; W_i is the weight of the suitability factor of item i; C_i is the score of the ith item suitability factor; n is the number of index factors.

3. Results and Analysis

3.1. Spatial Distribution Characteristics of Land Development Suitability for Elderly Facilities

The spatial distribution characteristics can show the spatial difference and the non-equilibrium between regions in the suitability of land use for elderly care facilities. The evaluation scores and spatial distribution characteristics of the suitability of land development for elderly care facilities were obtained by the ArcGIS platform. In Changsha, the geographical distribution of land suitability for elderly facilities verified a pattern of "high in the middle and low in the surrounding region, with high-value clusters in the city center and distributed outward, and infrequent scattered clusters in the suburbs." Due to its favorable environment, convenient transportation, obvious advantages in medical resources and high quality of economic development, the core area of Changsha shows the spatial distribution characteristics of high-value agglomeration. Furong District is highly suitable for the whole area. As the political and cultural center of Changsha in the early stage, resources and environment are at a high level. Gaotangling, Yujiapo, Dingzi Bay and Moon Island are sporadically distributed with the high-value areas of Xianglong, Xingsha, and Quantang in Changsha County, because the former is generally dependent on better medical facilities and transportation conditions, while the latter is closely connected with the municipal district. Low-value areas were most concentrated in Wangcheng District and Changsha County, with a small number distributed at the edges of the city-administered districts, such as Ladou River, Qingzhu Lake, and Shaping Street in the northern part of Kaifu District; Dato in the southern part of Tianxin District; the southeastern part of Yuhua District; and over half the southwestern part of Yuelu District. In comparison to other districts, Yuelu district has the highest proportion in low-value area due to its low development intensity, short construction period, and inadequate supporting facilities.

3.2. Comprehensive Evaluation Grade Analysis of Land Suitability for Elderly Facilities

As depicted in Figure 4, the GIS natural breakpoint approach was used to reclassify the evaluation results of the community home and institutional aged sites into four categories: inappropriate, less suitable, more suitable, and suitable.

(1) Suitable area. The land use characteristics in this type of area are optimal for senior facilities. The index range of community home suitable area is 2.95 to 4.82, accounting for 9.99% of the study area; the index range of institutional suitable area is 2.73 to 4.42, accounting for 10.06%, and the distribution of the two is very similar, concentrated in the municipal district and distributed in a row, and clustered sporadically in the north and south direction of the suburbs and the township centers on the outskirts of the city.

(2) More suitable area. Its construction and development conditions are second only to the suitable construction area, but certain indicators have disadvantages and

the development is constrained to some extent. The community home range is 1.55–2.73, accounting for 13.76%; the institutional range is 1.63–2.73, accounting for 12.37%. Spatially, they are all concentrated around the municipal district, extending in the north-south direction, and clustered in a point-like manner in the townships of Changsha County and Wangcheng District.

(3) Less suitable areas. There are more negative characteristics in this category, which hinders the total development of property for senior amenities. The range of the community house index is 0.92 to 1.78, representing 30.25%; the range of the institution index is 0.79 to 1.63, representing 26.7%. Overall, the distribution is comparable, with residential land use and transportation route direction accounting for the majority of the dispersion outside of the municipal area.

(4) Unsuitable area. This category is spread in regions where all indicators are deficient, and it is nearly inappropriate for development and construction usage as land for senior care institutions. The range for community homes is between 0.28 and 0.92, accounting for 45.88% of the total area, whereas the range for institutions is between 0.11 and 0.79, accounting for 50.14%of the total area, indicating that the land requirements for institutional elderly facilities are higher and more stringent for all indicators.

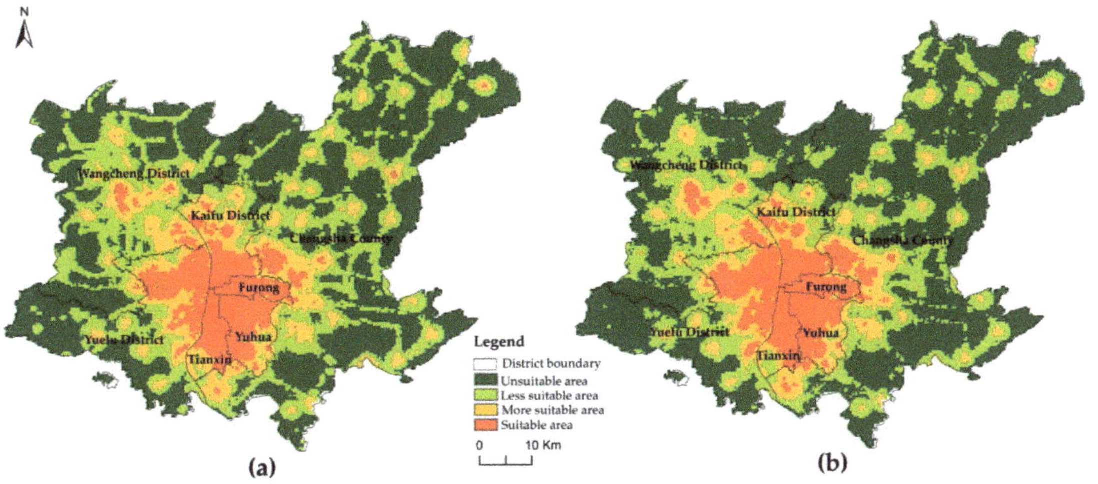

Figure 4. Evaluation of land suitability of community home and institutional endowment facilities. (**a**) Community Home; (**b**) Institution.

The more suitable and suitable areas in the study area are mainly concentrated in the central part of the city, accounting for a smaller proportion, but both transportation conditions and medical facilities and other resources occupy a considerable advantage: Superior transportation conditions can save time costs for children visiting the elderly, and are also convenient for the elderly to travel, reducing the sense of isolation of the elderly; also, sound medical facilities are the primary factor in the selection of land for elderly facilities. More than half of the total is comprised of less suitable and unsuitable regions, which are primarily located in the city's suburbs. We should expand road investment and medical facility building to establish the groundwork for conserving backup resources for land for senior care facilities.

The area ratio of different grades among administrative districts is shown in Table 5, Furong District is mainly suitable, and more suitable areas and the majority of suitable areas, which is the most suitable for the development of elderly care business; Tianxin District and Yuhua District are mainly suitable and more suitable areas, there are no unsuitable areas, and there are obvious advantages for the development of elderly care business. In Kaifu

District, there are some unsuitable areas and a few unsuitable areas, and the advantage of the northern area is weak. Wangcheng District and Changsha County have more than half of the unsuitable and less suitable areas, and most of the areas in them are less suitable.

Table 5. Distribution of area share of suitability class.

Type	County Grade / Districtc	Furong District	Kaifu District	Tianxin District	Yuhua District	Yuelu District	Changsha County	Wangcheng District
Land for institutional elderly facilities	Unsuitable area	0.00%	7.89%	0.00%	0.00%	40.46%	60.56%	49.70%
	Less suitable area	0.00%	29.76%	14.13%	5.08%	23.51%	27.97%	34.25%
	More suitable area	9.16%	29.39%	18.62%	23.73%	15.25%	9.20%	12.67%
	Suitable area	90.84%	32.96%	67.26%	71.19%	20.78%	2.26%	3.38%
Land for community home care facilities	Unsuitable area	0.00%	7.83%	0.00%	0.00%	38.65%	56.16%	46.02%
	Less suitable area	0.00%	27.76%	6.26%	5.28%	26.13%	31.50%	38.02%
	More suitable area	6.74%	32.39%	24.88%	22.80%	16.80%	10.67%	13.72%
	Suitable area	93.26%	32.02%	68.86%	71.92%	18.42%	1.67%	2.25%

3.3. Suggestions

3.3.1. Optimization of Layout Suggestions

To further verify the matching degree between the suitability evaluation of land use for elderly facilities and elderly facilities, this research visualized the results of the overlay suitability evaluation of the current elderly facility points and constructed ideal Thiessen polygon with the current elderly facility points as the center, and the intersection of Thiessen polygon had the characteristics of maximum coverage and facility balance, i.e., fairness, as shown in Figure 5. It was observed that the fit between the current elderly facilities and the suitability evaluation results was high, and the regional differences and unevenness of spatial distribution were very significant. The Thiessen polygon of both models shows the characteristics of a small geometric area in the urban center and gradually increasing or even multiplying towards the periphery, which indicated that the demand for elderly care in the urban center can be well met, while the supply in the periphery was insufficient, and the single elderly care facility in the periphery radiates too large an area and serves people even across several towns, so the pressure on the elderly care facilities was huge and it is difficult to meet the demand for elderly care facilities of the elderly population in the urban periphery. It was difficult to meet the needs of the elderly population in the urban periphery, and it also reflects the unbalanced distribution of resources among towns.

Urban centers and suburbs, with a high density of elderly population and obvious advantages of resources within the city, should lay out community home care facilities in each street as much as possible, and their best location should be in the Thiessen polygon intersection as well as in the suitable construction area, such as Xingsha, Quan Tang, Mapoling, Dongan, Yuehu, Qingshui Tang, Xinhe, Guanshaling, Moon Island, Liuyang River, Xianglong, Pioneer, Yanghu, etc.; for institutions, suburbs can be laid out in the southeastern part of Lukou Town, the eastern part of Quan Tang and the junction of Huanghua Town in the suburbs, Xianglong, and the area of Moon Island. The distribution of the elderly population in the urban fringe towns is more scattered and less dense, plus the traffic and medical conditions are far less than those in the urban center, so the demand for community home care facilities is smaller, and it is more suitable to add large elderly institutions to meet the needs of a larger range of elderly people, such as the lack of elderly resources in the northeastern part of Changsha County, and the addition of Junction Town, Fulin Town, Qingshanpu Town, and Baisha Town can greatly alleviate the pressure of existing elderly institutions. Therefore, giving full play to the role of institutional senior care facility sites in distant suburbs is crucial to improving the senior care problem in suburban villages.

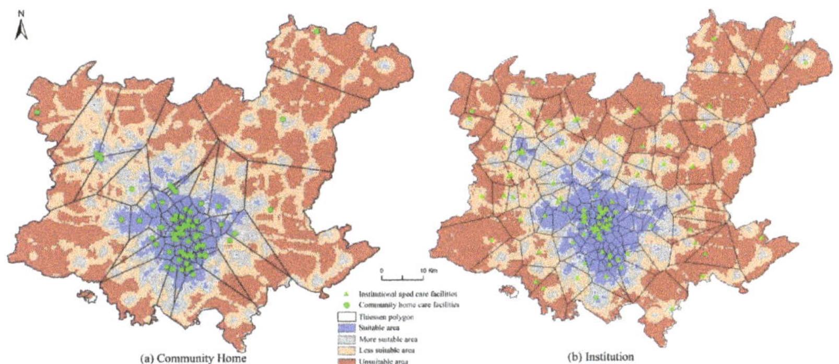

Figure 5. Optimization of land use layout for elderly care facilities.

3.3.2. Exploring Strategies for Redeveloping the Stock of Land

This study explores the potential of urban stock land from the perspective of proposing an optimization technique for land usage in senior care. To obtain the appropriate urban construction area, the suitable area of suitability evaluation is extracted and superimposed with the urban construction area retrieved from the land use status map. The visualization of community home care facilities in Yuhua District and Tianxin District and the construction of the Thiessen polygon, combined with the stock of residential land in the first quarter of 2022 in Changsha City, as an alternative location for future supplementary or reserved land for elderly care facilities. Since most of the stock land in the Changsha urban area is concentrated in Yuhua District and Tianxin District, and there is a relative lack of pension facilities in the border areas of these two districts, there is a large area of Thiessen geometry and Theisen geometry, indicating that a pension facility must serve a larger area. Since the service range of community home pension facilities typically includes the street level, the new points should be picked as much as possible in areas with big Thiessen polygons. In the interim, the new points should be satisfied in the city's acceptable building area and stock land, and alternative land locations for the new elderly care facilities should be provided, as depicted in Figure 6.

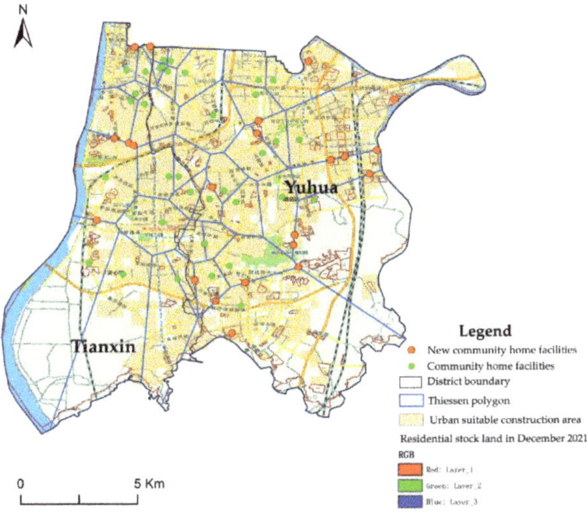

Figure 6. Additional sites for new community senior facilities. (The Chinese characters in the picture come from the Autonavi map and show the names of the city's main roads and green parks).

4. Conclusions and Discussion

4.1. Discussion

This paper takes Changsha urban area as the research object and evaluates the suitability of its land use for elderly facilities by using multi-source data such as POI data, OSM road network data and current land use data, and further discusses the layout strategy of land use for elderly facilities by interviewing the Elderly Section of Changsha Municipal Government, which can help alleviate the aging phenomenon in Changsha as well as provide reference for the planning of land use for elderly facilities in Changsha. The research results show that the suitable land for elderly care facilities in Changsha City is mainly concentrated in the central part of the city. The spatial distribution of elderly care facilities is uneven, and it is necessary to implement a stepped layout of elderly care facilities of different scales. For the source of land for elderly care facilities, the transformation of urban stock land can be emphatically considered. To ensure the effective supply of land for elderly care facilities, it is necessary to gradually strengthen institutional guarantees.

This study also has certain limitations. Due to the confidential nature of the data on the old population at the block level, the density of the elderly population can only be represented at the district level, so limiting the scope of the study and affecting its depth. In terms of evaluation factors, the selected buffer zone analysis approach is insufficiently unique and fails to account for the real route, which needs to be further optimized. Due to changes in the age composition, ethnic composition, and customs of the population, the special requirements for elderly care facilities and associated land use environment, such as self-care and non-self-care, and disregard for the preferences of the elderly, vary. The evaluation of the suitability of aged care land for various age groups, customs, and habits can be further refined and maximized through the use of extensive questionnaires and future research. In addition, due to data restrictions, this research only conducts empirical analysis from the standpoint of residential land stock, although the actual area that can be utilized for senior care is far more. Future studies should include field considerations for crucial locations, and data sources should be broadened to ensure quality. The conclusions of the analysis are more objective and rational.

4.2. Conclusions

(1) The spatial distribution characteristics of "strong in the middle and weak in the surroundings" are shown in the evaluation of the suitability of land for elderly facilities. The suitable areas are mainly concentrated in the central part of the city, while the less suitable and unsuitable areas are mainly concentrated in the suburbs and account for more than half of the area. In order to meet the expanding demand for land for elderly facilities, we should also continuously improve the construction of medical and transportation infrastructure in the suburbs to improve the suitability of land for elderly facilities and reserve backup resources.

(2) The Thiessen model centered on senior care facilities shows that the spatial distribution of senior care facilities in different modes varies significantly, with community home senior care facilities concentrated in urban centers and institutional senior care facilities spread throughout urban areas, but both have in common that the number of facilities in central cities is significantly higher than that in suburban areas, and the allocation of senior care facilities in urban and rural areas is less balanced.

(3) To optimize the spatial layout of land for elderly facilities, the following strategies are proposed. ① Enhance the hierarchical structure of senior care institutions in accordance with Changsha Civil Affairs Bureau's design in order to establish a three-tiered senior care service structure of "district-county-community-street (township)" that connects urban and rural areas effectively. For different types of land proposed for the elderly: the land for the elderly under the jurisdiction of the city due to its high suitability and dense elderly population, to the streets as a unit, with an emphasis on the layout of community home care facilities to provide basic protection for the elderly. In the suburban areas, the elderly population is relatively dense, the

traffic and medical facilities are complete, and the overall suitability is good, so the investment in community home and institutional elderly care facilities should be increased; in the peripheral areas of the city, more medium- and high-end elderly institutions can be built; in the distant suburbs, combined with the characteristics of its extensive and sparse population, large elderly institutions should be set up around the township centers, the more suitable areas with developed traffic networks and better medical facilities to expand the scope of services should be expanded. ② Based on the evaluation results of suitability of senior citizen land constructed in this paper, and from the perspective of urban stock land, we actively explore the potential senior citizen land, and take Tianxin District of Yuhua District as a case study, and propose alternative sites for new senior citizen facilities to effectively relieve the pressure on community senior citizens.

Author Contributions: Conceptualization, J.Y. and Z.L.; methodology, J.Y.; software, Z.L.; validation, J.Y., Z.L., X.T. and Y.S.; formal analysis, J.Y.; investigation, Z.L.; resources, J.Y.; data curation, Z.L., X.T. and Y.S.; writing—original draft preparation, J.Y.; writing—review and editing, Z.L.; visualization, Z.L.; supervision, J.Y.; project administration, J.Y.; funding acquisition, Z.L. All authors have read and agreed to the published version of the manuscript.

Funding: This work was supported by Research Project of Hunan Provincial Department of Natural Resources "Research on the Optimization of Land Use for Elderly Service Facilities in Urban Spaces under the Background of Population Aging" (2021G03).

Institutional Review Board Statement: Not applicable.

Informed Consent Statement: Not applicable.

Data Availability Statement: The data presented in this study are available on request from the corresponding author.

Acknowledgments: We sincerely thank Hunan Provincial Department of Natural Resources for their support of this research.

Conflicts of Interest: The authors declare no conflict of interest.

References

1. Liao, Y.; Lei, A.; Huang, Q.; Zhou, T. Research on pension land policy. *China Land* **2013**, *3*, 8–12.
2. Chen, X. Current situation and regional differences of disability in the elderly population in China. *Chin. J. Gerontol.* **2022**, *42*, 1197–1201.
3. Feng, J.; Hong, G.; Qian, W.; Hu, R.; Shi, G. Aging in China: An International and Domestic Comparative Study. *Sustainability* **2020**, *12*, 5086. [CrossRef]
4. Guo, Y.; Zhou, Y.; Han, Y. Spatio-temporal evolution of china's rural population aging and rural revitalization countermeasures. *Geogr. Res.* **2019**, *38*, 667–683.
5. Central Committee of the Communist Party of China State Council. Outline of the Tenth Five-Year Plan for the Development of China's Aging Career (2001–2005). Available online: http://www.gov.cn/zhengce/content/2016-09/23/content_5111148.htm?from=timeline&isappinstalled=0 (accessed on 28 December 2022). (In Chinese)
6. Central Committee of the Communist Party of China State Counci. National Plan for the Development of the Aged Cause and the Elderly Care Service System during the "Fourteenth Five Year Plan". Available online: https://www.mca.gov.cn/article/xw/mtbd/202202/20220200039833.shtml (accessed on 28 December 2022). (In Chinese)
7. Dizon, L.; Wiles, J.; Peiris-John, R. What Is Meaningful Participation for Older People? An Analysis of Aging Policies. *Gerontologist* **2020**, *60*, 396–405. [CrossRef]
8. Haines Terry Robinson Andrew, L.; Palmer Andrew, J. A new model of care and in-house general practitioners for residential aged care facilities. *Med. J. Aust.* **2021**, *215*, 44. [CrossRef]
9. Gan, D. Study on the Dilemma and Suggestions of Rural Pension Model in China. *World Sci. Res. J.* **2022**, *8*, 97–101.
10. Xia, M.; Wu, B.; Li, T. Analysis on the Development Status and Prospects of the Intelligent Pension Model in Tianjin. *Humanit. Soc. Sci.* **2019**, *7*, 34–48. [CrossRef]
11. Shuang, G.; Lu, Y.; Tao, P. Research Progress on the Model of Combining Medical and Elderly Care in Foreign Countries. *J. Nurs.* **2021**, *36*, 17–20.
12. Yi, H.; Xiaoguang, T. Analysis on the current situation of aging population in Chinese. *Chin. J. Gerontol.* **2012**, *32*, 4853–4855.
13. Peng, X.; Hu, Z. Aging of Chinese population from the perspective of public policy. *Chin. Soc. Sci.* **2011**, *32*, 106–124.

14. Hu, A.; Liu, S.; Ma, Z. Population Aging, Population Growth and Economic Growth: Empirical Evidence from China's Interprovincial Panel Data. *Popul. Res.* **2012**, *36*, 14–26.
15. Gao, X.; Wu, D.; Xu, Z. A Review of Aging Geography in China and the Construction of a Research Framework. *Prog. Geogr.* **2015**, *34*, 1480–1494.
16. Wang, H.; Liu, J.; Fang, Y. Multi-scale spatio-temporal evolution and influencing factors of population aging in Northeast China. *Reg. Res. Dev.* **2021**, *40*, 147–153.
17. Jiang, B.; Huang, B.; Zeng, B. Demand and suitability evaluation of "six-in-one" senior living circle facilities: Taking Gulou District of Nanjing as an example. *Sci. Technol. Eng.* **2022**, *22*, 5860–5870.
18. Zhang, S.; Liu, H.; Wang, H. Research on the Equilibrium of Institutional Pension Facilities in Tianjin Downtown. *Mod. Urban Res.* **2022**, *1*, 38–44.
19. Xiao, C.; Zhang, Y.; Lei, X.; Luo, J.; Zhao, G. Research on the Needs and Influencing Factors of Embedded Elderly Care Services in the Elderly Community. *J. Nurs.* **2022**, *37*, 91–93.
20. Peng, J.; Xin, L.; Yang, H. Research on Planning and Layout of Elderly Service Facilities Based on Supply and Demand Matching. *J. Geo-Inf. Sci.* **2022**, *24*, 1349–1362.
21. Ling, Q. Research on the Improvement and Implementation of Land Use Policy for Old-age Service Facilities in China. *Contemp. Econ. Manag.* **2016**, *38*, 52–55.
22. Lu, J.; Guo, F.; Chen, J. Rural Pension System Design in the New Era: Historical Context, Practical Dilemma and Development Path. *J. China Agric. Univ.* **2021**, *38*, 113–122.
23. Xuerui, Q. On the supply optimization and supervision of pension land. *China Land Resour. Econ.* **2019**, *32*, 58–62.
24. Mu, H. From "Pyramid" to "Olive Type": The Design and Optimization of the New Three-Pillar Old-age Security System. *Soc. Sci.* **2022**, *1*, 82–93. [CrossRef]
25. Li, B.; Ji, Y.; Zhao, X.; Qiu, W. Research on spatial accessibility of elderly care institutions in Xuzhou based on GIS. *J. Suzhou Univ. Sci. Technol.* **2021**, *38*, 65–69+75.
26. Yong, T. Study on the key factors and characteristics of the future elderly care model. *Basic Clin. Pharmacol. Toxicol.* **2020**, *127*, 291.
27. Amini-Behbahani, P.; Meng, L.; Gu, N. Walking distances from services and destinations for residential aged-care centres in Australian cities. *J. Transp. Geogr.* **2020**, *85*, 102707. [CrossRef]
28. Hao, Z. Spatial Matching and Policy-Planning Evaluation of Urban Elderly Care Facilities Based on Multi-Agent Simulation: Evidence from Shanghai, China. *Sustainability* **2022**, *14*, 16183.
29. Rahouti, A.; Lovreglio, R.; Nilsson, D.; Kuligowski, E.; Jackson, P.; Rothas, F. Investigating Evacuation Behaviour in Retirement Facilities: Case Studies from New Zealand. *Fire Technol.* **2021**, *57*, 1015–1039. [CrossRef]
30. Wang, H.; Liu, H.; Zhang, S. Research on Spatial Layout Evaluation of Institutional Elderly Care Facilities in Downtown Tianjin. *Urban Hous.* **2020**, *27*, 123–127.
31. Crooks, V.A.; Finlay, J.; Widener, M. Introduction to the special section: The changing geographies of aging. *Can. Geogr.* **2022**, *66*, 48–51. [CrossRef]
32. Xin, Y. How China Responds to the Challenge of Population Aging. *Natl. Gov.* **2014**, *21*, 13–18.
33. Hunan Provincial Bureau of Statistics. Bulletin of the Seventh National Census of Hunan Province (No. 4). Available online: http://tjj.hunan.gov.cn/hntj/wzzl/hnsdqcqgrkpc/qrptzgg/202105/t20210519_19054787.html (accessed on 30 December 2022).
34. Hunan Provincial Bureau of Statistics. *Hunan Statistical Yearbook (1998–2018)*; China Statistical Publishing House: Beijing, China, 2018.
35. Li, S.; Song, Q.; Yuan, S. Research on Site selection of elderly care real estate projects based on GIS: A case study of Tianjin. *J. Tianjin Univ.* **2017**, *19*, 204–209.
36. Zhu, C.; Li, Y.; Sun, X.; Xu, J.; Fu, Z. Urban public bicycle demand forecast considering land use. *J. South China Univ. Technol.* **2022**, *50*, 9–20+37.
37. Zhou, C.; Zhang, D.; He, X. Transportation Accessibility Evaluation of Educational Institutions Conducting Field Environmental Education Activities in Ecological Protection Areas: A Case Study of Zhuhai City. *Sustainability* **2021**, *13*, 9392. [CrossRef]
38. Hu, M.; Yong, X.; Li, X.; Zhou, X. Spatial layout optimization of rural residential areas under the guidance of urban-rural integration: A case study of Tongshan District, Xuzhou City. *Geogr. Geo-Inf. Sci.* **2022**, *5*, 104–110.

Article

Study on Green Utilization Efficiency of Urban Land in Yangtze River Delta

Qiaowen Lin * and Huiting Ling

School of Economics and Management, China University of Geosciences, Wuhan 430074, China; linghuiting125@cug.edu.cn
* Correspondence: qiaowen.lin@hotmail.com; Tel.: +86-151-7233-4036

Abstract: Based on the panel data of 41 cities in the Yangtze River Delta region from 2006–2018, this paper constructed an index system of measuring green urban land use efficiency (green land use efficiency), including input-expected output-unexpected output (input-expected output-unexpected output) and used the unexpected output-SBM model to calculate the green land use efficiency of the region. The spatial and temporal evolution characteristics of urban land green use efficiency in this area were studied by the Dagum Gini coefficient, decomposition, and exploratory spatial data analysis. The results show that (1) the temporal evolution characteristics of green land use efficiency in the whole region of the Yangtze River Delta, provinces (municipalities directly under the central government), and cities all show an upward trend of fluctuation. Among them, the green land use efficiency of the whole region is between 0.258 and 0.377, the gap in the green land use efficiency of the provinces (municipalities directly under the central government) is accelerating, and the green land use efficiency of the cities is gradually advancing to the middle and high efficiency areas. (2) Spatial evolution: On the one hand, the spatial difference in the green use efficiency of urban land in this region is gradually expanding. On the other hand, the green use efficiency of urban land in this region has a significant positive spatial correlation, the agglomeration pattern is dominated by high agglomeration and low agglomeration, and the low high agglomeration is supplemented.

Keywords: land green use efficiency; unexpected SBM model; spatiotemporal evolution characteristics; Yangtze River Delta

Citation: Lin, Q.; Ling, H. Study on Green Utilization Efficiency of Urban Land in Yangtze River Delta. *Sustainability* **2021**, *13*, 11907. https://doi.org/10.3390/su132111907

Academic Editors: Qingsong He, Jiayu Wu, Chen Zeng and Linzi Zheng

Received: 22 September 2021
Accepted: 18 October 2021
Published: 28 October 2021

1. Introduction

China's economy is in the transition period from "high-speed" development to "high-quality" development. The Fifth Plenary Session of the 19th CPC Central Committee proposed "promoting green development and promoting harmonious coexistence between human and nature", in which "green development" is an indispensable part of "high-quality development". The so-called green development refers to sustainable development in line with the requirements of ecological protection and environmental friendliness. It is required to realize low-carbon, environmental protection and the efficient utilization of production factors in each development link. As one of the most important means of production, land is the material basis of all economic and social elements and also the space carrier of economic and social development. It should play an important role in green development. This requires the implementation of the concept of harmonious coexistence and sustainable development between man and nature in the process of land use and the realization of the unity of economic and ecological benefits of land use. This process is defined as green land use [1]. Along with the process of urbanization in China, urban construction land is rapidly expanding. At the same time, it is facing the dilemma of total shortage, disordered utilization planning, high pollution, and low efficiency, which have become the shackles of the comprehensive green transformation of economic society.

The Yangtze River Delta is one of the regions with the most developed economy in China, the highest level of urbanization, the highest concentration of high-end talents, and the strongest innovation ability, but the contradiction between people and land is particularly serious. The shortage of land resources, serious pollution, low utilization efficiency, and multiple other problems restrict the pace of green development in the Yangtze River Delta. In order to change the current situation of development and pollution, the region actively seeks the practice road of coordinated development of ecological environment protection and economic and social progress. In order to improve the efficiency of land use, we should develop the leading area of ecological and green integration in the Yangtze River Delta, promote space sharing, and so on. However, with the development of practice, the traditional method of measuring land use efficiency only from the economic point of view cannot meet the demand, and it is necessary to increase the ecological perspective, including the concept of green development. Therefore, it is necessary to measure the green efficiency of urban land in Yangtze River Delta and explore its change trend. Therefore, we must use a reasonable measurement system to explore the efficiency of urban land green use in the Yangtze River Delta in order to promote the efficient and green use of urban land in the Yangtze River Delta.

Integrating the concept of green development into the process of urban land use and its quantitative analysis is the premise and basis for improving the efficiency of urban land green use and realizing the coordinated development of economic growth and environmental protection. It is also a hot topic in economics, management, and land science. The research of land use efficiency starts from theoretical research and gradually develops to empirical analysis. In the theoretical research stage, scholars mainly focus on the influencing factors of land use efficiency [2,3], index systems [4], path selection [5], and management systems [6]. In the empirical research stage, with the deepening of the research, the calculation method of land use efficiency is more scientific, the connotation is constantly enriched, and the research angle is more extensive. The calculation method of land use efficiency mainly experienced three stages: the single index method, the multi-index method, and the DEA model [7–11]. At present, the improved non-radial, non-angle, and non-expectation SBM model based on the traditional DEA model has become the mainstream method to measure multi-objective land use efficiency. The connotation of land use efficiency is further enriched, and the concept of green land use efficiency emerges as the times require and has become the research focus of relevant scholars [12–14]. As the concept of land green use efficiency is put forward and the calculation method is improved, the spatial-temporal differences and the sources of differences of green land use efficiency have become the focus of further research. Scholars use the Dagum Gini coefficient [15], exploratory spatial data analysis [16], the Theil index [17], kernel density estimation [18], and other methods to analyze the spatial-temporal evolution and difference degree of urban land use efficiency at different spatial scales.

At present, some scholars have carried out relevant research on the urban land use efficiency in the Yangtze River Delta [19–21], but no research has been carried out on the green land use efficiency of all 41 cities in the Yangtze River Delta, and no scholars have used the Dagum Gini coefficient and decomposition method to analyze the differences and sources of green land use efficiency in the Yangtze River Delta. Therefore, this paper constructs the measurement index system of "input expected output unexpected output" of green urban land use efficiency, calculates the land use efficiency of all cities in the Yangtze River Delta, and studies its spatial-temporal evolution characteristics by using the Dagum Gini coefficient, decomposition, and exploratory spatial data analysis.

2. Research Area, Indices Selection, and Methods

2.1. Research Area

The Yangtze River Delta includes 41 cities in Shanghai, Zhejiang, Jiangsu, and Anhui provinces (municipalities directly under the central government, the same below), according to the 2019 Outline of the Yangtze River Delta Regional Integrated Development Plan, which was recently approved by the State Council.

In order to effectively investigate the green use efficiency of urban land in the Yangtze River Delta, in this paper, the 27 central districts of Shanghai, Nanjing, Hangzhou, Suzhou, Wuxi, Changzhou, Nantong, Yangzhou, Zhenjiang, Yancheng, Taizhou, Ningbo, Wenzhou, Huzhou, Jiaxing, Shaoxing, Jinhua, Zhoushan, Hefei, Taizhou, Wuhu, Ma 'anshan, Tongling, Chuzhou, Anqing, Chizhou, Xuancheng, as well as Xuzhou, Huai 'an, Quzhou, Lishui, Lianyungang, Suqian, Bengbu, Huangshan, Huaibei, Huainan, Suzhou, Lu 'an, Haozhou, Fuyang and 14 non-central city districts as the study area.

2.2. Indices Selection and Data Sources

The term "green development" was first put forward by the British economist David, emphasizing a model of economic development that does not cost the depletion of natural resources. At present, the academic community has reached a basic consensus on the connotation of green development mode, that is, "green development is a sustainable development mode that not only pursues greater economic benefits but also requires the maximum reduction of consumption". Therefore, according to the connotation of green development and the experience of existing literature, the green efficiency of urban land is calculated, which should not only consider the economic output but also consider the pollution emission. Therefore, this paper takes land, capital, and labor as input indicators. Gross regional product is an indicator of expected output; the discharge of three wastes is an undesired output index; and the measurement index system of "input-expected output-unexpected output" is constructed. This system takes into account the ecological environment capacity and resource carrying capacity while considering the economic development and can measure the green use efficiency of urban land in the Yangtze River Delta more comprehensively and scientifically. Therefore, the following indices were selected for calculation in this paper:

(1) Input index. According to the theory of "three factors of production" put forward by the economist Say, the input of urban production system mainly includes the characteristics of land, capital, and labor. It is of more practical significance to consider the input index from the three aspects of land, capital, and labor [22]. The specific situation of input indicators is as follows: ① Land input (urban construction land area /km^2). ② Capital factor investment (fixed investment in municipal districts/CNY 10,000). ③ Labor input factors (total number of employees in urban units and urban private and self-employed employees/person). In addition, in order to eliminate the impact of inflation and increase the comparability of data, this paper uses the GDP deflator based on the year 2000 to treat the fixed investment of municipal districts. Similarly, the "gross regional product of municipal districts", the expected output index below, is treated in the same way.

(2) Expected output index. This index measures the "benefits" generated in the process of urban land use from the perspective of economic output, and the gross regional product is the most representative and comprehensive economic index. In addition, since the green use efficiency of land within a city is calculated, the "gross regional product of a municipal district" is chosen to represent the expected output.

(3) Unexpected output index. This indicator measures the "load" caused by the use of urban land from the perspective of green development. Industrial pollution is the most important pollution to the city, so the three indicators of "industrial sulfur dioxide emissions/ton, industrial dust emissions/ton, industrial wastewater emissions/ton" are chosen to represent the undesired output.

The data span of this paper is 13 years, which is determined by the degree of practice of the green development concept in China and the speed of data update. The data in this paper are based on China Statistical Yearbook, China Urban Statistical Yearbook, and China Construction Statistical Yearbook from 2007 to 2019. Urban geospatial location information is extracted with GeoDA software. Among them, the missing data of some cities are obtained by deductive calculation. Finally, the 13-year panel data of 41 cities are sorted out.

2.3. Research Methods

2.3.1. Unexpected Output SBM Model

The non-radial and non-angle undesired SBM model was proposed by Tone. The model puts the slack variable directly into the objective function [23,24], on the one hand, it solves the problem of input–output slackness, and on the other hand, it also solves the problem of efficiency evaluation in the presence of undesired output [25]. In the process of land use, there are expected outputs and undesired outputs at the same time. Therefore, this paper adopts the undesired output SBM model to calculate the green use efficiency of urban land in the Yangtze River Delta. In this way, a more comprehensive evaluation of the green land use efficiency of the cities in the region can be achieved.

The unexpected output SBM model assumes that the production system has n decision units, and each unit can calculate the green use efficiency of urban land in a certain region in a certain year. Each decision-making unit has m type input, s_1 type expected output and s_2 type unexpected output. This unit can be expressed as: $x \in R^m, x \in R^m, y^g \in R^{S_1}, y^b \in R^{S_2}$. The matrix that defines input, expected output, and non-expected output is expressed as:

$$X = [x_1, \ldots, x_n] \in R^{m \times n} \tag{1}$$

$$Y^g = \left[y_1^g, \ldots, y_n^g\right] \in R^{S_1 \times n} \tag{2}$$

$$Y^b = \left[y_1^b, \ldots, y_n^b\right] \in R^{S_2 \times n} \tag{3}$$

$$X > 0, Y^g > 0, Y^b > 0 \tag{4}$$

So, the production set can be expressed as:

$$\rho = \left\{(x, y^g, y^b)\middle| x \geq X\lambda, y^g \leq Y^g\lambda, y^b \geq Y^b\lambda, \lambda > 0\right\} \tag{5}$$

where $\lambda \in R^n$ is a weight vector, and $\lambda > 0$ represents constant returns to scale. In this study, the undesired output SBM model of a specific decision unit (x_0, y^g, y_0^b) is expressed as:

$$\rho = \min \frac{1 - \frac{1}{m}\sum\limits_{i=1}^{m}\frac{s_i^-}{x_{i0}}}{1 + \frac{1}{s_1+s_2}\left(\sum\limits_{r=1}^{s_1}\frac{s_r^g}{y_{r0}^g} + \sum\limits_{r=1}^{s_2}\frac{s_r^b}{y_{r0}^b}\right)} \tag{6}$$

$$s.t. x_0 = X\lambda + s^- \tag{7}$$

$$y_0^g = Y^g - s^g \tag{8}$$

$$y_0^b = Y^b\lambda - s^b \tag{9}$$

$$s^- \geq 0, s^g \geq 0, s^b \geq 0, \lambda \geq 0 \tag{10}$$

In the formula, s^-, s^g and s^b represent the relaxation variables of input, expected output, and non-expected output, respectively. The objective function ρ represents the urban land green use efficiency to be calculated, which is strictly decreasing with respect to s^-, s^g and s^b, the value is between 0 and 1. When ρ is 1, it is proven that the green use of urban land in this year is efficient, and the rest is lack of efficiency, which still needs to be improved.

2.3.2. Dagum Gini Coefficient and Decomposition Method

The Dagum Gini coefficient and its decomposition are the Gini coefficient proposed by Dagum and its subgroup decomposition method [26]. Using this method, the differences between regions and within regions can be explored, and the overall Gini coefficient can be divided into three types: intra-regional difference contribution (G_w), inter-regional difference contribution (G_{nb}), and supervariable density contribution (G_t). At present,

Dagum Gini coefficient and its decomposition method have been used to analyze various differences [27,28]. Therefore, this method is suitable for preliminary exploration of urban land use efficiency differences in the Yangtze River Delta as a whole, between provinces, and within provinces [29,30]. Dagum Gini coefficient and its decomposition formula are as follows:

$$G = \frac{\sum_{j=1}^{k}\sum_{h=1}^{k}\sum_{i=1}^{n_j}\sum_{r=1}^{n_k} | y_{ji} - y_{hr} |}{2n^2\overline{y}} \tag{11}$$

$$\overline{\mathrm{Y_h}} \leq \ldots \overline{\mathrm{Y_j}} \leq \ldots \overline{\mathrm{Y_k}} \tag{12}$$

$$G_{jj} = \frac{\frac{1}{2\overline{Y}}\sum_{i=1}^{n_j}\sum_{r=1}^{n_j} | y_{ji} - y_{jr} |}{n_j^2} \tag{13}$$

$$G_{jh} = \frac{\sum_{i=1}^{n_j}\sum_{r=1}^{n_j} | y_{ji} - y_{hr} |}{n_j n_h(\overline{Y_j} - \overline{Y_h})} \tag{14}$$

$$G_w = \sum_{j=1}^{k} G_{jj} p_j s_j \tag{15}$$

$$G_{nb} = \sum_{j=2}^{k}\sum_{h=1}^{j-1} G_{jh}(p_j s_h + p_h s_j) D_{jh} \tag{16}$$

$$G_t = \sum_{j=2}^{k}\sum_{h=1}^{j-1} G_{jh}(p_j s_h + p_h s_j)\left(1 - D_{jh}\right) \tag{17}$$

$$D_{jh} = \frac{d_{jh} - p_{jh}}{d_{jh} + p_{jh}} \tag{18}$$

$$d_{jh} = \int_0^{\infty} dF_j(y) \int_0^{y} (y - x) dF_h(x) \tag{19}$$

$$p_{jh} = \int_0^{\infty} dF_h(y) \int_0^{y} (y - x) dF_j(y) \tag{20}$$

Formula (3) is the formula for the calculation of the Gini coefficient. $y_{ij}(y_{hr})$ is the green urban land use efficiency in province $j(h)$. $\overline{y}$ cities in Yangtze River Delta is the average of the green land use efficiency, n is the number of cities, n is the number of the Yangtze River Delta provinces. $n_j(n_h)$ is the number of cities in the province of $j(h)$. Formula (4) is the ranking of the provinces according to the mean value of the green use efficiency of urban land in the provinces before the decomposition of the Gini coefficient. Equations (5) and (6) represent the intra-regional Gini index of j province and the inter-regional Gini coefficient of G_{jj}, j and h provinces, respectively. Equations (7)–(9) calculate intra-regional difference contribution, inter-regional difference contribution, and super-variable density contribution, respectively. According to the decomposition method of Gini coefficient, the following formula can be obtained as follows: $G = G_w + G_{nb} + G_t$. The influence of provinces on the relative green use efficiency of urban land is defined as Equation (10), in which the calculation of d_{jh} and p_{jh} are shown in Equations (11) and (12).

2.3.3. Exploratory Spatial Data Analysis

In this paper, the exploratory spatial data analysis (ESDA) method of global spatial autocorrelation and local spatial autocorrelation was used to measure the spatial evolution of urban land green use efficiency in the Yangtze River Delta

(1) Global spatial autocorrelation analysis. Global spatial autocorrelation can describe the global spatial relationship of all units in the study space. In this paper, it is used to analyze the overall spatial change of urban land green use efficiency in the Yangtze River Delta region. At present, Moran's I(Moran index) statistic is used to analyze the spatial autocorrelation of land green use efficiency. Its calculation formula is as follows:

$$I = \frac{n \sum_{i=1}^{n} \sum_{j=1}^{n} w_{ij}(x_i - \overline{x})(x_j - \overline{x})}{\left(\sum_{i=1}^{n} \sum_{j=1}^{n} w_{ij}\right) \sum_{i=1}^{n}(x_i - \overline{x})^2} \tag{21}$$

where $x_i = \frac{1}{n}\sum_{i=1}^{n} x_i$ represents the observed value of city i, n is the number of cities, and w is the spatial weight matrix. Moran's I is between -1 and 1. When $I > 0$, it indicates a positive correlation between cities, and the closer it is to 1, the stronger the positive correlation is. When $I < 0$, it indicates a positive correlation between cities, and the closer it is to -1, the stronger the negative correlation will be. When $I = 0$, it means that there is no spatial autocorrelation between regions.

(2) Local space autocorrelation analysis. Local spatial autocorrelation can distinguish different correlation patterns that may exist due to spatial location changes. In this way, we can investigate the local unevenness of space and explore the spatial heterogeneity between data. Due to spatial instability, global autocorrelation analysis is difficult to accurately reflect local spatial correlation, so local Moran's I is used to explore the local characteristics of urban land green use efficiency in the Yangtze River Delta region. Its formula is as follows:

$$I = \sum w_{ij} z_i z_j \tag{22}$$

where z_i and z_j are the observed values after standardization, and w_{ij} is the elements of the spatial weight matrix after standardization. The range of local Moran's I is not limited to -1 to 1. In this study, when I_i is significantly positive and $z_i > 0$, it means that the green land use efficiency of the city i and its neighbors is higher than the average level, which belongs to high agglomeration. When I_i is significantly positive and $z_i < 0$, it means that the green land use efficiency of i and its neighbors is lower than the average level, which belongs to low agglomeration. When I_i is significantly negative and $z_i > 0$, it means that the green land use efficiency of neighboring cities is much lower than that of cities i, which belongs to high agglomeration. When I_i is significantly negative and, it means that the green land use efficiency of neighboring cities is much higher than that of cities i, which belongs to low–high agglomeration.

3. Result Analysis

This part is based on the calculation of the SBM undesirable model for the green use efficiency of urban land in the Yangtze River Delta from 2006–2018 and analyzes its change characteristics from the perspectives of time and space. The time series analysis is carried out from the three dimensions of the region as a whole, the province, and the city; the spatial evolution analysis is divided into two parts: the spatial difference and the correlation of the urban land green use efficiency.

3.1. Analysis of Time Series Change

First of all, on the whole, the average value of green urban land use efficiency in the Yangtze River Delta presents a fluctuating upward trend, with a range from 0.2 to 0.4 (Figure 1). On the one hand, this shows that with the deepening of the green concept, the land green use efficiency within the urban area has been steadily improving; on the other hand, it also reveals the problem that the overall value of land green use efficiency is low and needs to be greatly improved. In addition, the change in urban land green use efficiency in the Yangtze River Delta can be divided into three stages. The first stage was from 2006–2009, when China put forward and vigorously implemented the "energy saving" emission reduction measures for the first time. During this period, due to the rapid reduction of "three wastes" emissions, the overall green use efficiency of urban land in the region rose rapidly. The second stage is from 2010 to 2016. In this stage, the green use efficiency of urban land maintained at about 0.3 after a rapid decline. Due to the prominent problems such as the slowdown of economic growth and the reduction in production benefits of enterprises, the green land use efficiency has returned to a more stable mode. The third stage is from 2017 to 2018. With the improvement of the "energy conservation and emission reduction" system, the market environment and the optimization of the economic development structure, the green use efficiency of urban land increases rapidly once again after energy storage.

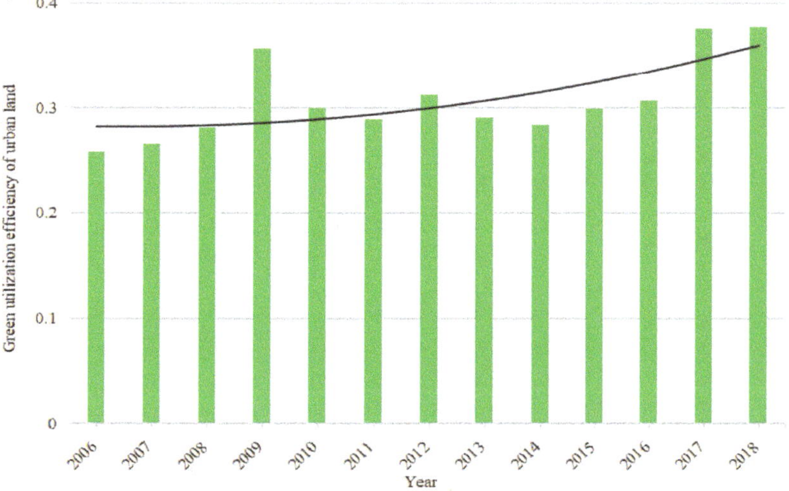

Figure 1. Urban land green use efficiency in the Yangtze River Delta region from 2006–2018.

Secondly, from the dimension of each province, its change trend follows the change trend of the whole region, showing a fluctuating upward trend (Figure 2). The green use efficiency of urban land in Shanghai has been in the leading position for a long time. Although it declined rapidly from 2010–2011, it rose rapidly in the following years and surpassed the previous level, completely opening up a gap with other provinces in the Yangtze River Delta. In the past three years, the efficiency value has been maintained above 0.7. The green use efficiency of urban land in Anhui Province is the lowest in the Yangtze River Delta region, fluctuating between 0.2 and 0.3 and even decreasing in the recent two years. The change trend of green land use efficiency of cities in Zhejiang Province and Jiangsu Province is roughly the same. Its change trend can be divided into three stages as the whole change trend. The first stage was from 2006–2009. Both provinces grew at a rapid rate and reached a small peak in 2009. The green land use efficiency of Zhejiang Province exceeded 0.45, while that of Jiangsu Province approached 0.43. In the second stage, from 2010–2016, the two provinces experienced a rapid rise after a lack of momentum, resulting in the efficiency growth being reversed and entering a long period of decline. The third stage is from 2017 to 2018. In these two years, the rising channel of green urban land use efficiency in the two provinces was opened, and it quickly recovered to or exceeded the highest state.

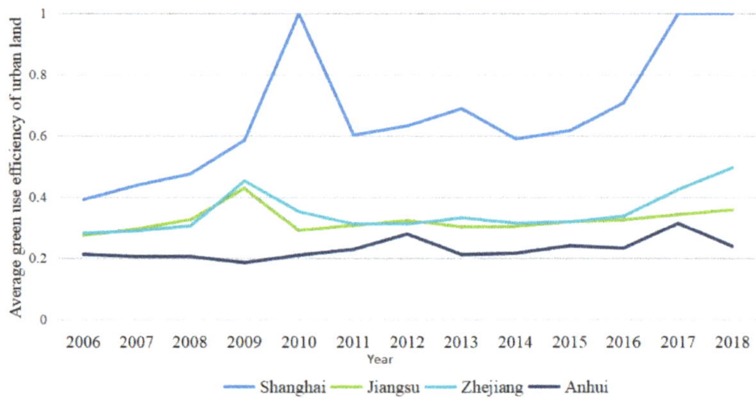

Figure 2. Average value of urban land green use efficiency of each province in the Yangtze River Delta from 2006–2018.

Finally, the results are analyzed from the perspective of city. Here, this paper presents the urban land green use efficiency of 41 cities in the Yangtze River Delta region in 2006, 2012, and 2018 through tables to analyze the results. In order to facilitate comparison, the natural breakpoint method was adopted to divide the table into a low efficiency zone (0.07–0.326), a medium efficiency zone (0.326–0.632), and a high efficiency zone (0.632–1). It can be seen from the following table (Table 1) that the green land use efficiency of all cities in the Yangtze River Delta region is less and less in the low-efficiency zone, and more and more in the medium-efficiency zone.

Table 1. Green land use efficiency of cities in Yangtze River Delta in 2006, 2012–2018.

Efficiency Range	2006	2012	2018
0.07–0.326	Changzhou, Lu'an, Haozhou, Huangshan, Huai'an, Fuyang, Huaibei, Chizhou, Bengbu, Lianyungang, Suqian, Lishui, Bengbu, Chuzhou, Suzhou, Anqing, Jiaxing, Shaoxing, Huaian, Yancheng, Jinhua, Hefei, Wuhu, Huzhou, Nanjing, Zhenjiang, Wenzhou, Taizhou, Ningbo, Nantong, Xuancheng, Xuzhou, Yangzhou,	Changzhou, Lu'an, Chuzhou, Lianyungang, Fuyang, Huaibei, Bengbu, Huangshan, Mizhou, Xuancheng, Shaoxing, Huainan, Chizhou, Hefei, Huai'an, Jinhua, Suqian, Anqing, Wenzhou, Yancheng, Quzhou, Wuhu, Taizhou, Lishui, Huzhou, Taizhou, Ma'anshan, Jiaxing	Changzhou, Fuyang, Lu'an, Huangshan, Mizhou, Chuzhou, Huaibei, Chizhou, Huainan, Anqing, Lianyungang, Bengbu, Jinhua, Xuancheng, Quzhou, Suzhou, Huzhou, Suqian, Ma'anshan, Huai'an, Wuhu,
0.326–0.632	Yangzhou, Zhoushan, Maanshan, Taizhou, Tongling, Hangzhou, Shanghai, Wuxi, Suzhou	Zhenjiang, Nantong, Tongling, Nanjing, Ningbo, Zhoushan, Hangzhou, Xuzhou, Wuxi, Yangzhou, Suzhou, Suzhou	Hefei, Tongling, Yancheng, Wenzhou, Nantong, Taizhou, Xuzhou, Zhenjiang, Shaoxing, Taizhou, Lishui, Nanjing, Jiaxing, Suzhou, Wuxi, Hangzhou
0.632–1	-	Shanghai	Ningbo, Shanghai, Zhoushan Yangzhou

3.2. Analysis of Spatial Evolution

3.2.1. Dagum Gini Index and Decomposition Results

In order to investigate the spatial differences and sources of urban land green use efficiency in the Yangtze River Delta region, a province was first divided into a subgroup. Based on the Dagum Gini coefficient and decomposition method, the Gini coefficient of green urban land use efficiency from 2006–2018 was calculated and decomposed. The results are as follows (Table 2).

1. The overall regional difference and evolution trend of the spatial distribution of urban land green use efficiency in the Yangtze River Delta. As can be seen from Figure 3, the Gini coefficient of urban land use efficiency in the Yangtze River Delta region shows an upward trend, fluctuating between 0.17 and 0.35. Based on 2006 and 2012, the difference of urban land use efficiency increased by 4.98% and 3.21% annually, respectively. At the same time, in the process of the overall rise, local fluctuations are larger.

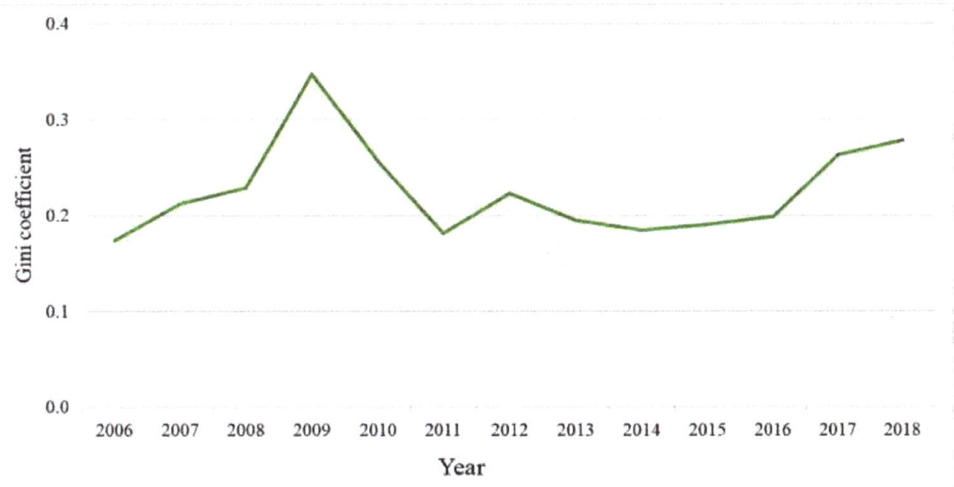

Figure 3. Overall Gini Coefficient in the Yangtze River Delta Region from 2006–2018.

2. Decomposition of regional differences in green use efficiency of urban land.

(1) Intra-regional differences in green use efficiency of urban land. Figure 4 describes the intra-provincial differences in green urban land use efficiency in the provinces of the Yangtze River Delta (except Shanghai). On the whole, the Gini coefficient within each province is not high, indicating that there is little difference in the green use efficiency of urban land within each province. In terms of subdivision, the fluctuation of regional Gini coefficient in Zhejiang Province is the most severe, showing a pattern of increasing first, decreasing, and then increasing again. The intra-regional Gini coefficient of Jiangsu Province showed a trend of increasing at first and then decreasing. After the increase from 2006–2009, the intra-regional Gini coefficient of Jiangsu Province gradually decreased at an average annual rate of 5.7%. The overall value of Gini coefficient in Anhui Province is the lowest among the three provinces, but the fluctuation is the most frequent, which indicates that the difference in urban land use efficiency among cities in Anhui Province is small and the change range is narrow.

Table 2. The Dagum Gini coefficient and its decomposition results of land green use efficiency in the Yangtze River Delta region from 2006–2018.

Year	Population Gini Coefficient	Intra-Regional Gini Coefficient				Interzone Gini Coefficient						Contribution		
		Shanghai	Zhejiang	Jiangsu	Anhui	Zhejiang-Shanghai	Jiangsu-Shanghai	Jiangsu-Zhejiang	Anhui-Shanghai	Anhui-Zhejiang	Anhui-Jiangsu	Gw	Gnb	Gt
2006	0.174		0.110	0.180	0.169	0.171	0.188	0.150	0.274	0.168	0.206	28.84%	34.80%	36.36%
2007	0.212		0.123	0.237	0.187	0.216	0.229	0.192	0.331	0.198	0.256	28.08%	37.33%	34.59%
2008	0.228		0.078	0.310	0.156	0.224	0.288	0.223	0.367	0.188	0.292	26.55%	44.04%	29.41%
2009	0.347		0.293	0.325	0.135	0.269	0.280	0.316	0.500	0.432	0.419	22.83%	56.55%	20.62%
2010	0.256		0.219	0.203	0.138	0.465	0.549	0.230	0.639	0.268	0.226	21.22%	61.66%	17.13%
2011	0.181		0.090	0.180	0.148	0.320	0.325	0.143	0.430	0.170	0.213	24.80%	44.99%	30.21%
2012	0.223	-	0.093	0.192	0.253	0.335	0.322	0.154	0.435	0.218	0.269	27.59%	25.18%	47.23%
2013	0.195		0.102	0.178	0.128	0.345	0.390	0.147	0.506	0.210	0.219	21.71%	59.08%	19.21%
2014	0.184		0.097	0.164	0.157	0.299	0.320	0.136	0.444	0.193	0.216	24.09%	52.84%	23.07%
2015	0.190		0.158	0.164	0.141	0.307	0.317	0.169	0.431	0.191	0.204	24.99%	45.68%	29.33%
2016	0.198		0.167	0.167	0.126	0.345	0.369	0.174	0.491	0.201	0.209	23.41%	53.46%	23.14%
2017	0.263		0.275	0.145	0.236	0.387	0.489	0.250	0.519	0.290	0.226	25.51%	40.58%	33.91%
2018	0.278		0.284	0.158	0.117	0.309	0.472	0.280	0.601	0.369	0.232	19.99%	69.07%	10.94%

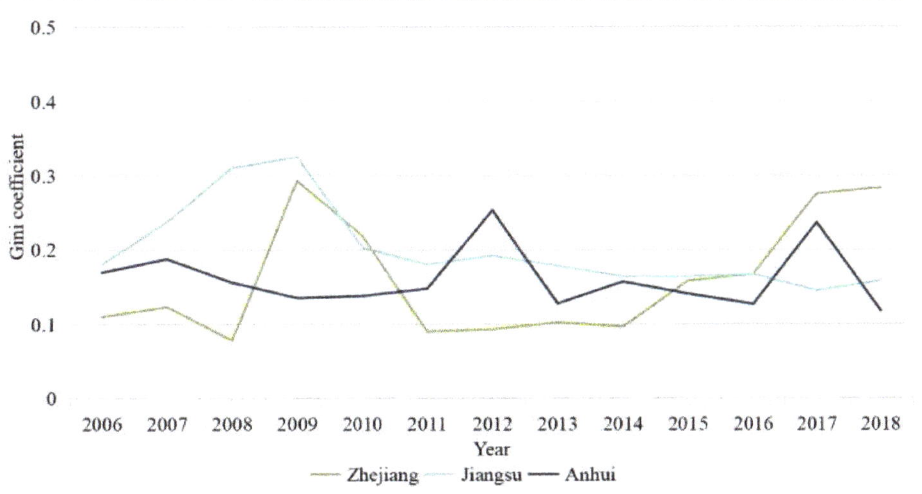

Figure 4. Intra-Gini Coefficient of Yangtze River Delta Region from 2006–2018.

(2) Regional differences in green use efficiency of urban land. Figure 5 describes the differences of urban land green use efficiency among provinces in the Yangtze River Delta. From the perspective of overall evolution trend, although the values of Gini coefficient in different regions are different, the trends are similar. The first wave peaks between provinces and regions concentrated in 2009 and 2010. After a collective rapid decline, small peaks appeared in 2012 and 2013, and then divergent trends emerged. In addition, compared with 2006, the gap in the green urban land use efficiency in all provinces is widening, in which the gap between Anhui and Jiangsu is the smallest and that between Anhui and Shanghai is the largest. From the point of view of the value, the Gini coefficient between Shanghai and other provinces in the region is high, among which the Gini coefficient between Anhui Province and Shanghai is the highest. Except Shanghai, the Gini coefficient among the other three provinces is relatively small, among which Jiangsu and Zhejiang have the lowest Gini coefficients.

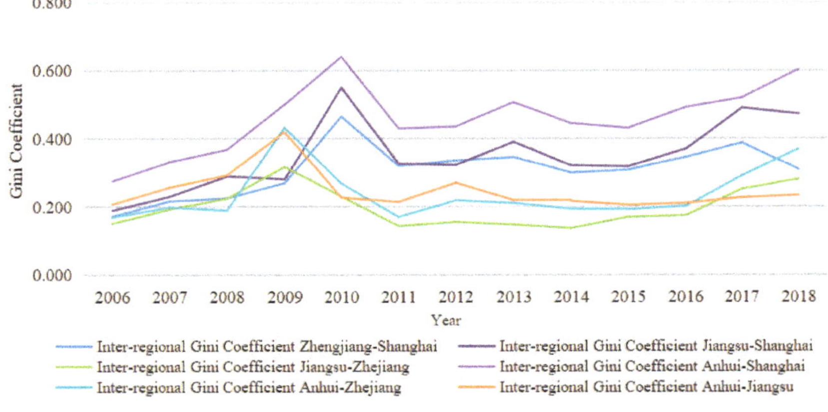

Figure 5. Inter-regional Gini Coefficient in the Yangtze River Delta from 2006–2018.

(3) Regional differences and contribution rates of green urban land use efficiency. As can be seen from Figure 6, during the research period, the difference in the urban land use efficiency in the Yangtze River Delta region was mainly caused by the differences among provinces, with an average contribution rate of 48.09%. The second is the contribution of superdensity, and the average contribution rate is 27.32%. Finally, the difference between cities within the province, the average contribution rate is 24.59%. Among these three types of contributions, the contribution rate of intra-regional differences is the most stable, while the contribution rate of inter-regional differences and the contribution rate of superdensity fluctuate greatly and tend to run in opposite directions.

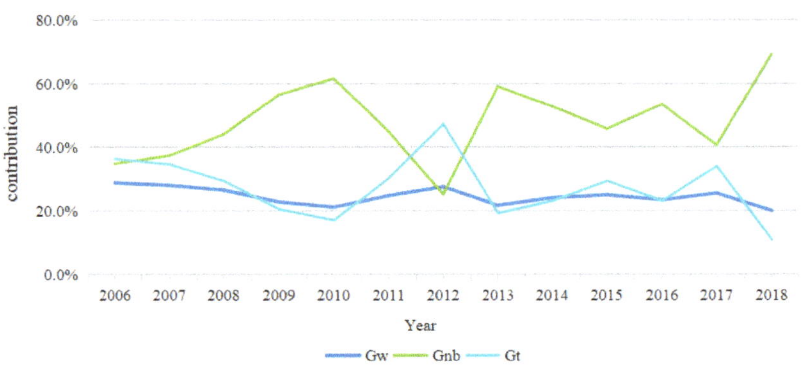

Figure 6. Contribution rate of green use efficiency difference of urban land in the Yangtze River Delta region from 2006–2018.

3.2.2. Exploratory Spatial Data Analysis Results

Using GeoDA software, the regional and regional Moran's I of the green use efficiency of urban land in the Yangtze River Delta region from 2006–2018 are calculated and analyzed, and the results are as follows.

(1) Analysis of global spatial evolution. The regional Moran's I passed the test at the significance level of 5%, and its value ranged from 0.070–0.354, indicating that there was a significant and positive spatial autocorrelation in the green use efficiency of urban land in the Yangtze River Delta. It can be seen from Figure 7 that the change trend of Moran's I is an overall increase and a local fluctuation, which indicates that there is not only spatial dependence but also spatial heterogeneity among the green use efficiency of land in different regions. During 2006–2009 and 2011–2012, Moran's I decreased at a relatively fast rate, indicating that the spatial difference and correlation of urban land green use efficiency in the Yangtze River Delta were increasing and weakening in these periods. Moran's I increased at a fast rate in three periods of 2008–2011, 2012–2013, and 2016–2018, indicating that the spatial correlation of the green urban land use efficiency in the Yangtze River Delta region was continuously enhanced, while the differences were gradually weakened during these periods. In particular, after the release of the "Yangtze River Delta Urban Agglomeration Development Plan" in 2016, Anhui Province was included in the Yangtze River Delta region, and the integration degree of the Yangtze River Delta region continued to deepen. Moran's I also rose rapidly and continued to break new highs, and the mutual influence of the green land use efficiency between cities was strengthened.

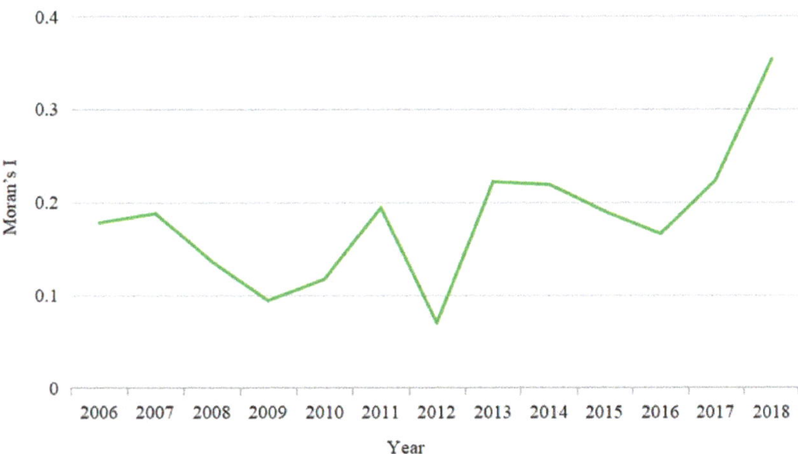

Figure 7. Moran's I index of the Yangtze River Delta region from 2006–2018.

(2) Analysis of local spatial evolution. Since global Moran's I is difficult to reflect the specific characteristics of local space, local Moran's I is used to study the spatial agglomeration pattern in the Yangtze River Delta. The figure below shows the agglomeration pattern of the Yangtze River Delta in 2006, 2010, 2014, and 2018.

It can be seen from Figure 8 that there are three types of agglomeration in the Yangtze River Delta: high agglomeration, low–low agglomeration, and low–high agglomeration. The specific characteristics are high agglomeration, low agglomeration into block distribution, and low and high agglomeration sporadic distribution. During the study period, the number of high-concentration cities stabilized between 3 and 5, mainly Shanghai, Suzhou, Jiaxing, and other economically developed cities. These cities not only have a high level of green land use efficiency but also have a strong positive influence on the green land use efficiency of the surrounding cities, and the spatial spillover effect is significant. The low–low agglomeration is the most widely distributed among the four agglomeration types and has an upward trend. Low and low agglomerations are mainly distributed in the northwest of the Yangtze River Delta, and most of these cities are marginal cities not included in the Yangtze River Delta urban agglomeration. Although the low–high agglomeration pattern of cities is not fixed, most of them are central cities with higher economic levels. The lack of the ability to receive a diffusion effect is the main reason why these cities cannot catch up with the green land use efficiency of surrounding cities.

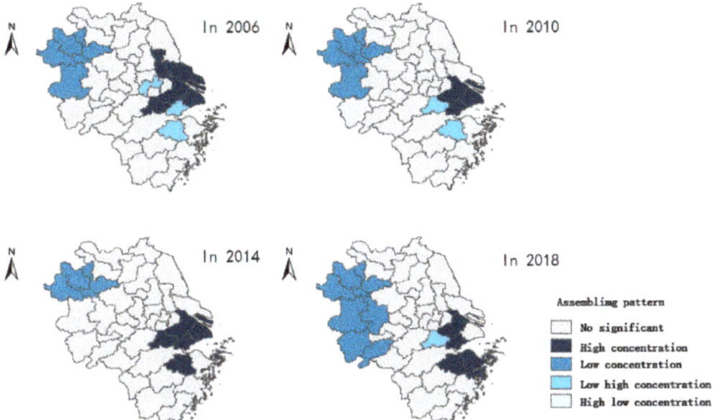

Figure 8. LISA cluster map in the Yangtze River Delta.

4. Conclusions

Based on the calculated green use efficiency of urban land in the Yangtze River Delta region, this paper analyzes the spatiotemporal evolution characteristics of green use efficiency of urban land in this region by using Dagum Gini coefficient and decomposition method and exploratory spatial data analysis method and comes to the following conclusions:

(1) From the perspective of overall temporal evolution, the overall green use efficiency of urban land in the Yangtze River Delta increased from 0.258 in 2006 to 0.377 in 2018, with a slight fluctuation during the period. As of 2018, it is still at a low level and has great potential for improvement.

(2) From the perspective of local time series evolution, the evolution characteristics of the green urban land use efficiency of all provinces in the Yangtze River Delta region are different. However, the overall trend is consistent with the regional trend. However, the overall trend is consistent with the regional trend.

① The efficiency of green land use in Shanghai increased rapidly and widened the gap with that of other provinces. As a super first-tier city in China, relying on the advantages of its economy, systems, and technology, Shanghai has continuously improved its green urban land use efficiency and has become the "leader" in improving urban land green use efficiency in the Yangtze River Delta region [31–33].

② The evolution characteristics of the green land use efficiency in Zhejiang Province and Jiangsu Province are similar, that is, the green land use efficiency in Zhejiang Province fluctuates at a slightly higher level than the overall level, and the increase rate is higher than the overall increase rate. Jiangsu and Zhejiang provinces are both located in the core area of the Yangtze River Delta, and both are representatives of the developed regions in China, with similar industrial structure [34]. The "two wings" of the improvement of the overall green urban land use efficiency in the Yangtze River Delta region dominate the development speed of the green urban land use efficiency in this region.

③ The green use efficiency of urban land in Anhui province increased slightly under the overall growth level. In the short term, this change has become an obstacle to the rapid improvement of urban land green use efficiency in this region. However, in the long term, Anhui Province, on the one hand, has undertaken relatively low-end industries in Jiangsu, Zhejiang, and Shanghai, and optimized the overall industrial layout and resource allocation of the region. On the other hand, it increases the development hinterland of the Yangtze River Delta region, which is the future growth point of the green urban land use efficiency in the Yangtze River Delta region [35,36]. From the inter-city level, more and more cities such as Hefei and Lishui have got rid of the low-efficiency state of urban land green use efficiency and entered the medium-efficient area. The inter-city difference is narrowing continuously, which makes the pattern of land green use efficiency in the Yangtze River Delta region optimize to some extent. From the perspective of inter-city level, more and more cities in the Yangtze River Delta region have gotten rid of the low efficiency of the green urban land use efficiency and entered the medium-high efficiency area. The inter-city differences are narrowing, and the pattern is constantly optimized.

(3) From the perspective of the difference of spatial evolution, the difference of the overall green urban land use efficiency in the Yangtze River Delta region is gradually expanding. The Dagum Gini coefficient and decomposition results show that the gap between developed and relatively backward provinces is getting bigger and bigger, which is the main source of the overall regional differences, while the intra-provincial differences are small and trend, and the contribution is less. Economically developed regions have absolute advantages in "hardware" construction (infrastructure construction, public service facilities, etc.) and "software" construction (institutional environment and public service level, etc.). These advantages are the acceleration of the ability to transform the green development concept into green development, which can enable the cities with advantages to realize the improvement of green land use efficiency more quickly, thus opening a gap with other regions. Cities in the same province are more likely to have the same economic development and policy environment, so the efficiency of green land use is more similar.

(4) From the perspective of correlation of spatial evolution, the green use efficiency of urban land in the Yangtze River Delta region has a significant and positive global spatial autocorrelation, and this correlation increases gradually with the deepening of the integration degree of the Yangtze River Delta. In the Yangtze River Delta, there are two main agglomeration patterns: high agglomeration and low agglomeration, which are respectively distributed in the regions with the strongest comprehensive strength and the weakest comprehensive strength. Developed cities such as Shanghai and Suzhou have worked together to promote industrial upgrading, eliminate backward production capacity, learn from effective policies, protect the ecological environment, bring positive externalities to each other, and form a high concentration of urban land green utilization efficiency. However, less developed areas such as Lu'an City and Fuyang City in the northwest of the Yangtze River Delta region have attracted a number of enterprises with high energy consumption and high pollution due to relatively loose environmental management policies, which promotes the improvement of local economic development level, but at the same time, leads to the low efficiency of land green use, thus forming a low–low agglomeration pattern. The spatial distribution of low–high agglomeration is not obvious, and these cities often find it difficult to accept the positive benefits of efficient cities and even suffer negative effects.

5. Policy Recommendations

(1) We should actively promote the integration of the Yangtze River Delta and comprehensively improve the green utilization efficiency of urban land. The Yangtze River Delta region is a community of economy, society, and ecological environment. The ecological environments are dependent on each other, and the ecological services are shared. Improving the green urban land use efficiency of the region as a whole will give full play to the "holistic thinking", together with the concept of "green water castle peak is the jinshan silver" as a guide, planning to break apart, overlapping conflict situation, through the methods of environment pollution co-government, backward infrastructure co-construction and sharing, overall upgrade to industry to promote the efficiency of the whole green urban land use.

(2) It is necessary to fully tap the potential of each region, make up for its strengths, and realize the optimization of the green utilization efficiency of urban land in the region. The development level of the three provinces and one city in the Yangtze River Delta region is different, so it is necessary to effectively tap the potential of each region, give full play to their advantages, and jointly improve the efficiency of green land use in the Yangtze River Delta region. Shanghai is open to the outside world at a high level, is capital and technology intensive, mainly develops high-yield, low-pollution happy technology industries, and its green land utilization efficiency has been at a high level for a long time. Therefore, we should give full play to the leading role and guide the direction of regional development. Both Zhejiang and Jiangsu provinces are big manufacturing provinces with the developed economy. While high-tech industries and strategic emerging industries are developing vigorously, the problem of insufficient development quality is still prominent. Therefore, we should take innovation as the first priority of the manufacturing industry, actively explore a new mode of innovation-driven manufacturing development, give full play to the pulling and lifting role of the two wings, and exchange higher technology for lower pollution. With abundant labor resources and urbanization in Anhui province, there is huge potential to develop land resources are relatively more, should give full play to the advantage of backwardness, actively undertake the industrial transfer of the Jiangsu region but also to strengthen the construction of urban infrastructure and public service facilities and pay attention to improving the management of urban land use and environmental pollution. We must not ignore environmental protection for the sake of pursuing economic benefits, which will lead to the low efficiency of land green benefits.

(3) It is necessary to explore the development mode of high-efficiency areas driving low-efficiency areas and promote the common growth of green urban land use efficiency.

At present, there are obvious differences in the efficiency of urban land green use among provinces in the Yangtze River Delta. Therefore, the association among provinces should be strengthened. The efficient provinces can help the inefficient ones improve their efficiency through industrial cooperation, policy sharing, technology, and talent support. On the one hand, this can speed up the efficiency improvement in the low-efficiency areas, solve the "pain point" of the low level of urban land green use efficiency in the Yangtze River Delta, and on the other hand, it can promote the coordinated development of green land use efficiency in the whole region.

Author Contributions: Conceptualization, Q.L. and H.L.; methodology, Q.L. and H.L.; software, Q.L. and H.L.; validation, Q.L., and H.L.; formal analysis, Q.L. and H.L.; investigation, Q.L. and H.L.; resources, Q.L. and H.L.; data curation, Q.L. and H.L.; writing—original draft preparation, Q.L. and H.L.; writing—review and editing, Q.L. and H.L.; visualization, Q.L. and H.L.; supervision, Q.L. and H.L.; project administration, Q.L. and H.L.; funding acquisition, Q.L. and H.L. All authors have read and agreed to the published version of the manuscript.

Funding: The project was supported by the Fundamental Research Funds for the Central Universities, China University of Geosciences (Wuhan) (No. CUG190268).

Institutional Review Board Statement: Not applicable.

Informed Consent Statement: Not applicable.

Data Availability Statement: The datasets used during the current study are available from the corresponding author on reasonable request.

Conflicts of Interest: The authors declare that they have no competing interest.

References

1. Tao, L.L.; Jun, Y.Y.; Guang, Y.C. Measurement of urban land green use efficiency and its spatial differentiation characteristics: An Empirical Study Based on 284 cities above prefecture level. *China Land Sci.* **2019**, *33*, 80–87.
2. Verburg, P.H.; Berkel, D.B.V.; Doorn, A.M.V.; van Eupen, M.; van den Heiligenberg, H.A.R.M. Trajectories of land use change in Europe: A model—Base exploration of coral futures. *Landsc. Ecol.* **2010**, *25*, 217–232. [CrossRef]
3. Chen, R. Theory of urban land use efficiency. *Urban Plan. Trans.* **1995**, *4*, 28–33+63.
4. Fang, X. Study on index system and method of land use efficiency measurement. *Syst. Eng.* **2004**, *12*, 22–26.
5. Sun, Y.; Yao, X. Path selection for improving urban land use efficiency. *Economist* **2002**, *12*, 50–51.
6. Ma, Y. Discussion on the efficiency of land use control system. *J. Shandong Univ. Technol. (Soc. Sci. Ed.)* **2002**, *6*, 26–31.
7. Li, Y.; Shu, B.; Wu, Q. Urban land use efficiency in China: Spatial and temporal characteristics, regional disparity and influencing factors. *Econ. Geogr.* **2014**, *34*, 133–139.
8. Wang, Y.; Song, G. Comprehensive benefit evaluation and case study of urban land use. *Geogr. Sci.* **2006**, *6*, 743–748.
9. Bao, X.; Liu, C.; Zhang, J. Comprehensive evaluation of urban land use efficiency. *Urban Issues* **2009**, *4*, 46–50.
10. Yang, K.; Wen, Q.; Zhong, T. Evaluation of urban land use efficiency in the Yangtze river economic belt. *Resour. Sci.* **2018**, *40*, 2048–2059.
11. Zhang, Z. Study on urban land use efficiency in China. *J. Quant. Tech. Econ.* **2014**, *31*, 134–149.
12. Li, C.; Hu, J. Spatial temporal difference of urban land use efficiency and its influencing factors based on DEA: A case study of nine cities in Jilin Province. *Resour. Environ. Yangtze River Basin* **2020**, *29*, 678–686.
13. Wang, W.; Song, Y.; Pang, X. Research on China's regional land use efficiency based on data envelopment analysis. *Explor. Econ. Probl.* **2011**, *8*, 60–65.
14. Ge, K.; Zou, S.; Lu, X.; Chen, D.; Kuang, B. Spatial convergence analysis of urban land green use efficiency under the background of regional integration: A case study of Yangtze River economic belt. *East China Econ. Manag.* **2021**, *35*, 31–41.
15. Ji, Z.; Zhang, P. Spatial difference and driving mechanism of urban land use efficiency under environmental constraints in China: A study based on 285 cities at prefecture level and above. *China Land Sci.* **2020**, *34*, 72–79.
16. Wang, D.; Pang, X. Study on green land use efficiency of Beijing Tianjin Hebei Urban Agglomeration. *China Popul. Resour. Environ.* **2019**, *29*, 68–76.
17. Li, L.; Dong, J.; Zhang, J. Regional differences and formation mechanism of urban land use efficiency in Yangtze River economic belt. *Resour. Environ. Yangtze River Basin* **2018**, *27*, 1665–1675.
18. Hu, B.; Li, J.; Kuang, B. Evolution characteristics and influencing factors of urban land use efficiency differences under the concept of green development. *Econ. Geogr.* **2018**, *38*, 183–189.
19. Zhang, Y.; Chen, J.; Gao, J.; Jiang, W. Impact mechanism of urban land use efficiency in Yangtze River Delta from the perspective of economic transformation. *J. Nat. Resour.* **2019**, *34*, 1157–1170.

20. Yang, Q.; Duan, X.; Ye, L.; Zhang, W. Evaluation of urban land use efficiency based on SBM undesirable model: A case study of 16 cities in Yangtze River Delta. *Resour. Sci.* **2014**, *36*, 712–721.
21. Yu, Y.; Du, X. Research on threshold effect of economic development, government incentives and constraints and energy conservation and emission reduction efficiency. *China Popul. Resour. Environ.* **2013**, *23*, 93–99.
22. Nie, L.; Guo, Z.; Peng, C. Study on urban construction land use efficiency based on SBM undesirable and meta frontier models. *Resour. Sci.* **2017**, *39*, 836–845.
23. Tone, K. A slacks-based measure of efficiency in data envelopment analysis. *Eur. J. Oper. Res.* **2001**, *130*, 498–509. [CrossRef]
24. Tone, K. A slacks-based measure of super-efficiency in data envelopment analysis. *Eur. J. Oper. Res.* **2002**, *143*, 694–697. [CrossRef]
25. Li, J. Study on the differences and influencing factors of regional environmental efficiency in China. *Nanfang Ji* **2009**, *12*, 24–35.
26. Dagum, C. A new approach to the decomposition of the Gini income inequality ratio. *Empir. Econ.* **1997**, *22*, 47–63. [CrossRef]
27. Liu, H.; Zhao, H. Analysis on regional differences of carbon dioxide emission intensity in China. *Stat. Res.* **2012**, *29*, 46–50.
28. Hong, H.; Xie, D.; Guo, L.; Hu, R.; Liao, H. Characteristics and types of spatial function differentiation of mountainous rural areas from the perspective of multi-function. *Acta Ecol. Sin.* **2017**, *37*, 2415–2427.
29. Li, X.; Zhang, Q. AHP-based resources and environment efficiency evaluation index system construction about the west side of Taiwan Straits. *Ann. Oper. Res.* **2015**, *228*, 97–111. [CrossRef]
30. Irwin, E. New directions for urban economic models of land use change:incorporporating spatial dynamics and heterogeneity. *J. Reg. Sci.* **2010**, *50*, 65–91. [CrossRef]
31. Tan, N.; Zhou, X.; Lin, J. Research on the economic growth effect of Shanghai Free Trade Zone: Counterfactual analysis based on panel data. *Int. Trade Issues* **2015**, *10*, 14–24.
32. Yin, H.; Gao, W. Have pilot free trade zones produced "institutional dividends"? Evidence from Shanghai free trade zone. *Financ. Res.* **2017**, *43*, 48–59.
33. Wang, J.; Yan, Y.; Hu, S. Spatial characteristics and influencing factors of urban technological innovation capacity in China: Based on spatial panel data Model. *Sci. Geogr. Sin.* **2017**, *37*, 11–18.
34. Analysis and Prospect of Economic Development in Yangtze River Delta Region: 2017–2018. Wang, Z. (Ed.) Economic Development Report of Yangtze River Delta Region, Shanghai Academy of Social Sciences Press: Shanghai, China, 2018; pp. 1–7.
35. Zhong, M.; Li, M.; Du, W. Whether environmental regulation can force industrial structure adjustment: An empirical test based on Chinese provincial panel data. *China Popul. Resour. Environ.* **2015**, *25*, 107–115.
36. Liu, Y.J. Optimization of modern industrial system in undertaking industrial transfer in central and western Regions—Taking Anhui Province undertaking industrial transfer in Yangtze River Delta region as an example. *East China Econ. Manag.* **2013**, *27*, 28–32.

Article

Sustainable Planning and Design of Ocean City Spatial Forms Based on Space Syntax

Longlong Zhang [1], Jingwen Yuan [1] and Chulsoo Kim [2,*]

[1] Department of Marine Design Convergence Engineering, Pukyong National University, Busan 48513, Republic of Korea
[2] Department of Industrial Design, Pukyong National University, Busan 48513, Republic of Korea
* Correspondence: yinghua@pukyong.ac.kr

Abstract: The form of an ocean city, as a physical space, has an important impact on the city's social economy, environment, etc. Whether the internal composition of an ocean city is well organized determines whether its form is sustainable and whether it can better carry out a variety of functions. Considering this context, in this study, we adopted the theory of space syntax (SS) to interpret the sustainability of the ocean city form. This was carried out from the perspective of the composition relationship of the internal organization of the ocean city (OC) physical space. We judged whether the composition relationship of internal space could effectively support the sustainable and healthy functioning of different features of ocean cities through the interpretation of SS-related theories. It is extremely hard to give an accurate definition of the form of a sustainable city. At the same time, it is impossible to make conclusions about which urban form is sustainable. However, combined with the concept of sustainable development, we argue that urban forms that continue to facilitate the virtuous cycle of the society, economy, and environment of a given city and also to be highly habitable for urban residents are sustainable. Thus, based on the above viewpoint, the research object and scope in this study only involved the ontology of the physical space form and whether urban physical space could effectively support the sound and sustainable development of three core elements: urban society, the economy, and the environment. This was comprehensively evaluated through our exploration of the form of urban physical space. Here, space syntax was taken as an analytical theoretical and practical tool to summarize the problems that existed in Shenzhen Bay through data analysis, and corresponding development proposals were put forward. The concept and method behind the strategy analysis of the ocean city (OC) design framework based on SS-related theories were presented and applied to practical cases to perform an objective and rational analysis, guide the design of actual projects, and promote ocean city (OC) design in the transition period in a judicious way. In addition, we discuss how design and planning can promote sustainable urban development.

Keywords: space syntax; ocean city; space form; sustainable planning and design

Citation: Zhang, L.; Yuan, J.; Kim, C. Sustainable Planning and Design of Ocean City Spatial Forms Based on Space Syntax. *Sustainability* **2022**, *14*, 16620. https://doi.org/10.3390/su142416620

Academic Editors: Qingsong He, Jiayu Wu, Chen Zeng and Linzi Zheng

Received: 21 October 2022
Accepted: 9 December 2022
Published: 12 December 2022

1. Introduction

1.1. Research Background

The spatial layout of cities has always been a leading factor in human activities. Today, cities are characterized by gaps between people and automobiles, between people, and between buildings and public spaces. This seclusion and discrepancy was not the original intention of urban planners, going against their initial attempts to create a sound communication environment. However, when looking at several traditional cities, we find that they do not have such problems. On the contrary, there is little seclusion and instead harmonious social relations. The essence of urban learning is to investigate human beings and their living space. Over time, the relationships between people and their living spaces are becoming more complex. During the normal operation of a city, people readily attribute the driving force of urban development to social and economic factors and ignore its subjects and law. Theoretically, this simple judgment made by academic circles lays a foundation for the prerequisite of social and economic planning, reducing urban planning to a passive position in social development, economic construction, spatial layout, etc. Finally, the planning and design (PAD) of urban spaces fails to give full play to their due function and pay adequate attention to the rules of urban space, which leads to many misunderstandings in urban and rural construction. This argument must be explored in the context of the effect of economic and social forces on a city and how they come into being [1].

Cities are bound to face one particular challenge; that is, how to develop the population, resources, environment, and social economy in a coordinated way. Along with the development of the social economy and the acceleration of urbanization in China, cities—especially big cities and megacities—are faced with a grim situation [2]. For this reason, how to identify the nature, scale, and development direction of a city and create a pathway of sustainable development are currently issues of importance.

SS is a research method that combines space with society. Spatial ontology quantifies the complex spatial form, behaviors, and activities of humans, and discusses the internal relationship between the spatial organization, urban structures, and functional modes [3]. The OC form, which is a kind of physical space, is a significant carrier. Whether its internal structure is good or bad determines its sustainability and the variety of functions that it bears. In a nutshell, the OC design framework is a strategic and holistic OC design tool. The framework can be understood as the provision of a stable mechanical structure. This structure can be shared and replicated during operation and can be extended to satisfy the design requirements of different regions. When introducing the OC design framework, one should add the features of local elements, not only to ensure the integrity of the regional OC design but also to fully protect the local characteristics. The OC design framework should adapt to local and seasonal conditions and fulfill the requirement of diversity. In addition, the OC design framework is tantamount to supplying a framework or beginning for OC design, whose connotations and outcomes must vary from place to place [4].

To this end, by applying SS theory to the relationship between internal structure and composition in the OC physical space of marine cities, we determined their sustainability and judged whether their internal spatial structure can fully support the sustainable and healthy functioning of different features based on SS-related theories.

1.2. Research Purpose

The importance of oceans runs through the whole process of urban development, and the planning and design of marine cities are also crucial. Due to the "urban fringe" characteristic of coastal space, the planning and design methods used for marine cities are completely different from those of inland cities. However, most of the existing planning and design strategies for marine cities focus only on the location of the coast and fail to view the construction of marine cities from multiple levels. How to conform to and utilize the spatial relations of marine cities to create a better urban space is an important research question and was the focus of this study.

Since they are not incisive enough, the previous case studies on space syntax cannot offer guidance on construction problems in actual projects and cannot be satisfactorily used to develop design strategies for urban construction. Therefore, based on SS analysis, we merged spatial units as the marine space of "three bays and one estuary" via zoning and performed a quantitative evaluation and analysis of the marine ecological economy in each sea area by building an evaluation index system for the sustainable development of the marine ecological economy. We classified and divided sea areas according to their main features and the spatial autocorrelation analysis structure, as well as planned and designed a sustainable OC spatial form for marine cities to provide a reference to better guide the development of the marine economy and protect and utilize the marine resources and environment in each sea area. Here, backed by the SS research theory, we summarized strategies for the OC design framework to solve the problems existing in the construction of OCs and guide OC design during the transition period [5]. The main aim of this study was to obtain plans and proposals for sustainable spatial development based on a case study of Shenzhen Bay, under the common premise of SS analysis and a sustainable viewpoint, and to guide the spatial planning of other marine cities.

1.3. Research Significance

A city is both a complex and complicated internal system. The spatial structure of a city has a great impact on human activities. First, a reasonable spatial layout can boost the healthy development of a city, while a reasonable spatial structure can create a vibrant and safe environment and promote the development of society, the economy, and the environment. If a space is ill-planned, its urban function will fail and the whole society may even lose balance. The discussion of methods for urban function and form is becoming increasingly common. In this study, the analytical approach of SS provided a scientific explanation for the genesis and operation mode of urban space, delivered accurate information and rigorous data for future urban problems, and gave suggestions and optimal solutions to ensure normal operation and promote sustainable development of cities [6].

Based on SS theory, this study undertook a rational analysis of the characteristics of current OC spaces, guided the layout of various OC design elements, guaranteed the rationality of OC design, discussed the compatibility between OC design and OC planning systems, and ensured the enforceability of OC design. The significance of this research is that it summarized strategies for an OC design framework during the transition period through qualitative and quantitative analysis, which can help to shape a good OC living atmosphere; i.e., an environment where humans and nature can coexist in harmony in a vibrant OC space [7]. Most importantly, using the current spatial form of Shenzhen Bay, our study analyzed the prominent problems existing in the current spatial development of Shenzhen Bay and put forward targeted development proposals for these problems, to promote the economic development of Shenzhen Bay and provide experience to draw from for the development of other regions.

1.4. Research Methods

In this study, OC planning exhibition halls with different combinations of internal space were selected and analyzed, and research was conducted based on the acquired text resources, photos, etc. We adopted the following methods:

Literature review: Through the consultation of a large number of online and offline data sources, the literature was systematically sorted and analyzed. The classification of the internal spatial forms of OC planning exhibition halls was summarized and the relevant theories underpinning this article were clarified.

Field survey: The survey included surveying and mapping of the internal dimensions of the case investigated in this study using a laser measuring instrument.

Software analysis: To begin, according to the theory and technical platform of SS, the quantitative parameters of the internal space of the planning hall were calculated using the DepthMap software; then, the maps were read. The spatial differences in the different planar organizations were analyzed in accordance with the indicated colors and data, and the quantitative results of the abstract space were extracted.

Comparative analysis: After a quantitative and qualitative comparative evaluation of different spatial organizations, we provide reasonable advice on planning and design in this paper.

1.5. Domestic and Foreign Literature Review

SS was introduced in the 1970s. By Bill Hillier and Julienne Hanson is an integral part of SS theory. In the late 1990s, J. Hanson conducted a large empirical analysis on the relationship between spatial structure and social culture in the book. It systematically explains the relationship between different combinations of architectural space and social culture, and pointed out the relationship between social cognition, the mode of spatial organization, and space [8].

In the book, B. Hillier sets forth the central position of spatial structure in research on spatial relations. In addition, based on an empirical analysis [9], he gives an in-depth demonstration. During the 21st century, SS has been widely applied in urban planning, structural design, architectural space design, landscape design, etc. Using SS, A. H. Mahmou analyzed the influence of the tree planting mode in gardens on tourists' line of sight and found that its spatial layout had a certain guiding effect on pedestrians' walking propensity. In addition to the theoretical basis, analytical method, and modeling techniques of SS, K. Karimi stated that urban layout should be combined with the characteristics of "configuration" [10]. Y. Lerman et al. built a walking model based on pedestrianism and discussed the impact of walking paths on the siting of basic bus stops [11].

Jin Dongsheng published an article entitled "Space Syntax" in a journal called in 1985. This article first appeared in China, and its reach was gradually expanded with the development of SS theory [12]. In 2007, Duan Jin and B. Hillier jointly published and developed it in the application field in China [13]. Since then, research on SS by Chinese scholars has gradually increased and the theory has been widely applied across different disciplines. Tan Chengzhong investigated architectural space and put forward an evaluation criterion for spatial social stratification from the perspective of society. Wang Hongchao extracted and encapsulated different grid types from different cities and comprehensively evaluated the cohesion between social attributes in each stage using the SS method [14].

From the perspective of SS, Shen Peng examined the spatial form of Zhengdong New District in two historical stages; quantitatively evaluated the spatial structure regarding several aspects, such as integration, synergy, and intelligence; and determined development trends, including the expansion of the scope of the regional integration core, a reduction in spatial intelligence, and a decrease in the accessibility of external traffic [15]. By carrying out an analysis of the spatial form of Zhukou Ancient Village in Qimen County, Anhui Province, Chen Dandan compared it with the planned spatial form and verified the feasibility and effectiveness of the project in terms of the protection and renewal of the spatial form of Zhukou Ancient Village [16].

1.6. Research Review

In this work, research findings on SS published at home and abroad were collected; in terms of spatial ontology, it was found that the studies on SS were full-fledged and widely applied at several macro levels, such as for urban spatial structures, road networks, urban form evolution, the accessibility of urban parks, and the internal streamlining of buildings. The relationship between their internal spaces can be accurately reflected through an SS-related analysis and studied in conjunction with various perspectives, including sustainability planning. On this basis, a rational optimization strategy for a spatial layout is presented.

2. Overview of Related Theories

2.1. Ocean City (OC)

An OC (ocean city) generally refers to a coastal city with abundant marine resources and a large total marine economic output. OCs are usually cities with a long coastline and a developed marine economy. An OC has an "ocean" attribute, which is not only reflected in the urban location but is also embodied in distinctive "ocean" features and outstanding "ocean" advantages. From the denotation of these features and advantages, modern marine cities are comprehensive coastal cities with international influence, e.g., a famous international marine city and a gulf metropolis [17]. Such cities have prominent urban attributes, such as openness, security, connectivity, and flexibility, as well as high internationalization. They have an advanced modern industrial system, play a critical pivotal or nodal role in the global industrial and supply chains, and have very powerful innovation and service functions. Distinctive marine features are found in every aspect of these cities, which include marine science cities, port shipping hubs, seashore tourist cities, and maritime service centers.

2.2. Planning and Design (PAD)

Planning and design (PAD) refer to the specific planning or overall design of the project, taking into account the political, economic, historical, cultural, and folk customs, as well as geography, climate, transportation, and other factors, to further improve the design scheme and propose planning expectations, vision and development modes, development directions, control indicators, and other policies [18]. The content of PAD, as the outline for construction, is later used to guide architectural design, landscape design, and other design aspects (including building volume, form, architectural color, interface, structure, spatial layout, and landscape form).

In this study, the design included urban planning, landscape architecture, and architectural design. The PAD used in this study was concrete and data-specific.

2.3. Sustainable Planning

The concept of sustainable development was first raised by the International Union for Conservation of Nature (IUCN) in 1980 in the World Conservation Strategy. Sustainable planning is established based on sustainable development. Sustainable urban planning not only refers to sustainability at the technical level but, more importantly, sustainability of the environment, culture, and psychology. Sustainable planning is not limited to the mechanical layout and material construction of urban space but also includes the combination of urban development, natural ecology, and the principles of sustainable development in urban planning and their implementation at every level of planning; this not only satisfies the needs of contemporary people without jeopardizing the ability of future generations to meet their demands but also ensures their sound, sustainable, and coordinated development [19].

2.4. Evolution Mode of an Urban Form

Individual activities and space are interactive. The utilization of space by people entails a simple and clear straight line—that is, the space from one point to another—to reinforce ties with the outside world. At the same time, it is also necessary to examine the inner connectivity of cities; that is, the shortest distance between two points. The contradiction between these two factors leads to the realistic form of a city. The evolution law of urban spatial structure in China can be derived through an analysis of the evolution of global cities. In the beginning, most cities either develop linearly along the main street or form an irregular ring along a space with varying lengths and widths. With the maintenance of the linear space and the addition of extended and compact grids, the deformation of the city is generated; this reflects the basic morphological law of cities. In general, whether square or irregular, the network structure of a city is always dense in the center and sparse in the periphery. The spatial composition of a city proper is not important; instead, the key lies in the overall understanding of space and time.

3. The PAD of an OC Based on Space Syntax (SS)

3.1. Correlation Analysis of SS

3.1.1. Meaning of SS

Over the past forty years, SS research methods have been constantly improved and refined, and a well-established theoretical system has gradually been formed. Founded on spatial ontology, the basic idea of SS is to divide large-scale space and mathematically abstract and model it to explore the connection between spatial ontology and other non-spatial elements. The scope of SS research involves a large number of aspects, such as urban planning, urban design, architecture, and spatiography, gradually forming an integrated discipline with multiple ways of thinking [20].

SS is a theory and method that investigates physical space and the spatial composition of human cognition by quantitatively describing the spatial structure of a human settlement [21]. It highlights the research on the correlation between "space" as an independent element and architecture, society, and human cognition. According to SS, the OC space system consists of two parts: closed space and free space. Closed space generally refers to buildings, while free space is a place separated by objects where people can move around at will. In free space, any point can reach other points in space and demonstrate continuity; thus, it is also called open space. The space system created by people is produced in social development. Upon the completion of the space system, it has social attributes; these social attributes of OCs or architectural spaces are not determined artificially and subjectively by attaching functional labels.

The SS method is employed to understand and utilize the effect of a spatial configuration in urban systems or complex buildings. It can be seamlessly linked to GIS network analysis, superposition analysis, data connection, and other functions, and has great superiority regarding data modification, sorting, and mapping. ENVI focuses on image processing, such as correction, stitching, index calculation, and classification; in contrast, ArcGIS emphasizes the construction, analysis, and calculation of geographical information to provide statistical and regional analyses based on geographic space, and is more suited for the analysis and processing functions of spatial information.

According to the new space theory, which states that space is a part of social life, human settlements, such as the layout of OCs or architectural spaces, will influence society, the economy, and culture and innovatively show the social logic of spatial structure. SS is different from traditional research on space theory in that SS takes space as the starting point to dissect the relationship between the OC and social and cognitive areas (e.g., architectural space). The general research relations of SS are shown in Figure 1.

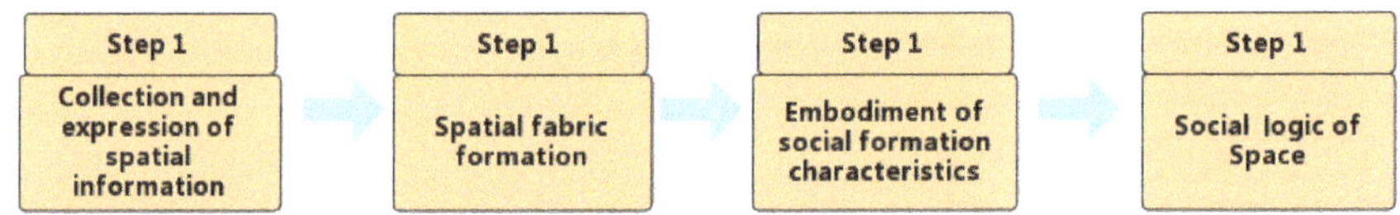

Figure 1. General research relations of SS.

The division of spatial scale and the segmentation of space is an important foundation of SS research.

Division of spatial scale: The division of spatial scale is the basis of SS and also the primary step in SS research. SS divides space into two space scales: large and small. The division criterion is whether individuals in a space can fully perceive the space. If so, then the space is a small-scale space; if not, or if they can only perceive a part of the space, then the space is a large-scale space [22].

Segmentation of space: SS assumes that it is difficult for individuals to feel the entirety of a large-scale space. In studies on large-scale spaces, such as cities, SS divides them into three types of small-scale space. Axial segmentation refers to the farthest distance that can be seen from a given point in space. Convex space segmentation refers to the ability of any point in space to see another point in space; that is, the space can be fully perceived. FOV (field of view) segmentation refers to the scope of horizontal space that can be perceived at a given point. At present, axial segmentation is most widely used in research utilizing SS theory.

3.1.2. Basic Principle of SS

The basic principle of SS is spatial segmentation; i.e., transforming the whole space system into a relational diagram made up of nodes and interconnections. The process in which the whole space system is divided into single components is called spatial segmentation. A walking route is usually considered a straight line, while interpersonal space can be regarded as a convex space. People generally move around in a visible range within which they can feel safe. Beyond this range, people usually lack a sense of security as a result of an unknown space. Thus, axial, convex space, and FOV segmentations come about through an investigation of people's mental activities.

3.1.3. Core Theory of SS

SS is a theory and method that quantitatively analyzes the form of a surrounding space and examines the relationship between spatial structure and human society. Founded on the internal structural characteristics of the architectural form itself, it draws support from the logic of internal structure; analyses with software according to certain quantitative criteria allow for the elucidation of the deep logical relations between spatial combinations and the optimization of the spatial structure of architecture. In popular terms, the effect of the method with which a space inside a building is established based on people's behaviors, such as walking tendency, staying, and gathering, is analyzed from a quantitative perspective [23].

3.2. Brief Analysis of SS Steps

SS theory aims to transform a perceptual space into a rational space by using SS methods and techniques and understanding spatial characteristics more objectively. The major steps are shown in Figure 2.

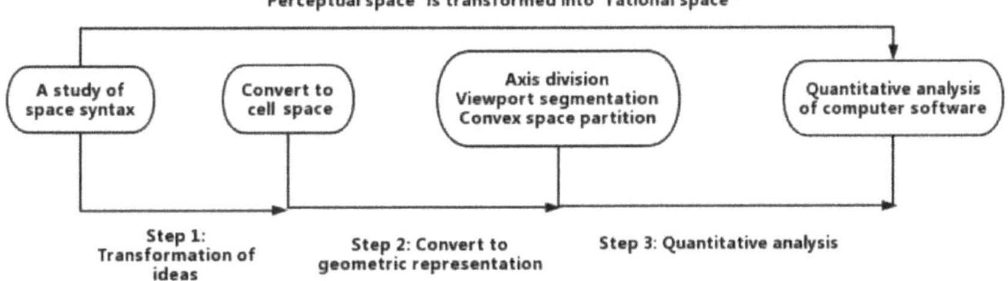

Figure 2. Brief analysis of SS steps.

First, based on SS theory, the space is reasonably converted; that is, the space system is divided into individual spatial units via internal logic. These spatial units are interconnected and mutually influenced. Each spatial unit has a different spatial form variable in the space system. Second, the most crucial step for SS to convert a perceptual space to a reasonable space is to resort to the three kinds of segmentation methods; i.e., axial, convex space, and FOV, to transform spatial configuration to understandable geometric representations and space activities that can barely be perceived in reality into rational geometric representations [24]. Finally, the geometric representations are imported into computer software and numerical calculations and analyses are performed. The value of the morphological variable is obtained, and the spatial characteristics can be understood more defensibly and completely. The details are as follows.

1. Analysis of the vector space

The analysis of the vector space mainly exploits the potential information of spatial targets via a conjoint analysis of spatial data and a spatial model. The basic information about these spatial targets includes their spatial position, distribution, morphology, distance, azimuth, and topological relations. Among them, the distance, azimuth, and topological relations constitute the spatial relations of spatial targets. This is a spatial characteristic between geographical entities and can be adopted as the ground of data organization, query, analysis, and inference. By dividing the geospatial targets into different types—that is, points, lines, and planes—we can acquire the morphological structure of these types of targets. The spatial calculation and analysis of many specific tasks can be carried out by combining the spatial data and attribute data of spatial targets [25].

- Primitive merging
- Primitive merging is the aggregation of the vector space, which merges or transforms data types according to spatial adjacency and categorical attribute fields to achieve the merging of spatial regions (data synthesis). Due to spatial aggregation, complex categories are often transformed into simpler ones; this analytical treatment is often needed during the graphical transformation from places and areas to large regions.
- Spatial query

A spatial query is meant to discriminate the spatial topology (the inclusion, separation, intersection, encompassment, and intersection of rectangles) of the elements of the input layer and query layer or the query scope of the interactive input and extract primitives that satisfy the topological discrimination conditions from the input layer.

- Superposition analysis

Overlay and superposition analysis is an operation that superimposes two or more layers of map elements and generates a new element layer. The result is that the original elements are segmented to generate new elements. The new elements synthesize the attributes of the original two or more layers of elements, where the overlay and superposition analysis not only produces new spatial relations but also associates the attributes of the input data layer to produce new attribute relations. Overlay and superposition analysis calculates and analyzes attributes of new elements based on a given mathematical model, thus producing the results required by users or answering the questions raised by users.

2. Grid space analysis

Grid-data-based spatial analysis is the foundation of GIS spatial analysis, which incorporates distance mapping, density mapping, surface analysis, statistical analysis, reclassification, grid calculation, visibility analysis, topographic factor analysis, hydrological analysis, etc.

- Distance mapping

Distance mapping analyzes and maps in accordance with the distance between each grid and its nearest element (also known as the "source") to reflect the interrelation between each grid and its nearest neighbor source. A large amount of relevant information can be obtained through distance mapping, which can be used to guide people to plan and utilize resources rationally [26].

- Density mapping

Density mapping mainly calculates the data distribution of the whole area in accordance with the value and distribution of known point elements, thereby producing a continuous surface. It is mainly generated based on point data and searches the ring area of each grid point taken as the center to calculate the density value of each grid point.

- Surface analysis

Surface analysis is mainly conducted by generating new datasets, such as isoline, slope, aspect, and hillside data; the process is used to acquire more information that reflects the spatial characteristics, spatial pattern, etc., implied in the original dataset.

3.3. *Advantages of the Introduction of SS*

By investigating its background, the problems with traditional OC design can be clearly defined. During the transition to urbanization, quantitative analysis can effectively avoid the subjective tendency of traditional design. Currently, all domestic quantitative analysis methods use geometry, such as the analysis of maritime city OCs or architectural drawings. Although such a quantitative analysis method is simple, it has limitations. That is, it cannot analyze the spatial structure based on people's characteristic behaviors, and thus, it is difficult to understand the essential characteristics of the OC space [27].

On the other hand, the spatial environment also influences people's activities, and people's activities, in turn, change the spatial structure. Through a quantitative analysis of the spatial environments and structures where people reside, play, and work, actual quantified data can be used to guide the functions in maritime city OCs, the establishment of the ecological environment and humanistic environment, etc. They play a significant guiding role in building an environmental space suitable for residing, playing, and working based on space research.

3.3.1. Necessity of Spatial Form Planning for OCs under SS

SS plays a considerable role in the analysis of the structural form of OCs; the most fundamental principle on which this is built is that an OC is formed through the interaction between self-organization and other organization and a complex spatial self-organization system formed by the interaction and interplay of multiple factors, such as social factors, economic factors, and environmental factors, over a long time. These elements will inevitably exert a coupling effect on long-term development, thus generating a series of complex internal connections and interactions between the physical space of the OC and its corresponding social and economic functions [28].

Just as the forms of OCs differ considerably, the spatial forms that they describe from the axis model are also dramatically different. However, by using topological logic to induce and deduce the relationship between these axes, we can see that the internal structures of these axes are highly unified, and it is this common structural relation that forms the internal development law of OC space.

3.3.2. Sustainability of OC Spatial Form Planning under SS

When exploring the form of sustainable OCs, SS theory considers form and function as inseparable, and the organizational structure of OC not only reflects form but also embodies function, which can explain the interaction between form and function [29]. For this reason, SS theory argues that the organization and composition form of space are crucial and essentially determine whether a space is sustainable. SS theory examines the OC spatial form in terms of three aspects: the natural law of space, the formation mechanism of the network, and the evolution of the OC [30]. For guidelines on the sustainable form of the OC, such as compactness, high density, diversification, mixed land, and green transportation, SS no longer evaluates in advance but examines the OC spatial structure or spatial "fabric" first to see how each local space is established and organized at various scales [31].

3.4. Sustainable PAD of Marine City Spatial Form

3.4.1. Compact Spatial Layout

A compact spatial layout can exert an important effect on the whole form of an OC, mainly based on shape, scale, distance, accessibility, centrality, and agglomeration. Taken together, the compact layout of all kinds of spaces in marine cities, such as production spaces, living spaces, and social interaction spaces, can help to provide more diversified OC functions and services for OC residents within different service radii so that residents in different regions of marine cities can be closer to the main functions and service centers, and the speed at which residents can access and be involved in the functional operation of the OC increases. A compact layout also delivers a major guarantee and support for more intensive development in marine urban areas than in suburban and rural areas [32].

3.4.2. Reasonably Built Density Distribution

Generally, high-density areas are those in the marine urban center or subdistrict center. These areas are typically the most concentrated places in terms of the variety of functions and services of marine cities. Whether these high-density areas can be rationally distributed in different regions of an OC has a direct bearing on whether residents in different regions of the OC can be closer to these functions and service centers. Meanwhile, if high-density areas in a built-up region are too concentrated in a given area, the overall development intensity in this area will be too high and traffic jams in the central area will become inevitable [33].

Many experts and scholars contend that a multi-center marine urban form is more sustainable than a single-center marine urban form; one of the important reasons for this is that the distribution of high-density areas in built-up regions is more well-founded. On this account, the rational density and distribution of built-up regions play an important role in the overall benign development of marine cities.

3.4.3. Mixed and Diversified Land Use

The mixed use of land is always an important principle that is followed by OC planning. The mixed use of the OC form has striking diversity and selectivity in functions and can deliver more OC functions and services to residents in the whole OC and region. While still offering residents higher diversity and selectivity, the mixed use of land can also bring residents closer to the function and service centers of the OC; it no longer separates marine urban functions mechanically but cuts the time and distance for residents to access these functional centers [34].

3.4.4. High-Density Marine Urban Road Network

As the most important framework for an OC, the network density of marine urban roads has always been an important indicator for measuring the effectiveness of the planning and implementation of marine cities. High-density marine urban roads offer more choices for the development of plots with different functions, thus providing space for the mixture of different functions and increasing the ability of residents to approach the function and service centers of marine cities [35]. Generally, the higher the density of the road network in marine cities, the easier it is for residents to quickly connect and access the various subdistrict centers of marine cities to become involved in the division of labor.

4. Case Analysis

4.1. Evaluation and Analysis of Regional Marine Eco-Economic Zoning in Shenzhen

4.1.1. Analysis of Spatial Heterogeneity

Considering the sub-indexes of the three subsystems in the Shenzhen marine eco-economic system—namely, marine ecology, marine economy, and marine society—an evaluation index system for marine eco-economic zoning was built and the marine ecological economy of Shenzhen was evaluated via zoning using comprehensive factor evaluation and analysis. Briefly, this process was divided into three steps: First, a zoning evaluation index system was built. Second, the weight of each index was calculated and determined by issuing questionnaires and scoring by experts. Third, rational values were assigned to concrete indexes for each region according to the evaluation criteria. Finally, the evaluation index was obtained via calculations and the zoning evaluation was made.

4.1.2. Data Source

The research data used in this study primarily included Google satellite images, the previous urban master plans for Shenzhen, other specialized plans, and social and economic development plans. Amid studies on the urban form of Shenzhen City, our work mainly referred to the compilation year of the latest version of Shenzhen's urban master plan, and we processed data using three main reference pictures that depended on the data collection time of the Google satellite map; that is, the status quo map of construction land in 1994, the historical satellite image from 2004, and the status quo image from 2014 (Figures 3 and 4). The image from 2004 represented the year before the start of the "great construction" period. In 2004, Shenzhen entered a period of "great construction" and the changes in urban form were very dramatic. For this reason, in this study, we analyzed and investigated the basic evolution characteristics of Shenzhen's spatial form in the past and present by comparing the urban form structure data of Shenzhen City from three periods over the past 20 years to lay a foundation to determine the sustainability of the urban spatial form of Shenzhen City [35].

Figure 3. Shenzhen regional map.

Figure 4. Picture of present-day Shenzhen.

4.1.3. Data Preprocessing

According to the basic principle of the establishment of a central axis map in SS theory and based on the status quo map of construction land in 1994 and the satellite images of Shenzhen in 2004 and 2014, we created the urban spatial structure axis maps of Shenzhen for these three periods in history and at present in CAD, as shown in Figure 5 [36]. The definition of the central axis in SS theory involves taking the street as the convex space. If this space is penetrated by the least and longest ray, then this ray is defined as the axis of the space. In the whole city, the axis of each road is taken as a node in the topology and linked by referring to the interconnection between nodes; the resulting relationship diagram between nodes represents the axis diagram of the city. According to this basic principle, in our study, we took urban roads above the urban branch network as the basic reference elements to draw the axis map and adjusted some axis data based on social attributes, spatial transition, and the visibility relationship implied in the axis map to guarantee the comparability of the data. After the axis diagram was processed in CAD, it was imported into the DepthMap software and analyzed via the axis and line segment models; then, the relevant data were read into the SPSS software for comparative analysis.

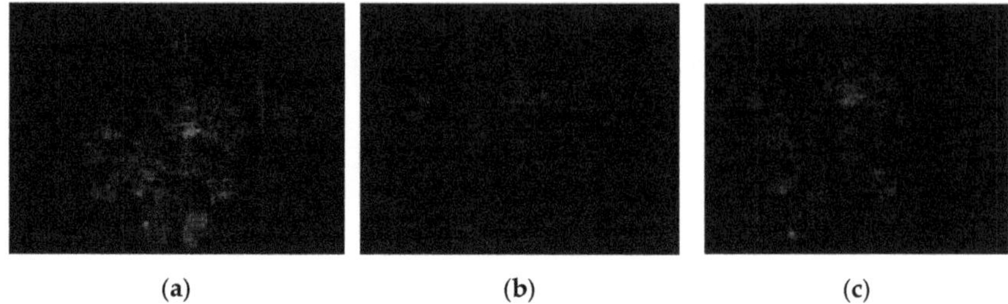

(a) (b) (c)

Figure 5. Satellite images of Shenzhen in (**a**) 1994, (**b**) 2004, and (**c**) 2014.

5. Research Methods

A Shenzhen marine eco-economic database was built and the spatial unit map of marine zoning in Shenzhen was drawn. Each zone contained information at three levels: marine natural environment, marine industrial economy, and marine social culture. A standard scoring system was created for the marine eco-economic zoning evaluation and the factors, weights, and grading were determined. The marine eco-economic zoning evaluation system and digital maps were completed using the Shenzhen marine eco-economic database.

The criteria and principles for the selection of evaluation factors in the Shenzhen marine eco-economic zoning evaluation model were as follows. The evaluation factors had a certain hierarchy and were evaluated in terms of natural, humanistic, economic, social, and other dimensions. They could be divided into evaluation categories, classes, and levels, and had some inclusive relationships with one another. The selected evaluation factors represented the main characteristics of Shenzhen's marine ecological economy, offered guidance for the next evaluation, and generated corresponding data values to lay a foundation of quantitative analysis for evaluation. Relying on the above principles and the actual situation of marine eco-economic development in Shenzhen, the evaluation factors and hierarchical structure of zoning were determined.

5.1. *Establishing Evaluation Units*

In terms of natural and geographical conditions, Shenzhen coastal waters were divided into two parts: the Pearl River Estuary and Shenzhen Bay in the west by Nantou Peninsula, and Dapeng Bay and Daya Bay in the east by Dapeng Peninsula. In our study, combined with the management, development, and exploitation direction of Shenzhen, the marine ecological economy of the coastal waters was analyzed.

5.2. *Determining Weights*

In this study, the analytic hierarchy process (AHP) was adopted to rank several levels from high to low according to the subordination relationship among natural, economic, and social marine factors. Using the objective reality of the actual development of the marine ecological economy, the relative importance of each level was expressed quantitatively, and the weight of the relative importance order of all elements at each level was identified using a mathematical method.

5.2.1. Data Processing

Since the selected indicators had different units, the data were first standardized. In our study, a normalization method was applied to eliminate the differences between variables in magnitude and dimension so that they could be calculated uniformly and a longitudinal comparison and analysis could finally be carried out.

Profitability indicator:

$$z_i = \frac{x_i - \overline{x}}{\sqrt{\frac{(x_i-\overline{x})^2}{n}}} \quad (1)$$

Cost indicator:

$$z_i = -\frac{x_i - \overline{x}}{\sqrt{\frac{(x_i - \overline{x})^2}{n}}} \tag{2}$$

In the equations above, x_i represents the value of the indicator in year i and z_i denotes the value of the standardized indicator in year i. An increase in the profitability indicator signifies better performance, while an increase in the cost indicator signifies worse performance.

5.2.2. Indicator Weighting

When determining the weight of each level and factor, we generally used the consistent matrix method. According to the scale method shown in the following table, the (pairwise) importance of factors was scored via subjective evaluation to obtain a judgment matrix.

To guarantee the credibility of the hierarchical ranking, it was important to check the consistency of the judgment matrix; that is, to calculate the random consistency ratio. The formula is as follows:

$$CR = \frac{CI}{RI} = \frac{\lambda_{\max} - n}{RI \cdot (n-1)} \tag{3}$$

The result of a single hierarchical ranking was considered satisfactory only when CR < 0.1. Otherwise, the value of the judgment matrix element was adjusted. The hierarchical ranking and consistency check were performed on the index system using professional AHP software. The results are shown in Table 1.

Table 1. Results of the consistency check of the judgment matrix.

Judgment Matrix	CR	Result
Sustainable development index of marine economy	0.0000	Adopt
Natural resources	0.0362	Adopt
Economic development	0.0379	Adopt
Social resources	0.0000	Adopt

The test results showed that the marine economy sustainable development index, natural environment, industrial economy, and sociocultural judgment matrices all passed the hierarchical ranking and consistency check.

5.2.3. Determining the Index Weight

If the pairwise comparison matrix was a consistency matrix, then a normalized eigenvector W corresponding to eigenvalue n was taken as the weight vector of the factors C1, C2, . . . , Cn of the criterion layer C to the target layer. If the comparison matrix A was not a consistent matrix but within the allowable range of inconsistency, then the eigenvector (after normalization) corresponding to its maximum eigenvalue was adopted as the weight vector of the factors of criterion layer C to target layer O.

The indexes were derived through the following equations:

$$E_t = \sum_{i=} a_i z_i \tag{4}$$

where E_t is the marine ecological and economic sustainable development index; $i = 1, \ldots, 17$; a_i represents the total weight of each index; and z_i denotes the treated observation value of the ith index.

$$S_i = \sum_{i=}^{k} \beta_i z_i \tag{5}$$

Here, S_i is the marine ecological and economic sustainable development index; i represents the sub-item weight of each index; z_i refers to the treated observation value of the ith index; β_i denotes the weight value of a single index; and k is the marine pollution coefficient.

6. Research on Sustainable PAD for the Spatial Form of OCs Based on SS

6.1. Eco-Economic Accounting of Ocean Cities

The marine eco-economic accounting value was calculated by correcting the marine economic accounting indexes based on marine economic accounting by considering the consumption of marine resources, the cost of marine environmental pollution, and the value of marine ecological adjustment in the process of the economic activities of the marine ecosystem. The marine eco-economic accounting framework is shown in Figure 6.

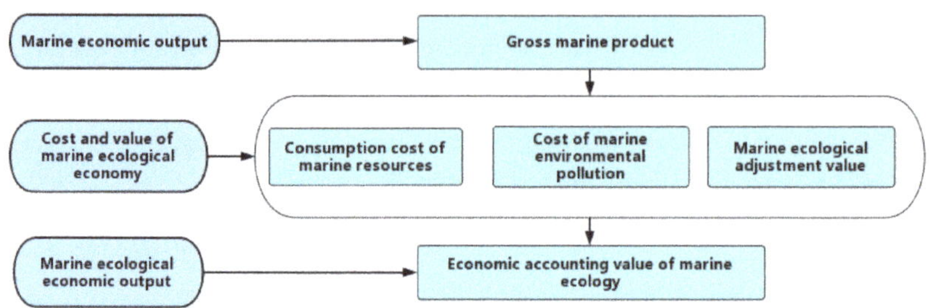

Figure 6. Marine eco-economic accounting framework.

The marine eco-economic accounting was performed by considering the consumption of marine resources, the cost of marine environmental pollution, and the value of marine ecological adjustment, thus furthering the current research on China's marine economic accounting system [37]. To enable the accounting work to be carried out smoothly, the scope of the cost accounting of the consumption of marine resources according to the actual situation of marine industrial development included marine fishery resources, marine oil and gas resources, and marine resources. In this study, based on the average land transfer price of each year from 2011 to 2015, the use cost of marine resources per unit area was derived by deducting the development cost and the added value of a sea area attribute change. The results are shown in Table 2 and Figure 7.

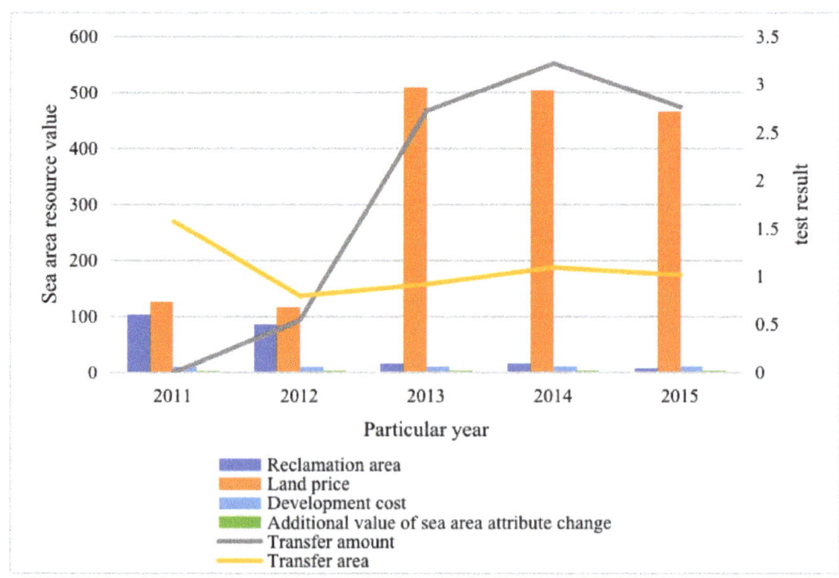

Figure 7. Analysis of the value of marine resources.

Table 2. Value of marine resources formed by reclamation in Shenzhen.

Year	Reclamation Area	Land Price	Transfer Amount	Transfer Area	Development Cost	Additional Value of Sea Area Attribute Change
2011	0.6	0.73	-	269.97	0.06	0.0015
2012	0.5	0.68	93.14	136.82	0.06	0.0015
2013	0.1	2.97	467	157.50	0.06	0.0015
2014	0.1	2.94	552	187.50	0.06	0.0015
2015	0.04	2.72	473.78	173.93	0.06	0.0015

The reclamation area of each part of Shenzhen's "three bays and one estuary" from 2011 to 2015 was divided by the sea area to obtain the proportion of reclamation per unit area of sea area and then multiplied by the sea area within each cell grid and the use cost of marine resources per unit area to calculate the environmental cost for the marine resources of each cell grid. The results are shown in Table 3 and Figure 8.

Table 3. Use cost of marine resources in the "three bays and one estuary".

Year	Pearl River Estuary	Shenzhen Bay	Dapeng Bay	Daya Bay	Total
2011	31.87	8.02	0.00	0.45	40.33
2012	24.46	6.16	0.00	0.34	30.96
2013	22.94	5.77	0.00	0.32	29.04
2014	22.77	5.73	0.00	0.32	28.83
2015	9.40	2.37	0.00	0.13	11.90

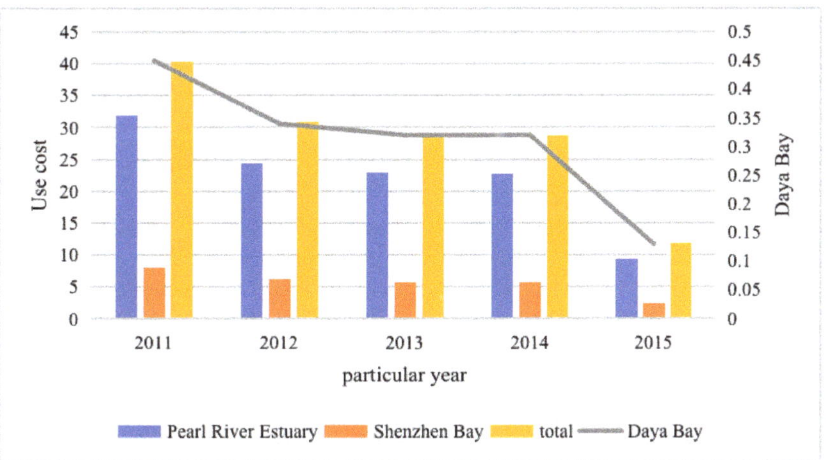

Figure 8. Data chart of use cost of marine resources.

Regarding the marine eco-economic center of Shenzhen, the overall situation basically remained unchanged from 2011 to 2014; however, in 2015, there was an obvious trend of individuals moving to the northeast. Since the offshore oil and gas industry, marine shipping industry, and other weighted industries were mainly concentrated in Nanshan, the two industries shrank substantially from 2014 to 2015, affected by the downturn in the international market, which led to a dramatic decline in their proportion in Shenzhen's marine economy and placed emphasis on the shift eastward.

6.2. Results and Analysis

From the chart analysis above, we can see that the development of the marine economy was the main basis for the sustainable development of the marine ecological economy. On the one hand, the development of the marine economy required marine resources and marine space and imposed tremendous pressure on the ecological environment of offshore areas. On the other hand, the ecological environment and resource-carrying capacity of the oceans imposed objective requirements for the development of the marine industry. At different stages of economic development, it was essential to match the industrial type and economic model with the ecological environmental resources. Accordingly, the sustainable development of the marine ecological economy should have an economic foundation and industrial activities, protect the ecological environment while developing the economy, and promote harmonious coexistence and coordinated development between humans and the oceans.

According to the evaluation results of this study, Shenzhen Bay, having the highest level of sustainable development of its marine ecological economy, mainly relied on two major urban areas, Nanshan and Futian. Shenzhen Bay was the area with the highest land value, while key development areas, such as the Shekou Free Trade Zone, Houhai Headquarters Cluster, and Super Headquarters Base of Shenzhen Bay, had the most intensive high-end service industry and economic activities in Shenzhen [38].

A planning proposal for the Shenzhen marine eco-economic zone was put forward based on the results above:

- Optimization and adjustment zone for the marine ecological economy in Shenzhen Bay

Shenzhen Bay has an important coastal landscape coastline for an OC. Shenzhen Bay and the Houhai Headquarters Base were built in the rear. It is a densely populated area with high-end OC functions. In the future, the area of the marine urban greenbelt and forest will be woefully incompatible with high marine urbanization and high population density in this region. Thus, it is necessary to reserve a public coastline [39]. Futian Mangrove Reserve can be built into a marine environmental protection education base to increase pro-sea-space support and restore the coastline to an ecological coastline and nature reserve.

- Coordinated development zone for the marine eco-economy in Dapeng Bay

This area is the main container port area and coastal tourist resort in east Shenzhen [40]. Dapeng New District is a national marine ecological civilization demonstration zone and a core bearing zone of marine strategies in Shenzhen. Future development should focus on improving the pilot innovation of integrated coastal zone management; facilitating the sustainable development of coastal zone resources; exploring a market-oriented pricing mechanism for the right to use the beach and sea resources; promoting the market-oriented allocation of sea and island resources; exploring the establishment of an early warning index system for the environmental carrying capacity of marine resources within the jurisdiction; making use of Shenzhen's financial advantages; increasing green financial input; exploring the establishment of public welfare funds for the construction of marine ecological civilization, investment, and financing mechanisms for the protection of the marine ecological environment; unleashing the vitality of the bay-style city; and building a demonstration village for coastal ecological civilizations.

- Countermeasures for developing the marine ecological economy in Shenzhen

Build a grid planning spatial unit system; actively explore the further refinement of spatial functional units based on the planning of the main marine functional zones; establish a marine eco-economic planning evaluation unit based on a kilometer grid in coastal areas; and set up a comprehensive evaluation system for marine economic development accounting, marine ecological value evaluation, and the quantitative evaluation of marine resources and environment based on the kilometer grid space unit.

- Develop a differentiated development and exploitation direction for the marine urban space

Set up a differentiated development pattern of "West City and East Park": Extend the axis structure in space, intensify the agglomeration of the modern service industry, and accelerate industrial transformation and functional upgrades in the western region. The eastern region is positioned as a regional ecological stabilizer and a world tourist destination that is connected to the natural terrain in space via the implementation of decentralized development.

Boost the functional transformation of the coastal area: Adapt to the transformation of marine cities and make overall plans for the use of land and sea space with the coastal zone as the core to promote connectivity between land and sea functions. Further increase the proportion and balanced layout of living shorelines, improve the inventory of ecological shorelines, and make shoreline utilization more scientific and rational. Optimize and upgrade the functions of the coastal OC in Shenzhen, combined with the functional positioning in eastern and western regions and the functional adjustment of marine cities in coastal areas [41].

7. Discussion

Based on field surveys, SS draws on graph theory to form spatial relation structure diagrams. SS can be used to transform the spatial form relationship of a city into a mathematical model, which can be used to quantitatively analyze the influential factors of urban spatial form and calculate and compare invisible elements in a quantitative way to more clearly and accurately describe the spatial form characteristics of the research area. SS theory has evolved to help architects and planners find potential space in cities and develop targeted designs for specific planning and construction projects. In addition to the results of the above analysis, the following problems existed in Shenzhen Bay.

- Lack of connection between subdistricts

The degree of synergy in Shenzhen Bay in Zhengzhou was low, and the connection between the parts and the whole of the park was poor. There was a shortage of connection between different areas, accessibility was poor, spatial vitality was low, and the ability to attract people was weak. Thus, the movement of people was reduced.

- Poor accessibility of the local area
- The global integration degree of Shenzhen Bay was found to be on the low side at 0.7495 on average, indicating that the connection between different areas of Shenzhen Bay was not close, and the ability to travel between areas was reduced; thus it was inconvenient for tourists to commute between areas.
- Differences between the east and west
- The development levels of the terrestrial and marine eco-economies in central China were much higher than those in the eastern and western regions. The development level of the marine eco-economy in the sea area was influenced by the residents' living and production activities. The development, exploitation, and environmental pollution in central and western waters were much higher than that in eastern waters, making the marine eco-economy in central and western waters lower than that in eastern waters.
- Differences between the north and south
- Ocean-related enterprises tended to aggregate in areas near the coastline. As they extended inland, the aggregation of ocean-related enterprises fell sharply and the development level of the marine ecological economy also dropped rapidly.

8. Conclusions

SS is a theory and method that investigates physical space and the spatial composition of human cognition by quantitatively describing the spatial structure of human settlements. It highlights the study of space as an independent factor and the correlation between space and architecture, space and society, and space and human cognition. In this study, based on the SS theory, we measured the sustainability of Shenzhen's urban spatial form, analyzed and compared Shenzhen's spatial forms, and drew the following conclusions. Relying on its regional advantage, Shenzhen gathered some high-quality marine enterprises in the early stage of development. It had some scale merits in traditional industries, such as marine transportation, marine oil and gas, and coastal tourism, and strong innovation ability in emerging industries, such as marine electronic information, marine engineering equipment, and marine biology. However, Shenzhen also had distinct "short slabs" in terms of its marine resources and environment. From the perspective of resource conditions, Shenzhen had scarce marine resources, coastal resources, and island resources. The development space of its marine industry was limited to a large extent. It was difficult for the emerging marine industry to form agglomeration strength. Furthermore, in terms of the conditions of the marine environment, there were marked differences in environmental quality between the eastern and western waters of Shenzhen. Reinforcing the guidance on spatial planning; optimizing the spatial layout of the marine economy; and vigorously promoting the concentrated, agglomerative, and intensive development of the marine economy are important approaches for heightening the quality and efficiency of marine economic development in Shenzhen and improving Shenzhen's economy through sustainable spatial planning.

Regarding the coordination degree between the marine economy and the development of resources and environment, there were remarkable differences between the sea areas. The ecological environment of the Pearl River Estuary was under great pressure and short-term recovery was challenging. The socioeconomic intensity of offshore land is expected to continue to grow. In the years to come, it is imperative to control the scale of reclamation, change traditional shoreline use, avoid non-intensive land use and ecological destruction, and promote the sustainable growth of the city.

Shenzhen Bay is a densely populated area with high-end urban functions while also being an area experiencing great pressure in terms of ecological protection and sea environment management. In the future, it is necessary to optimize the marine economy and ecological environment concurrently, avoid non-intensive land use and ecological destruction, and promote the process of sustainable and smart urban growth. The marine ecological and environmental conditions in Dapeng Bay and Daya Bay are superior. For this reason, it is appropriate to adhere to the principle of ecological protection first and limit development; prudently plan and leverage coastal land resources; and develop coastal tourism and emerging marine industries with low pollution, low energy consumption, and high added value in a targeted way, to push forward the development of the marine ecological economy. According to the spatial development characteristics of Shenzhen Bay, development can proceed in terms of the following aspects. First, integrate the landscape resources of Shenzhen Bay, plan plants, build local habitats, connect and repair water systems, shape an open marine space, refine the landscape resistance zone, and build an open space system with sustainable planning in the park. Second, to relieve the development pressure in the downtown region, intensify the development of the public transport system and increase the connections between old and new urban road networks to form a smooth and sustainable road network. Regarding the active public space in the region, we should promote the construction of infrastructure, enliven the regional atmosphere, and promote the coordinated and sustainable development of the region.

According to SS theory, the space system of OCs consists of two parts: closed space and free space. The social attributes of OCs and architectural spaces are not determined artificially and subjectively by attaching functional labels. The space system created by people often acts on people's behavior after its completion, thus producing different social effects. Our study generated a sustainable PAD strategy for the spatial form of the OC based on SS. This is beneficial to the sustainable development of Shenzhen in all aspects. Through a general survey of findings published at home and abroad, we found that the research on the sustainable planning and design of ocean cities has aroused wide concern in theoretical and practical circles, and its theory and methods are constantly being improved. However, as a new research frontier and vibrant subject, the sustainable planning and design of ocean cities warrant further study in many regards. Therefore, the contribution of the results presented in this paper to the sustainability of spatial planning in Shenzhen will also be evident in future practice.

9. Research Limitations and Further Research Directions

Based on SS theory, we carried out a quantitative analysis of the spatial structure of Shenzhen Bay, summarized the existing problems in the spatial structure of two areas, and put forward concrete optimization suggestions. However, the following aspects can be improved in future research. When performing data analysis, we used the satellite map as the main reference. However, due to the large study area and limited personal competence, there were still some phenomena for which the relevant data were not complete enough. Although this had little effect on the analysis results, it should be avoided in future research as much as possible.

Further potential research directions are as follows.

- How to develop a compact strategy and sustainable development

Shenzhen has built a fully functional business and cultural area and its industry and logistics have also made considerable progress. However, part of the periphery of Shenzhen has developed in a scattered fashion along both sides of the transportation lines. Thus, on a larger regional scale, based on the existing urban structure and the research undertaken in this study, considering how to develop a compact strategy and sustainable development may be a further research direction.

- How to develop sustainable town traffic and spatial strategies

In our study, the traffic road system was investigated in an attempt to build a continuous and unified traffic relationship and improve the accessibility and visibility of the space. How to extend and modify the rail system and associate it with spatial design to expand the traffic system to the edge zone, break through natural limitations, and form an overall pattern of north–south extension and sustainable development along the main roads in the town are also important research directions.

Author Contributions: Conceptualization, L.Z. and C.K.; methodology, L.Z. and J.Y.; software, L.Z. and J.Y.; validation, L.Z. and C.K.; formal analysis, L.Z. and J.Y.; investigation, L.Z. and J.Y.; resources, L.Z. and J.Y.; data curation, L.Z. and J.Y.; writing—original draft preparation, L.Z. and J.Y.; writing—review and editing, L.Z. and C.K.; supervision, C.K. All authors have read and agreed to the published version of the manuscript.

Funding: This work was supported by a grant from Brain Korea 21 Program for Leading Universities and Students (BK21 FOUR) MADEC Marine Designeering Education Research Group.

Institutional Review Board Statement: Not applicable.

Informed Consent Statement: Not applicable.

Data Availability Statement: Not applicable.

Conflicts of Interest: The authors declare no conflict of interest.

References

1. Boeing, G. Spatial information and the legibility of urban form: Big data in urban morphology. *Int. J. Inf. Manag.* **2021**, *56*, 102013. [CrossRef]
2. Bai, R.S.; Jiang, Y.P.; Jiang, J.D. Multi-Scale Analysis on Functional Structure of Ecological-Production-Living Spaces of Jianghuai Urban Agglomeration. *Chin. Famous Cities* **2016**, *10*, 21–28.
3. Dalton, R.; Bafna, S. The syntactical image of the city: A reciprocal definition of spatial elements and space syntaxes. In Proceedings of the 4th International Space Syntax Symposium, London, UK, 17–19 June 2003.
4. Yang, L.; Tian, Y. Art Design of Urban Public Space Based on Marine Culture. *J. Coast. Res.* **2020**, *106*, 427–430. [CrossRef]
5. Xu, Y.; Rollo, J.; Esteban, Y. Evaluating Experiential Qualities of Historical Streets in Nanxun Canal Town through a Space Syntax Approach. *Buildings* **2021**, *11*, 544. [CrossRef]
6. Yu, H.S.; Yang, J.; Li, T.; Jin, Y.; Sun, D.Q. Morphological and functional polycentric structure assessment of megacity: An integrated approach with spatial distribution and interaction. Sustain. *Cities Soc.* **2022**, *80*, 103800. [CrossRef]
7. Guo, S.; Wang, S.; Chen, L. Research on the Regional Characteristics of Marine Coastal Architecture from the Perspective of Ecological Development. *J. Coast. Res.* **2020**, *108*, 221–225. [CrossRef]
8. Esposito, D.; Santoro, S.; Camarda, D. Agent-Based Analysis of Urban Spaces Using Space Syntax and Spatial Cognition Approaches: A Case Study in Bari, Italy. *Sustainability* **2020**, *12*, 4625. [CrossRef]
9. Hillier, B.; Leaman, A.; Stansall, P.; Bedford, M. Space syntax. *Environ. Plan. B Plan. Des.* **1976**, *3*, 147–185. [CrossRef]
10. Chun, H.; Wang, N.; Liu, Y. Disaster Prevention Route Planning of Fenglin Ancient Village Based on Space Syntax Analysis. *IOP Conf. Ser. Earth Environ. Sci.* **2021**, *781*, 032009. [CrossRef]

11. Chen, Y.; Chen, J. Analysis and Research on Spatial Nodes of Historical Villages based on Spatial Syntax Theory: Cuiwei Village in Qianshan Area of Zhuhai as an Example. *J. Comput. Sci. Technol. Stud.* **2022**, *4*, 7–34. [CrossRef]
12. Zheng, Z.; Zhu, Q.; Fu, J. A Research on Pingshan Village's Sustainable Development Based on Space Syntax Theory-Take the Improvement of Tourism Format Quality for Example. *Landsc. Archit.* **2019**, *2*. [CrossRef]
13. Qi, L.; Liang, W.; Xi, M. The Public Space Pattern research of Guangfu Traditional Villages Based on space syntax: A Case Study of Huangpu Village in Guangzhou City, China. *IOP Conf. Ser. Earth Environ. Sci.* **2019**, *2*, 59–70. [CrossRef]
14. Wang, F.; Chen, J.; Tong, S.; Zheng, X.; Ji, X. Construction and Optimization of Green Infrastructure Network Based on Space Syntax: A Case Study of Suining County, Jiangsu Province. *Sustainability* **2022**, *14*, 7732. [CrossRef]
15. Zheng, W.; Du, N.; Wang, X. Understanding the City-transport System of Urban Agglomeration through Improved Space Syntax Analysis. *Int. Reg. Sci. Rev.* **2022**, *45*, 161–187. [CrossRef]
16. Chen, Y.; Zhu, Q. Street network or functional attractors? Capturing pedestrian movement patterns and urban form with the integration of space syntax and MCDA. *Urban Des. Int.* 2022, *prepublish*. [CrossRef]
17. Qi, B. Interactive Landscape Design of Ocean City Wetland Park. *J. Coast. Res.* **2020**, *112*, 26–28.
18. Tian, Y.; Yang, L. Cultural and Artistic Design of Coastal Cities Based on Marine Landscape. *J. Coast. Res.* **2020**, *106*, 431–434. [CrossRef]
19. Russo, F.; Rindone, C.; Panuccio, P. External Interactions for a Third Generation Port: From Urban Sustainable Planning to Research Developments. *Int. J. Transp. Dev. Integr.* **2022**, *6*, 253–270. [CrossRef]
20. Penn, A. Space syntax and spatial cognition: Or why the axial line? *Environ. Behav.* **2003**, *35*, 30–65. [CrossRef]
21. Mora, R. The Cognitive Roots of Space Syntax. Ph.D. Thesis, Faculty of the Built Environment, The Bartlett, University College London, London, UK, 2009. Available online: http://discovery.ucl.ac.uk/18920/1/18920.pdf (accessed on 20 October 2022).
22. Geng, S.; Chau, H.-W.; Jamei, E.; Vrcelj, Z. Understanding the Street Layout of Melbourne's Chinatown as an Urban Heritage Precinct in a Grid System Using Space Syntax Methods and Field Observation. *Sustainability* **2022**, *14*, 12701. [CrossRef]
23. Li, Y.; Xiao, L.; Ye, Y.; Xu, W.; Law, A. Understanding tourist space at a historic site through space syntax analysis: The case of Gulangyu, China. *Tour. Manag.* **2016**, *52*, 30–43. [CrossRef]
24. Safari, H.; Nazidizaji, S.; Fakouri Moridani, F. Social Logic of Cities and Urban Tourism Accessibility; Space Syntax Analysis of Kuala Lumpur City Centre. *Space Ontol. Int. J.* **2018**, *7*, 35–46.
25. Ali, H.H.; Al-Hashimi, I.A.; Al-Samman, F. Investigating the applicability of sustainable urban form and design to traditional cities, case study: The old city of sana'a. *Int. J. Archit. Res. ArchNet-IJAR* **2018**, *12*, 57. [CrossRef]
26. Siregar, J.P.; Surjono; Rukmi, W.I.; Kurniawan, E.B. Evaluating accessibility to city parks utilizing a space syntax method. A case study: City parks in Malang city. *IOP Conf. Ser. Earth Environ. Sci.* **2021**, *916*, 012015. [CrossRef]
27. Yin, L.; Wang, T.; Adeyeye, K. A Comparative Study of Urban Spatial Characteristics of the Capitals of Tang and Song Dynasties Based on Space Syntax. *Urban Sci.* **2021**, *5*, 34. [CrossRef]
28. Yamu, C.; van Nes, A.; Garau, C. Bill Hillier's Legacy: Space Syntax—A Synopsis of Basic Concepts, Measures, and Empirical Application. *Sustainability* **2021**, *13*, 3394. [CrossRef]
29. Laura, B.; Alvarez. *A Beginner's Guide to Urban Design and Development: The ABC of Quality, Sustainable Design*; Taylor and Francis: Abingdon, UK, 2022.
30. Salama, S.W. Towards developing sustainable design standards for waterfront open spaces. *City Territ. Archit.* **2022**, *9*, 26. [CrossRef]
31. Arora, N.K.; Mishra, I. Ocean sustainability: Essential for blue planet. *Environ. Sustain.* **2020**, *3*, 1–3. [CrossRef]
32. Pierce, C.H. Boston—A Maritime City with a Continental Climate: A Part Time Air Conditioned City. *Weatherwise* **2010**, *14*, 138–141. [CrossRef]
33. Harris, P. Urban renewal of a Pacific maritime city. -Sydney's redevelopment. Paper presented to the Pan Pacific Congress of Real Estate Appraisers, Valuers and Counselors (18th: 1996: Sydney). *Valuer Land Econ.* **1996**, *34*, 164–168.
34. Hwang, Y.W.; Yhang, W.J. International Tourists' Attitude to Theme Park Development of Busan. *J. Navig. Port Res.* **2003**, *27*, 345–349. [CrossRef]
35. Sepe, M. Urban history and cultural resources in urban regeneration: A case of creative waterfront renewal. *Plan. Perspect.* **2013**, *28*, 595–613. [CrossRef]
36. Cheng, L.; Feng, R.; Wang, L. Fractal Characteristic Analysis of Urban Land-Cover Spatial Patterns with Spatiotemporal Remote Sensing Images in Shenzhen City (1988–2015). *Remote Sens.* **2021**, *13*, 4640. [CrossRef]
37. Liu, X.; Huang, B.; Li, R.; Zhang, J.; Gou, Q.; Zhou, T.; Huang, Z. Wind environment assessment and planning of urban natural ventilation corridors using GIS: Shenzhen as a case study. *Urban Clim.* **2022**, *42*, 101091. [CrossRef]
38. Liu, X.; Sang, X.; Chang, J.; Zheng, Y.; Han, Y. Sensitivity analysis and prediction of water supply and demand in Shenzhen based on an ELRF algorithm and a self-adaptive regression coupling model. *Water Supply* **2022**, *22*, 278–293. [CrossRef]
39. Cheng, H.; Lai, Y.; De, T. Decoding the decision-making in the new wave of urban redevelopment in China: A case study of a bottom-up industrial land redevelopment in Shenzhen. *Land Use Policy* **2021**, *111*, 105774. [CrossRef]
40. Deng, W.; Luo, W. Activation of Traditional Villages in Dapeng New District, Shenzhen, China—A Case Study on Wangmu walled-village. *IOP Conf. Ser. Earth Environ. Sci.* **2021**, *783*, 012115. [CrossRef]
41. Hsieh, C.M. Sustainable planning and design: Urban climate solutions for healthy, livable urban and rural areas. *J. Urban Manag.* 2021, *prepublish*. [CrossRef]

Article

Kerbside Parking Assessment Using a Simulation Modelling Approach for Infrastructure Planning—A Metropolitan City Case Study

Premaratne Samaranayake [1,*], Upul Gunawardana [2] and Michael Stokoe [3]

[1] School of Business, Western Sydney University, Penrith, NSW 2751, Australia
[2] School of Engineering, Design and Built Environment, Western Sydney University, Penrith, NSW 2751, Australia
[3] Freight and Servicing, Sydney Coordination Office, Transport for NSW, Sydney, NSW 2000, Australia
* Correspondence: p.samaranayake@westernsydney.edu.au

Abstract: The main purpose of this research is to investigate the effect of kerbside parking demand and provision on short-term parking (STP) and freight activity space (FAS) as a benchmark for infrastructure planning, considering the impacts of expected future growth and capacity changes. In this study, we adopted a mixed-methods approach of quantitative analysis including a spatial view of parking using manual and video-captured camera data from the majority of STP and FAS parking bays covering a diverse range of loads/tasks with different levels of elasticity and substitutes, as well as simulation of current demand influenced by various factors, as a basis for the development of strategies and prioritisation of the allocation of limited kerbside spaces in Parramatta, a rapidly transforming/growing CBD city centre environment. Parking demand consisted of a diverse range of FAS and STP categories. Spatial analysis showed a non-homogeneous distribution of parking demand and loads across several sections of the city. A large proportion of short-term parking spaces is attributed to two peak periods during the day and increased traffic volumes at peak times. Comparatively lower average parking times in the northern and western regions compared to those in the city centre indicate the potential to reduce peak parking periods and therefore traffic congestion in the city centre by changing parking limits. The presented simulation model can be used as a reliable benchmarking model for the simulation of future impact scenarios and to make recommendations with respect to infrastructure planning and to develop travel demand management strategies. This research is based on a case study and is therefore subject to limitations in its applications in other contexts. Extension of the baseline simulation with future impact scenarios is planned for the next stage of this research. A simulation model is presented and illustrated as a reliable benchmarking tool for the simulation of future impact scenarios through a case study of a rapidly changing city environment.

Keywords: kerbside parking; baseline scenario; short-term parking (STP); freight activity space (FAS)

Citation: Samaranayake, P.; Gunawardana, U.; Stokoe, M. Kerbside Parking Assessment Using a Simulation Modelling Approach for Infrastructure Planning—A Metropolitan City Case Study. *Sustainability* **2023**, *15*, 3301. https://doi.org/10.3390/su15043301

Academic Editors: Qingsong He, Jiayu Wu, Chen Zeng and Linzi Zheng

Received: 8 December 2022
Revised: 23 January 2023
Accepted: 3 February 2023
Published: 10 February 2023

1. Introduction

Parking is a broad topic that has been investigated from different perspectives, mainly considering parking assessment and infrastructure planning across a range of parking demands and facilities, including off-street and kerbside parking in urban city environments, policies, and parking limits. Parking infrastructure relates to current facilities, requirements and usage in urban cities and other large established areas such as university campuses, business parks and sporting venues. Parking infrastructure comprises a range of facilities, including kerbside parking in metropolitan city centres and private multistorey carparks. Furthermore, parking policy, as well as urban planning and space policy, is an integral parts of parking assessment. Therefore, the provision of parking and space for communities under urban planning is a main question for policymakers at the local government/council level, focusing on types of kerbside parking space/allocation, parking time limits, pricing of parking, and space for people [1].

Parking assessment is a continuing area of interest for various stakeholders due to the changing environment in many urban/metropolitan cities. For example, the difference between supply and demand has increased over the years, and most local/city governments are behind and institutionally unprepared for planning, regulation. and management of parking [1]. At the same time, parking assessment is becoming a complex and challenging task when cities are being planned with the concept of "car-free cities" or "planning for people, not for cars". Most local authorities (and states) are proposing a reduction in parking provisions in cities to provide more places for people (than vehicles). Main changes include increasing the level of infrastructure, commercial and residential developments, increasing level of demand for scarce space for different types of parking and daily temporal patterns, and changes to existing parking infrastructure.

Access and kerbside parking assessment are critical for any future infrastructure development, particularly when existing parking infrastructure is affected during the construction stage and given expected changes to existing parking facilities. Some of the changes impacting the availability of kerbside parking include (i) increasing public space and pedestrian prioritisation, (ii) possible closure of existing kerbside parking due to new public transport infrastructure such as light rail, (iii) increasing demand due to population growth, (iv) increasing freight and servicing activities, and (v) changes to other parking facilities (e.g., closure of existing multistorey carparks for the development of new residential and commercial buildings). Some studies have investigated the impact of changes in specific factors on overall parking capacity and needs/behaviour, including the provision of a minimum number of parking spaces, taking into consideration historical relationships between specific land uses and parking needs [2], as well as parking pricing and kerbside allocation when the private sector provides garage parking [3]. Although parking needs and capacity, which are impacted by several factors (e.g., population growth and public transport infrastructure) is a well-studied/-reported area, segmentation of kerbside assessment of FAS is less defined/studied. In this context, there is a lack of science supporting the understanding of different segments of demand and providing useful guidance for policymakers in making decisions on parking limits and schemes.

Thus, the main purpose of this research is to propose a simulation modelling approach for kerbside parking assessment. The key objectives of this research are to (i) assess the current kerbside demand during regular peak periods using a survey of parking events based on manual and camera data and (ii) model and simulate current kerbside parking demand (baseline scenario) as the basis for modelling and simulation of future kerbside parking impact scenarios. Camera data are collected using video of kerbside parking events. Simulation modelling of the baseline (current demand and capacity/supply) scenario based on comprehensive camera data and analysis of future parking demand growth forms the basis for simulation modelling of kerbside impact scenarios, which are influenced/generated by potential future changes to parking capacity and potential increases in demand due to construction activities, population growth, and increased freight and serving activities.

This study was carried out as a kerbside parking case study of a rapidly changing city centre within metropolitan Sydney. The remainder of the paper is structured as follows. First, the research background and an overview of the research methodology are presented. The case study background and data collection are outlined under the research methodology, followed by simulation modelling, results, and analysis. In this case, results and analysis include verification of camera data using manual survey data, simulation model-based analyses, and analysis of the growth of kerbside parking demand and future changes to parking capacity. Finally, findings, discussion, and conclusions are presented by outlining recommendations for policymakers regarding kerbside parking, limitations, and practical implications.

2. Research Background

Kerbside parking provision and management is a major area of focus of city planning for various stakeholders, including local councils and local government transport agencies. Key areas of consideration include kerbside parking infrastructure (e.g., capacity perspective), kerbside parking policy (e.g., pricing perspective), current practices (e.g., demand, utilisation, trends, etc.), future impacts, and policy objectives for urban centres [4]. All these areas are very broad and have been considered in various research projects, including the planning of infrastructure in different settings [2], setting of policies at local government and council levels, and operational issues and challenges from demand and utilisation perspectives [5]. Research studies on parking assessment cover a range of aspects and have investigated parking assessment from a range of viewpoints, including evaluation of parking facilities and infrastructure from the perspective of constraints and/or opportunities for urban development [2], energy management for a large-scale electric vehicle (EV) charging-enabled municipal parking facility [6], a cost–benefit analysis of inner-city parking using network optimisation [7], and the effectiveness of off-street parking pricing schemes under changing conditions [5]. Studies investigating kerbside parking demand from the perspective of influencing factors have considered changes in parking cost, shifting demand, and their interaction with cruising for parking [8,9]; a tradeoff between cruising and walking cost [10]; pricing elasticity of parking occupancy and associated parking supply restrictions [11]; and management of freight capacity in the context of a major CBD transformation while accommodating business growth with reduced kerbside capacity [12]. Furthermore, studies of demand modelling have considered latent demand (e.g., unrealised demand and drivers searching/cruising for an available spot), showing that it accounts for a significant proportion of overall demand [13,14]. In addition, factors affecting the level of satisfaction of commuters can also be used as a key input for the promotion of sustainable modes of transport to solve parking problems in rapidly growing urban cities [15].

Based on a range of research studies around these key areas, it is clear that most of studies cover long-term planning and short-term operations. These areas are related and interconnected through the planning/management process and involve many stakeholders, including local government for the planning of infrastructure, councils for setting of parking policies, and users from a demand and utilisation perspective. Because most research is driven by the parking assessment context, key areas of kerbside parking are discussed from both theoretical and practical perspectives herein.

2.1. Kerbside Parking Infrastructure—Capacity Perspective

Every metropolitan city is facing the challenge of balancing kerbside parking infrastructure with various spatial and usage demands. Demand is changing at a rapid rate due to various influencing factors, such as growth in e-commerce, increasing numbers of share economy providers, and shorter delivery windows [2,5,16], as well as changes to supply due to rapidly changing city infrastructure, particularly with popular light rail transit projects [17]. These changes are dynamic and need to be considered as part of a holistic approach when dealing with planning for parking infrastructure needs. These changes are common in high-growth cities such as the Sydney CBD and Parramatta [18] due to minimum development requirements [2]. This is exacerbated by the increasing rate of changes to existing infrastructure and changes to the mix of demand needs, among many factors.

Parking infrastructure in urban centres consists of a diverse range of parking spaces, including kerbside parking, available for a range of users under different parking categories, time limits, and pricing conditions. An assessment of paid parking in an urban city environment [1] indicated that off-street parking facilities are growing much faster than kerbside parking. According to [1], off-street parking spaces now account for more than 90% of the total parking spaces in Beijing—a more than 290% increase since 2008. However, parking infrastructure assessment studies are very limited, particularly with respect to changes to the composition of various parking facilities from the perspective of urban city settings. Although the growth of kerbside parking is decreasing in some urban settings [1], it represents critical infrastructure in an urban city environment and serves various users with increasing demand due to increasing residential and commercial activities. Thus, kerbside parking facilities are considered to be a significant part of the overall parking space, especially in urban centres. The rapid growth of residential and commercial activities in urban centres has led to the need for assessment of current parking supply, demand, location, and duration/limits, among many other factors, as the basis for improving current parking practices and setting appropriate parking policy to meet future demand under changing conditions. Several research studies have focused on various aspects of parking in urban centres, including parking infrastructure [2], parking policy [16,19,20], and environmental assessment of parking [21] and parking requirements [22].

Parking infrastructure is another area of investigation, mainly due to the direct connection to the source of the major problem of shortage of parking spaces and the increasing gap between supply and demand [1]; increasing demand for parking spaces due to growth in residential living in association with increasing car ownership and commercial activity in urban centres [23]; and the very important influence of infrastructure development, resulting in changes to existing parking infrastructure [2]. Parking infrastructure studies are mainly focused on the assessment of parking requirements based on zoning regulations and future developments, including the assessment of the extent and location of parking infrastructure within metropolitan areas [2], setting of minimum parking requirements using cost–benefit analysis [22], and policy shifts towards setting maximum limits for buildings (off-street) for general parking to encourage active and public transport. On the other hand, some scholars and planners suggest multiple ways to reduce parking requirements and even suggest additional parking for other users, such as cyclists, carpoolers, and transit users [22]. As evident from the literature, a reduction in parking requirements can address some of the issues caused by kerbside parking infrastructure (e.g., abundant "free" and low-cost parking), such as traffic congestion, poor air quality, more household spending on mobility, equity issues, and underused land [24–26]. This suggests that addressing parking infrastructure issues not only reduces the gap between demand and supply but also societal benefits, such as by promoting more space for people than cars in populated cities. Recently, addressing a key issue associated with minimum parking requirements set by historical relationships between specific land use and parking needs [27] has emphasised the importance of considering remaining incentives for auto use created by the existing parking infrastructure when reforming parking policies.

2.2. Kerbside Parking Policy

Parking policy can cover a range of perspectives, including barriers to the emergence of off-street parking markets [28], on-street parking pricing [29], non-residential parking policies [30], kerbside parking time limits [16], time-varying parking prices [31], and user-attitude-driven pricing [32]. There is a considerable literature on parking policy-related studies, mainly focusing on the investigation of the supply of parking and parking price [29,32,33]. At the same time, the supply of parking in urban centres covers a range of parking spaces including kerbside parking and public and private parking facilities. Thus, broader parking policy covers the pricing and supply of various parking spaces.

Recently, kerbside parking policy has received considerable interest, in particular from the perspective of parking time limits [16] and pricing schemes [16,33–35]. According to [16], underpricing of kerbside parking leads to wasteful cruising for parking. Similarly, based on the survey method of comparing parking occupancy, the objective of parking policy to increase the ease of finding a vacant parking place is evident from underutilised parking spaces in city centres [33]. However, these studies are limited to developing kerbside parking policies subject to only economic factors. Shortage of parking spaces is a major issues, particularly in urban centres, and has been exacerbated by an increase in traffic and associated demand in recent times [33]. It is emphasised that the increase in Internet shopping due to the COVID-19 pandemic has contributed to an increase in freight-related traffic, particularly in urban centres. According to [36]:

"We've entered an entirely new way of buying goods and services, but our infrastructure is only adapting incrementally."

Studies on parking policy can be divided into two categories: studies on the supply of parking and studies on parking price. Most studies on parking policy are case-study-based, as policies are area-specific, given the very specific nature of the supply of parking and its price. In this context, the authors of [28] emphasise the need for a more rigorous policy effort, taking into consideration both market fostering and regulation. Recently, [20] highlighted the importance of adopting new planning practices of "maximum provision codes", limited parking development, and demand pricing over traditional practices. Based on the case study approach, a few studies have suggested new policy frameworks, including spatial distribution and usage of parking in Melbourne, Australia [19], and the strong need for reform in urban parking management to promote urban transportation and maximise social welfare [1]. In the specific parking policy of pricing downtown parking, many research studies have reported different perspectives, [16] recommending underpricing of kerbside parking as a sound policy response to the free parking provided by suburban shopping centres, in particular with respect to heterogeneous individuals using the parking. However, this approach limits the parking policy to differences in individuals using parking but does not consider types of vehicles.

2.3. *Parking Usage and Behaviour*

Parking presents several challenges, particularly in metropolitan areas, due to limited supply and demand/usage by a range of users [37]. With limited parking in the metropolitan city environment, vehicles searching for parking create an environmental and economic impact [38]. Thus, parking usage is considered a central part of parking assessment studies, taking into consideration various parking demand and supply scenarios, including kerbside parking assessment, the impact of varying user and parking load categories, the impact of expected growth and changes in parking behaviour, and the variety of assessment situations in suburban cities and large centres such as university campuses [39]. The underlying principles of all these assessments are changing demand and supply situations, with a range of dynamic conditions impacting demand and supply [30] and various characteristics of supply and demand, including specific differences between commercial parking and other parking facilities [39]. From the perspective of optimum parking facilities, the authors of [40,41] proposed modelling approaches to determine the optimum number of park-and-ride facilities as a way to reduce traffic congestion in the considered urban cities.

Kerbside parking assessment studies mainly focus on the assessment of current demand for various purposes, including the impact of the parking price change on demand [42], parking slot allocation subject to dynamic conditions [37], the evaluation of parking demand for policy assessment [20], and assessment of pricing impact [43]. From the perspective of usage and impact, studies have focused on the source of knowledge for car parking strategies, emphasising the need to develop and implement staff car parking strategies [44], environmental and economic impact due to searching for parking [38], the concept of freight landscape, and requirements for different city logistics strategies [45], in addition to highlighting limited parking capacities with increased demand and intensity in metropolitan cities.

The impact of price changes on kerbside parking usage has shown mixed results in different contexts, including a reduction in double parking and cruising for parking and improved driver experience [43], determining optimal parking price, taking into consideration users' attitudes [32] and optimal parking rate/price to achieve the desired level of parking occupancy. The authors of [37,42] proposed an approach for public parking slot assignment using advances in parking sensing and communication technologies, which rely on eliciting truthful private information from drivers while maximising social welfare.

Furthermore, a few research studies on parking assessment and allocation of limited parking capacity have reported on the application of advanced technologies, smart parking management systems to increase effective capacity [46], and potential benefits of using intelligent parking guidance systems, mainly using electronic signage systems that direct drivers to vacant parking lots [20].

These studies have mainly focused on assessing current kerbside/off-street parking, taking into consideration pricing/rate changes and/or application of technologies to improve effective parking usage/behaviour. In this case, effective parking usage/behaviour is achieved by minimising the gap between demand/requirements and supply across a diverse range of parking spaces and loads and under different conditions, including parking limits and/or pricing schemes. It is also evident from these research studies that kerbside parking infrastructure, kerbside parking policy, and parking usage are inter-related. Rapidly changing cities are most likely to influence changes to all or most of these aspects. Therefore, the need for parking assessment taking into consideration all of these aspects is imperative, particularly when cities are subject to the rapid growth of infrastructure development, population growth, and changes to existing parking infrastructure. In this context, parking assessment taking into consideration all of these aspects is considered with respect to a case of the fastest-growing cities in Australia.

2.4. Background of Current Parking Infrastructure—Parramatta CBD

One of the rapidly changing urban cities in Australia is Parramatta CBD. The increase in development in Parramatta CBD aims to support more employment and residents [47]. Parramatta is recognised as one of several major cities in Sydney as part of A Metropolis of Three Cities—the Greater Sydney Region Plan prepared concurrently with Future Transport 2056 and the State Infrastructure Strategy [48]. Parramatta light rail and the Sydney Metro West rail line are two major infrastructure projects currently in progress. In the case of Parramatta CBD as the focus of this research, the proposed light rail project will have a major impact on the current kerbside parking facilities. Figure 1 shows a map of the planned route of light rail infrastructure and the parking study area in Parramatta CBD on a 2 km × 2 km map. The parking study area of short-term parking (STP) and freight activity space (FAS) defined by the boundaries of the area is about 0.78 km^2 (i.e., 0.91 km × 0.86 km) and is shown in Figure 1 (blue colour).

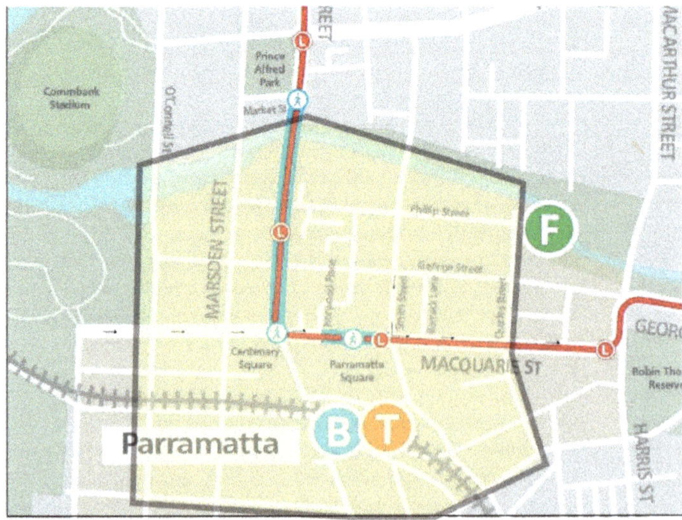

Figure 1. Planned route of light rail infrastructure (red) and the parking study area (shaded) of Parramatta CBD [49].

The second major change expected in the Parramatta CBD is ongoing infrastructure development, which also impacts the current kerbside parking infrastructure [50]. There are also several other residential and commercial infrastructure projects that are in the pipeline, including a large number of applications for commercial office development in the Parramatta CBD [51]. Impact on kerbside parking due to these projects are in two stages. First, there is an immediate impact on kerbside parking requirements due to increased demand by various vehicles such as construction workers' vehicles and freight services. Secondly, there would be additional demand and varied demand composition once the development is complete and buildings are occupied by residents and businesses.

The next major change is the reduction in existing parking facilities due to other developments in the Parramatta CBD. In Central Parramatta, there are three public multistorey carparks, two of which are flagged for demolition to make way for other developments. Commuter carparks on the fringe of the CBD are likely to replace this infrastructure. While this scenario can be approached from a land value perspective, a uniform profile of commuter parking may be required. However, tradespersons who used the carpark on Horwood Place (now demolished for future development) may be using kerbside parking due to its convenient proximity to the large city centre developments such as Parramatta Square, particularly if workers are carrying tools to the site [50]. The relationship between the Horwood carpark and Parramatta Square should be considered as one of the critical factors when parking behaviour is further investigated within the city infrastructure development planning framework, particularly in the context of the scale of change in Parramatta CBD. All of these expected changes and the resulting impact on overall kerbside parking behaviour from the perspective of usage (e.g., different loads/tasks, vehicles) with time and bottleneck are complex to visualise and need to be investigated to develop travel demand management strategies for all stakeholders. Therefore, the broader aim of this research is to provide a holistic kerbside parking assessment with a spatial view of parking distribution and an analysis/forecast of expected changes, taking into consideration current parking demand, non-homogeneous distribution of parking loads, and overall utilisation as the basis for investigation of expected future changes and developing travel demand management strategies for all stakeholders. The key research question investigated in this paper is: What is the current kerbside demand in terms of parking distribution and utilisation, and how can potential changes to kerbside parking demand be modelled as a basis for evaluating the impact on kerbside parking, taking into consideration of all possible supply and demand changes in the future? To answer this research question and achieve the main purpose of this study, the following aims/objectives are set:

- Investigate current kerbside parking capacity, demand, and utilisation patterns, including a spatial view of parking and behaviour;
- Model current kerbside parking demand and capacity using simulation modelling as a basis for modelling kerbside impact scenarios, taking into consideration the non-homogeneous distribution of parking loads, future demand increases, and changes to kerbside capacity due to infrastructure development;
- Illustrate, forecast and analyse expected future kerbside parking demand due to population and construction growth, as well as capacity reduction due to infrastructure development, as the basis for investigating various kerbside impact scenarios to be carried out in the next stage of the research;

Because the proposed light rail infrastructure and expected increased demand due to rapid growth in construction activities are expected to directly impact parking spaces in the Parramatta CBD, this study focuses on kerbside parking within Parramatta CBD. Thus, the scope of the research includes:

- Kerbside parking spaces for short-term parking (STP) and freight activity space (FAS) in the Parramatta CBD, as defined by the boundaries of the area of about 0.78 km^2 (i.e., 0.91 km $\times$ 0.86 km) (Figure 2);
- Video data collection of kerbside parking events at selected locations as a representation of Parramatta CBD over one week;

- Manual data collection of parking events using a survey at selected locations in Parramatta CBD during peak periods over two days. Each parking event is recorded with time in/out, type of vehicle/load, and the type of parking space the vehicle occupies;
- Analysis of manually captured data to verify the camera dataset first, followed by setting of input parameters for a baseline traffic/parking simulation model using the camera dataset.

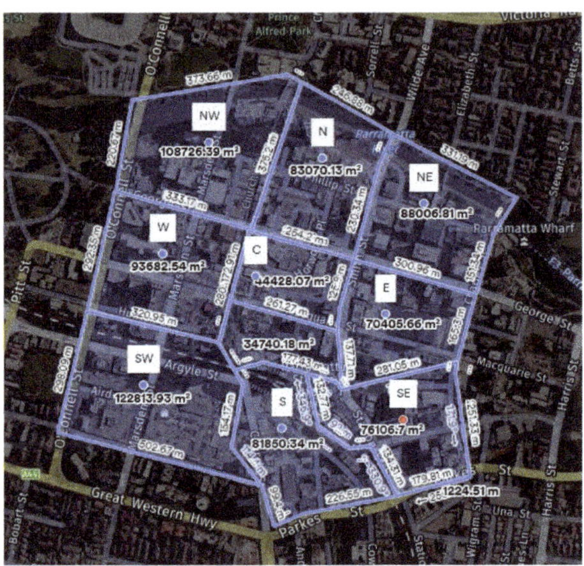

Figure 2. Area of Parramatta CBD (2021) kerbside parking spaces for assessment.

This research was carried out using a combination of experimental and simulation modelling approaches based on real manually collected and camera data. Details of the research methodology are presented next.

3. Research Methodology

In this study, we adopted a mixed-methods approach of (i) analysis of parking in terms of load distribution; (ii) a spatial view of parking in five regions/areas, including the centre of the CBD; and (iii) simulation modelling of the current parking scenario for prioritisation of the allocation of limited kerbside spaces in a rapidly changing city environment due to increasing population growth and rapid infrastructure development. The research methodology involves three stages: (i) analysis of current kerbside parking demand and a spatial view of parking loads using video camera data; (ii) simulation modelling of current parking demand, taking into consideration expected searching for parking; and (iii) estimation of key model input variables using expected population and construction growth as a benchmarking for modelling and simulation of four future kerbside impact scenarios. The research was carried out by assessing current kerbside parking demand/usage using real data, simulation modelling of current kerbside parking demand/usage, and illustration using a case of an urban centre/city in Sydney, Australia. The benchmarking of current parking demand using simulation modelling forms a basis for the development of strategies and prioritisation of the allocation of limited kerbside spaces in a rapidly changing city environment. The first stage of parking assessment involves data collection and associated analysis. This is followed by traffic/parking simulation modelling of the current demand/usage (baseline) scenario using input parameters from a comprehensive kerbside parking dataset. The baseline (benchmarking) traffic/parking simulation model using key parameters forms a basis for investigating the impact of future increased demand, changes to parking capacity, and infrastructure development Expected population growth and construction growth forecast are evaluated as key model inputs for the development of four kerbside impact scenarios for the next stage of this research project.

3.1. Data Collection

Kerbside parking data collection was carried out using a combination of manual survey data collection over two working days and video-captured camera data over one week at selected parking spaces in respective parking segments. Parameters of kerbside data collection are outlined in Table 1.

Table 1. Parameters of kerbside data collection in Parramatta CBD.

Data Collection Method	Number of Parking Spaces Used		Dates and Times of Data Collection	Parking Time Slots
	STP: 5M, 1/4P, 1/2P, 1P & 2P	**FAS:** LZ, TZ, DR		
Manual kerbside survey (8 segments)	26	20 (15 LZ, 3 TZ and 2 DR)	19–20 October 2017; 7 a.m.–4 p.m.	PT6 = before 0700, PT7 = 0700–0800, PT8 = 0800–0900, PT9 = 0900–1000, PT10 = 1000–1100, PT11 = 1100–1200, PT12 = 1200–1300, PT13 = 1300–1400, PT14 = 1400–1500, PT15 = 1500–1600
Camera dataset using video capture (19 segments)	84	28 (All LZ)	16–20 October 2017; 24 h	24 h; Hour of day: 0 (12 a.m. to 12:59 a.m.) to 23 (11 p.m. to 11:59 p.m.)

Total kerbside parking spaces in Parramatta CBD (Figure 2) include 501 STP and 46 FAS spaces. STP and FAS are the two main parking zone types; STP consists of five different types (5M (5 min), 1/4P (1/4 h), $\frac{1}{2}$ P (half an hour), 1P (1 h), and 2P (2 h)), and FAS consists of three categories (loading zones (LZ),truck zones (TZ), and driveways (DR)). LZ spaces have a maximum dwell time of 30 min. STP can be used by FAS vehicles. LZ spaces can legitimately be used by any vehicle for the purpose of pick-up and drop-off. Manual data collection was carried out for all parking types subject to different usage depending on the time (e.g., LZ spaces have a maximum dwell time of 30 min (parking time limit), and STP can also be used by FAS vehicles), whereas camera data capture was limited to STP and LZ parking zones.

3.2. Data Collection Methods

Manual kerbside data collection was carried out at eight predetermined segments within eight subsections of the CBD; each parking event was recorded in a spreadsheet, identifying key attributes such as bay number, parking zone at the start of parking, arrival time, vehicle type, load type, and departure time. Because data collection methods involve different locations and numbers of parking spaces, details of kerbside segments available and used for both types of data collection are outlined in Table 1. To overcome the limitations of manual kerbside data and the limited number of parking spaces surveyed, camera data collection was carried using video capture of parking events for selected parking segments over five workdays (16–20 October 2017).

The camera (video-captured) dataset was carefully checked for any missing data, overlapping data, and outliers (e.g., parking events with more than 24 h parking durations) and was cleaned before analysis for kerbside hourly arrival rates and parking duration of parking events for each hour over the selected period. The refined kerbside parking demand dataset was used as the main kerbside demand input for the baseline simulation model. The refined kerbside dataset for the assessment and baseline model is based on measures and parameters outlined in Table 1.

The main assumptions and limitations of kerbside parking using camera data include:

- Removing outliers, duplicate data, and events with overlapped parking times in the same parking space of camera data are assumed to be a good representation of all parking events [50];
- Limitations of kerbside camera data include (i) data captured in a small scope (61% of LZs and only 17% of STP, as most STP is in locations unlikely to be used for freight and servicing activity) of parking spaces (84 STP and 28 LZ) compared to the full scope of Parramatta CBD kerbside parking (501 STP and 46 LZ), (ii) the fact that FAS parking spaces are limited to the LZ category only (i.e., no data capture for TZ and DR parking spaces), and (ii) removal of some records due to missing data.

3.3. *Modelling and Simulation of Kerbside Baseline (Current) Scenario*

Because manual data are limited (only two days from 7 a.m. to 4 p.m.), the kerbside parking demand data captured using video cameras over 5 days were used. Hence, model inputs on arrival rates and parking durations for kerbside modelling were obtained from the camera dataset outlined in Table 1. The parking duration was modelled by evaluating all available weekday kerbside parking data according to the parking duration distribution for parking events in each hour. The resulting parking distributions determined and reported in [50] were used in the simulation of the baseline scenario.

Because kerbside video data collection had some gaps due to likely video recording errors, a "typical day" parking scenario was constructed by considering the highest hourly parking arrivals. Kerbside usage during the week (16–20 October 2017) and usage of a typical day are shown in Appendix A (Table A1). Using Flexsim [52], a simulation model was first constructed using the typical day scenario, which consisted of 31 STP and 8 LZ parking spaces. The typical day scenario was then verified against the expected parking usage and arrivals to verify the kerbside parking model and the procedure.

The typical day scenario was then extended to the baseline scenario simulation model by appropriately evaluating hourly arrival rates to reflect the baseline (current) kerbside scenario. The baseline scenario kerbside model takes into consideration a kerbside capacity of 501 STP and 46 LZ parking spaces. Details of data analysis, results, and findings are presented in the following section and categorized into the key areas of analysis of an intercept survey, kerbside and carpark data, and analysis/results of traffic/parking simulation models.

4. Results and Analysis

Kerbside parking assessment is presented in five stages: (i) assessment of manual kerbside data, (ii) assessment of camera data, (iii) spatial analysis of parking demands and loads, (iv) baseline modelling of current kerbside demand, and (v) evaluation of future impact scenarios as the basis for simulation modelling of kerbside impact scenarios.

4.1. *Analysis of Manual Kerbside Data*

The first stage of data collection involved a small-scale kerbside parking survey to verify the validity of camera data. The purpose of validation is to ensure no significant discrepancy between manual data and camera data before incorporating the camera data into simulation modelling. The preliminary analysis found that camera data closely aligns with the manual dataset. Therefore, the camera dataset is valid and reliable for use in the development of the traffic/parking baseline scenario simulation model.

Key analysis of manual data collected at several kerbside parking sections of Parramatta CBD over two days (19–20 October 2017) includes descriptive statistics of key measures (number of parking events and length of stay) and analysis of those measures using key variables/parameters, including (i) time slots/zones and (ii) parking zones/categories. Key analyses are illustrated using column/bar charts for visualization of peak/valley points and trends.

Based on the analysis of parking events in three parking time zones (morning: before 0700 to 1200; mid-day: 1200 to 1400; and afternoon: 1400 to 1600), the distribution of parking events in three categories of parking zones (STP, FAS, and other) (Figure 3) shows a fairly uniform distribution. For example, STP parking events are distributed in proportion of 44%, 42%, and 46% over the three parking time zones (morning, mid-day, and afternoon, respectively), whereas FAS parking events are distributed in proportions of 44%, 46%, and 45% over the three parking time zones. Furthermore, both distributions of STP and FAS parking events show similar percentages of parking events (44% and 45%, respectively), whereas the percentage of parking events in other parking zones show a very low percentage of 11% of the parking load during the considered period.

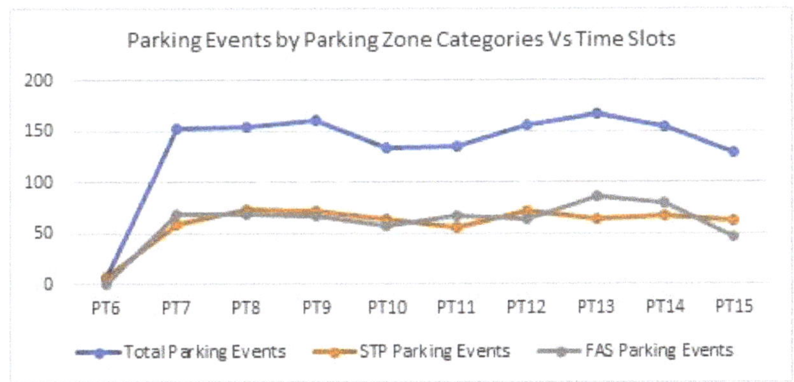

Figure 3. Number of parking events by parking zone category vs. time slots.

The above results show that time zone 2 (mid-day) is the busiest period across both parking zone categories (STP and FAS) when the number of parking events per parking zone is considered. For example, the average (mean) number of parking events per parking zone in STP parking zones is 55, 68, and 65 for time zones 1, 2, and 3, respectively, with the highest average number of parking events recorded for time zone 2. Similarly, the average number of parking events during time zone 2 (Mid-day) is 75 for FAS. This means that the demand for parking during mid-day is at its peak across both categories of parking zones.

Table 2 presents descriptive statistics of the length of stay, as well as a detailed profile of each time slot, using the number of parking events. Overall, the average length of stay of parking events over individual time slots (PT6 to PT15) is within a range of 15 min to 38 min, suggesting a fairly uniform distribution over the considered period, including a peak of an average of 38 min during the unrestricted period (before 0700) and a low average of 15 min during the last period/slot (1500–1600). Figure 4 shows a boxplot of the length of stay by time slot, and Figure 5 shows a 95% confidence interval of mean by time slot, indicating potential implications concerning the differences in length of stay across those time slots. It is interesting to note that between 0700 and 1400, the 95% confidence is tightly contained around the 30-minute mark, which is very close to the average length of stay (27 min). Furthermore, an Anderson–Darling normality test is presented to confirm the distribution of each category, leading to a better assessment of the length of stay of the studied sample and supporting these conclusions. It is evident from the results that none of the distributions of individual lengths of stay are normally distributed except for PT6 (before 0700).

Table 2. Descriptive statistics of length of stay (minutes) of parking events during two days (19–20 October 2017).

Parking Time Slot	Number (N)	Mean (Minutes)	Median (Minutes)	Standard Deviation (Minutes)	1st Quartile	3rd Quartile	95% Confidence Interval for the Population Mean (Lower; Upper)	Anderson–Darling (A-Squared Normality Test)	Skewness	Kurtosis
PT6	7	38.57	35.00	25.59	15.00	62.00	14.90; 62.24	0.39	0.31	−1.92
PT7	153	33.00	12.00	67.62	5.00	33.00	22.21; 43.81	26.11	4.49	22.95
PT8	155	31.40	7.00	71.75	3.00	25.00	21.02; 42.78	30.92	4.17	19.04
PT9	159	35.66	13.00	56.95	4.00	38.00	26.74; 44.58	19.35	3.38	15.91
PT10	133	23.31	10.00	32.72	5.00	27.00	17.70; 29.92	13.80	3.15	12.90
PT11	135	27.78	15.00	42.14	5.00	31.00	20.61; 34.95	15.17	3.77	17.78
PT12	156	30.20	14.00	41.44	5.25	37.75	23.65; 36.75	14.73	2.88	10.28
PT13	167	26.90	14.00	35.25	5.00	32	21.52; 32.29	17.06	2.13	4.14
PT14	154	19.15	12.50	21.34	3.75	30.0	15.75; 22.55	9.67	1.69	2.88
PT15	129	15.30	11.00	13.89	5.00	22.00	12.88;17.72	5.26	1.18	0.80
All	1348	27.32	12.00	47.21	5.00	30.75	45.49; 49.06	173.62	4.76	31.15

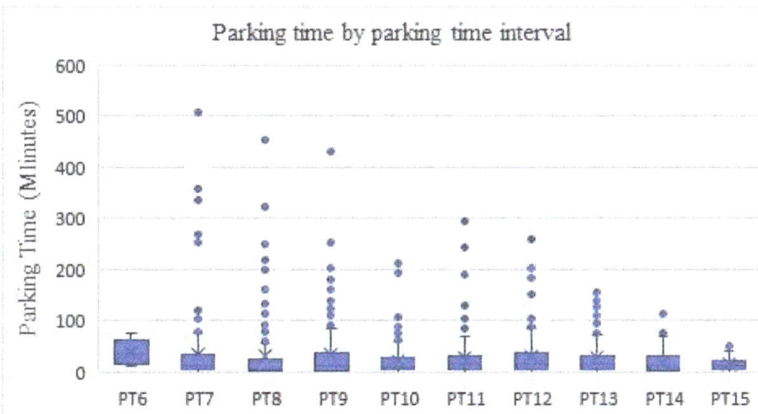

Figure 4. Boxplot of parking time (length of stay) of all parking events on FAS and STP spaces by parking time slot.

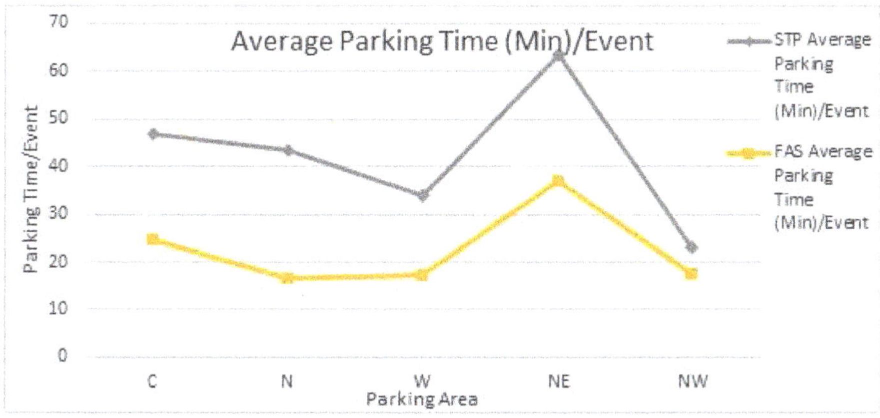

Figure 5. Average parking time in STP and LZ spaces across five areas/regions.

Although some parking events span two or more periods, in particular freight and services (FAS)-related parking events, the average length of stay of parking events in each time slot is less than 60 min (duration of the time slot). However, the range of length of stay (indicated by standard deviation) is significantly large across several parking time slots, including 72 min for time slot PT8 (0800–0900). Analysis of manual data includes a comprehensive analysis of the following aspects:

- Define the peak time of day activity and the average length of stay for parking events in short-term and loading zone bays;
- Define the use of short-term parking by the freight and servicing vehicle classifications;
- Examine the use of "freight activity spaces" by other types of vehicles;
- Define the current state minimum short-term or loading zone bays required to meet service demands.

In summary, the following results are noted from the comprehensive analysis of kerbside assessment using a manual dataset. The peak time of day activity across two categories of parking zones (STP and FAS) is the mid-day (1200 to 1400) period. The average length of stay for parking events is 27 min and within an average range of 15 min to 38 min, with a fairly uniform distribution over the studied period (0700–1600).

The analysis shows that there is a considerable amount of FAS loads/tasks (11% of all parking events) performed in STP zones. On the other hand, more than half of all parking events in STP (i.e., 52%) are associated with FAS loads/tasks. Additionally, passenger vehicles not only use STP zones but also FAS parking zones. This indicates that parking usage by various users is not fully matched with the parking zone categories.

A considerable (if not significant) difference was found between the average length of stay in FAS (23 min) and the average length of stay in STP zones (13 min). This emphasises the need for further research to determine whether the difference is significant, as well as practical implications for prioritising parking allocation in the future.

Kerbside assessment using analysis of current capacity and actual demand shows that there is spare capacity for short-term parking zones. In this case, the spare capacity of time is around 27% of total short-term parking time (i.e., only around 74% of the total short-term parking time available during the considered period was used). Because full capacity utilisation is assumed to be around 90% of total capacity, taking into consideration lost time between events and a large number of events with a short length of stay [18], the current spare capacity is only around 17%.

It can be noted from the above analysis of current kerbside parking that overall parking assessment is limited by (i) the scope of the dataset (parking events over two days of the week from 7 a.m. to 4 p.m.) and (ii) analysis using two key variables/parameters: (a) time slots/zones and (b) parking zones/categories. In order to use camera data for the development of a baseline traffic/simulation model, kerbside camera data were validated using a representative sample of data, details of which are presented in Appendix A (Table A1).

The kerbside usage shown based on the validation of camera data (Appendix A) is closely aligned with the results of the manual data analysis reported earlier. By considering 90% of utilisation as full-capacity usage, overall, kerbside STP parking reaches full capacity over three time periods (8:30 a.m.–9:30 a.m., 10:30 a.m.–3:30 p.m. and 7 p.m.–8 p.m.). Kerbside usage of around 77% between 10 a.m. and 11 a.m. is somewhat lower than expected. This could be due to (i) a small sample size of 39 spaces compared to a total of 547 kerbside (510 STP and 46 LZ) spaces, (iii) usage based on a simple method of hourly arrivals and departures, and (iii) possible errors when the recording camera data from video capture. Therefore, for simulation modelling of baseline kerbside parking (current demand and capacity) as the basis for the testing of kerbside impact scenarios, we adopted kerbside camera data gathered in 2017. In addition, key inputs of the model are hourly arrival rates and parking duration of hourly events. In this case, parking duration distributions of each hour based on a synchronised camera dataset are determined by best-fit distribution functions within simulation software (Flexsim) based on the approach adopted for distribution of kerbside parking manual data [50].

The kerbside usage/behaviour considered in this analysis is mainly focused on benchmarking current demand/requirements and supply under given economic conditions, such as parking limits and pricing schemes set by the local council. Because the parking demand for STP is at full capacity at three different time intervals and there could be unnoticed demand during these times, it is suggested that there is a need for further investigation into kerbside parking demand as part of promoting car-free city principles and providing more space for sustainable mobility in future studies.

4.2. Spatial View of Parking Spaces and Utilisation

Because the parking assessment is based on parking data (spaces and loads) across a relatively small area (0.78 km^2) but with a diverse range of parking spaces/limits, it is expected that parking utilisation is not only non-uniform but also non-homogeneously distributed. To understand this non-homogeneous distribution of parking utilisation as a basis for the development of guidelines for changes to parking infrastructure, the representative dataset of manual data collected over two days was further analysed in terms of average parking times in both STP and FAS spaces, as well as distance from the centre of the CBD. Details of parking utilisation from the perspective of distance from the centre of the CBD are presented in Table 3.

Table 3. Spatial view of kerbside parking using camera data.

Area	Total (FAS&STP) Spaces	Distance from Carpark (m)	Collected STP Parking Space Data	Collected FAS (Only LZ) Parking Space Data	STP Average Parking Time (Min)/Space	STP Average Parking Time (Min)/Event	FAS Average Parking Time (Min)/Space	FAS Average Parking Time (Min)/Event
NW	75	342.12	10	5	321.30	23.12	300.6	17.68
N	174	211.70	24	8	631.67	43.44	326.875	16.55
NE	116	325.75	6	2	740.33	63.46	631	37.12
W	86	294.97	18	3	666.72	33.90	366.33	17.17
C	25	0.00	9	6	474.89	46.97	493.5	24.88

As shown in Table 3, there is a non-homogeneous distribution of parking in the CBD in terms of parking limits and times across two major groups of parking spaces (STP and LZ of FAS). In this case, the centre section of the CBD has (i) the highest average FAS parking time per space and (ii) the second highest parking time per event for both FAS and STP parking. Although the northern section is the closest to the centre section of the CBD, it has the lowest FAS parking time per parking event and space. Furthermore, when the average parking time per space is compared across the range of several areas, the average STP parking time/event ranges from 23.12 min in the NW section to 63.46 min (Figure 4). Similarly, the highest average parking time of 37.12 min in FAS spaces (LZ) is also associated with the NE area/region.

As shown in Figure 5, it is interesting to note that both STP and FAS average parking times in the NW and N regions are shorter than those in the NE region. Based on this parking behaviour, increasing parking spaces and/or increasing limits on existing spaces in the northern section first, followed by the western section, could alleviate the current parking demand for the centre and thereby reduce congestion in the CBD.

Because the metropolitan city associated with this study is changing very rapidly with several infrastructure development projects and the planned demolition of two out of three multistorey carparks, parking assessment requires incorporation of these expected changes if the assessment is to be current and used to plan for future needs. To model various kerbside impact scenarios influenced and generated by these developments to plan for future kerbside demand, an analysis of kerbside parking demand using the baseline simulation model is presented next. Analysis of simulation modelling of impact scenarios is not presented here because it is beyond the scope of this research paper.

4.3. Analysis of Kerbside Parking Demand Using the Baseline Simulation Model

The baseline model captures the current (2019) kerbside parking demand in Parramatta CBD. Kerbside parking demand is modelled using model inputs and parameters including parking capacity, hourly arrival rates, and parking usage/dwell times of parking events derived from the camera dataset.

The traffic/parking baseline scenario simulation model is developed as a discrete-event simulation using hourly parking duration distributions, hourly parking arrival rates, and parking capacities as the main inputs. The kerbside parking logic of the simulation modelling of kerbside parking assumes two distinct routes of parking upon arrival and searching for parking for two minutes and leaving if a parking space is not available. The parking logic model with key events and associated paths is shown in Figure 6.

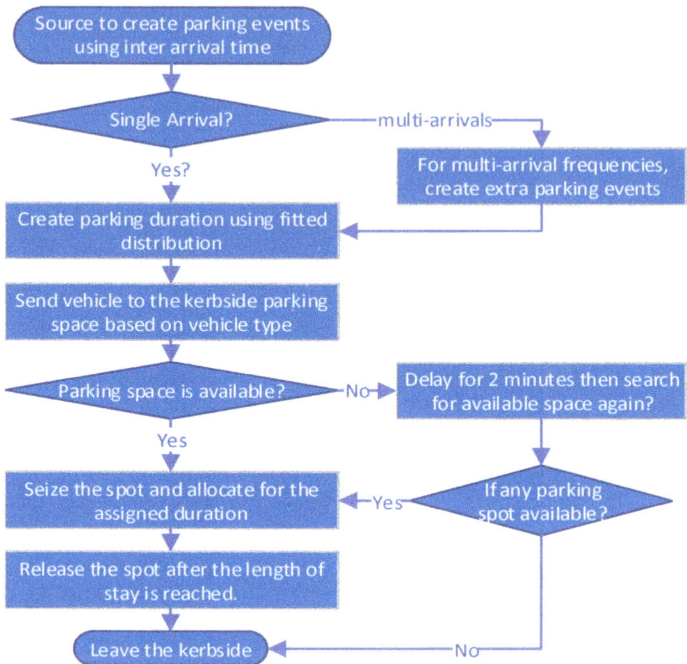

Figure 6. Kerbside parking logic model (Source: [50]).

The simulation modelling platform used is Flexsim [52], and the analysis of results was performed using MATLAB [53] and Microsoft Excel. The kerbside parking events and percentage of parking events reproduced by the baseline simulation model are shown in Figures 7 and 8, respectively.

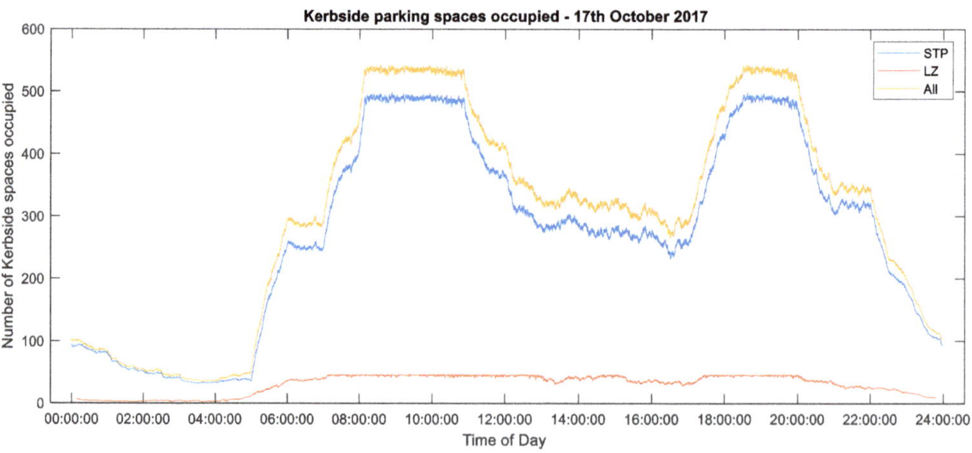

Figure 7. Kerbside parking events using the baseline simulation model (17 October 2017).

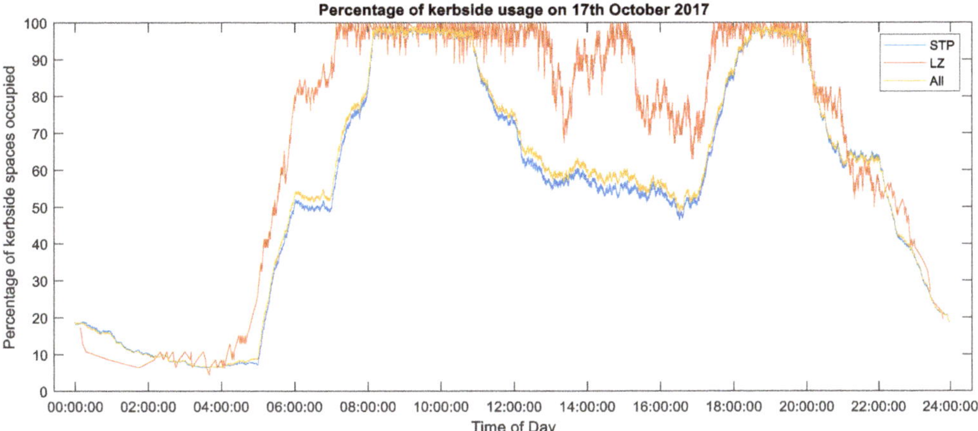

Figure 8. Percentage of kerbside parking events using the baseline model (17 October 2017).

As shown in Figures 7 and 8, kerbside parking reaches full capacity at two-time intervals (from around 8:08 a.m. to 10:54 a.m. and from 6:00 p.m. to 8:00 p.m. Because kerbside parking is expected to reach full usage over a considerable period during the peak period (7 a.m. to 4 p.m.) rather than spikes (just small peaks), the method used in the simulation modelling is more accurate. The baseline model was tested using data collected in 2019 to simulate impact scenarios (e.g., population growth and changes to parking infrastructure) and was found to be a reliable modelling approach to test current and expected changes to parking demand and supply [50].

Because the actual kerbside usage is closely reproduced by the baseline scenario simulation model using hourly arrivals and best-fit distributions of parking duration, with some variations (e.g., possible underestimation of usage) between parking usage using data analysis and simulation modelling, typical weekday usage is assumed, using the maximum usage of each hour. Kerbside usage during the week (16–20 October 2017) and usage of a typical day are shown in Appendix A. Because kerbside parking demand is closely reproduced by the baseline model using weekly data including a typical day, the baseline model can reliably be used to model future kerbside scenarios, incorporating various kerbside impact situations depending on assumed demand and capacity changes [50].

4.4. Analysis of Growth of Kerbside Parking Demand and Future Changes to Parking Capacity

Because expected changes are critical for the development of travel demand management strategies, all the expected changes are investigated as the basis for modelling future kerbside impact scenarios.

Key areas of changes to kerbside parking capacity and requirements are outlined in Section 2.4. Because expected future changes to capacity and requirements are critical inputs for the development of traffic/parking simulation modelling and the development of travel demand management strategies for stakeholders, a preliminary analysis of those changes is presented here. In this case, changes include increased demand due to population and construction growth and reduction in kerbside capacity due to light rail infrastructure development. The forecasted population growth in Parramatta CBD is shown in Table 4. The year-to-year demand growth for kerbside parking is projected using the population growth rate.

Table 4. Predicted population growth for Parramatta.

Year	2019	2020	2021	2022	2023	2024	2025
Population	260,130	266,763	273,565	280,541	287,695	295,031	302,555
Cumulative Growth	0%	2.55%	5.16%	7.85%	10.60%	13.42%	16.31%
Year-to-Year Demand Growth		2.55%	2.55%	2.55%	2.55%	2.55%	2.55%

The increase in freight and servicing activity is assumed to be 10–15% annually from 2019 to 2025, which is based on the actual growth in Sydney CBD from 2017 to 2019 [4]. Therefore, a 12.5% increase in demand due to freight and serving is assumed for each year. Although this assumed increase in demand could be adjusted for COVID, no adjustments were made to the future parking demand in the modelling/simulation, considering that construction largely continued through COVID.

Year-to-year parking demand growth due to construction worker growth for 2020–2025 is based on the information provided by Transport for NSW supported by the following guidelines and forecast growth forecast using historical data as reported in [50] and outlined below.

i. It is assumed that the current estimation of construction workers at one of the major infrastructure developments in Parramatta CBD labelled "Parramatta Square" (PSQ) using a ratio of workers to the construction area is reliable for estimating future construction worker demand growth, using development application (DA) approvals for 2020–2025;
ii. It is assumed that the construction worker forecast for 2018–2020 can reliably be estimated using DA approvals during the corresponding period;
iii. The construction worker forecast for 2021–2025 is based on exponential smoothing of historical Parramatta construction activities during the period of 2010–2018 and estimated values of construction workers during the period of 2019–2020 (available from Transport for NSW).

Based on the above assumptions, the construction worker forecast was developed for the period 2020–2025 and is shown in Table 5 and Figure 9. The developed year-to-year growth rate was used to predict the number of construction workers who may use the carpark if available.

Apart from changes to parking demand, it is expected that existing kerbside parking spaces will be reduced from 501 to 436 for STP and from 46 to 44 for LZ in 2020. Four kerbside impact scenarios taking into consideration all of the above changes of kerbside capacity and demand over the next 6-year period are difficult to comprehend without visualization of those impacts using a time scale for better understanding of key points (e.g., peak period(s), percentage of utilisation, and the transition to full usage with changes). Therefore, the impacts outlined above are modelled in the next stage of the research project based on the baseline model, the modelling details of which are not presented here because they are beyond the scope of this paper.

Table 5. Forecasted construction worker growth.

Year	2019	2020	2021	2022	2023	2024	2025
Construction Workers (Forecasted)	6690	5890	6135	6304	6452	6749	7037
Year-to-Year Growth	N/A	−11.96%	4.17%	2.75%	2.33%	4.61%	4.27%

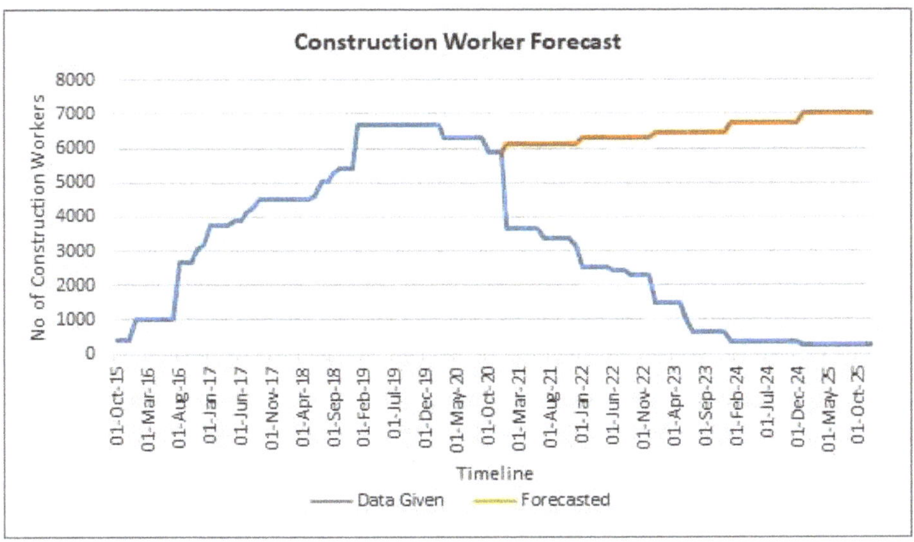

Figure 9. Construction worker forecast for 2021–2025. (Source: [50]).

5. Research Findings

5.1. Stage 1—Kerbside Assessment

The assessment is focused on providing evidence-based information on the current state of demand for kerbside parking for various activities for the planning of parking allocation in the future.

Analysis (estimation) of the capacity of parking zones (measured by both time availability and the allowable number of parking events) and evaluation of the actual number of events and length of stays during the consider period show that there is a spare capacity for short-term parking zones. In this case, the spare time capacity is around 27% of total short-term parking time (i.e., only around 74% of the total short-term parking time available during the considered period was used) based on the data for 2017. However, this spare capacity is expected to be reduced, given an expected increase in demand and changes to parking capacity, as evidenced by the testing of impact scenarios [50].

5.2. Stage 2: Results of the Baseline Simulation Model Replicating the Refined Camera Data

It was found that the baseline simulation model using camera data provides a reliable representation of current demand over the entire kerbside scope using a selected set of kerbside spaces based on camera data. Kerbside impact can easily be modelled using parking usage duration and hourly arrival distributions determined from the camera dataset.

5.3. Stage 3: Results of Future Kerbside Capacity and Demand Changes

The predicted growth of kerbside demand due to infrastructure development to support more employment and residents shows that overall parking behaviour will be significantly different, resulting in daily temporal patterns, patterns of freight-related kerbside utilisation, and changing peak/saturation periods. These expected demand patterns could impact business and the community broadly due to limited parking access, leading to loose demand among several users. Furthermore, COVID interruption exacerbate parking demands due office workers preferring to drive a private car rather than use public transport. Kerbside impacts over the next six years are expected be widespread in terms of overall kerbside utilisation, peak periods, and saturation periods. This suggests that kerbside impact scenarios influenced/generated by such changes need to be investigated from the point of view of planning for future travel demand management strategies within a broader urban city infrastructure development planning framework. Therefore, the baseline model can be used as the basis to simulate all these impact scenarios by incorporating the growth and expected changes using key measures.

6. Findings, Discussion, and Conclusions

Kerbside parking assessment using a comprehensive camera dataset provides an opportunity for a better understanding of current parking behaviour in terms of peak demand and distribution of parking utilisation in different parking categories on weekdays. Results show that current kerbside parking capacity can meet the overall captured current demand with different utilisation of demand in a range of capacities of FAS and STP categories. However, the captured demand does not include the potential demand (e.g., searching for parking) not captured during full-capacity utilisation (e.g., peak periods) and potential demand increase due to the increasing construction workforce. Furthermore, peak periods, the profile of capacity utilisation by various loads and kerbside categories, and saturation periods are identified as key indicators of kerbside parking assessment for both the baseline and impact scenarios, which transport infrastructure planners can use to develop policies and travel demand management strategies. Peak periods, segmenting of demand, capacity utilisation profile, and saturation periods are critical measures from the perspective of economic analysis of parking assessment towards the development of policies, as reported in similar studies [3,7,20].

Modelling and simulation of current demand/utilisation using a comprehensive camera dataset were found to be a reliable approach for modelling and simulation of future impact scenarios, in particular kerbside impact from the perspective of saturation times over a 6-year period, taking into consideration all the factors affecting kerbside parking capacity and utilisation. Furthermore, kerbside parking assessment can be used to develop recommendations for travel demand management strategies, providing guidelines for several stakeholders in the metropolitan city considered in this research.

Modelling and simulation of kerbside parking demand and supply under dynamic conditions is a very effective approach for urban parking evaluation and policy assessment, as evidenced in some recent studies, including estimating the effectiveness of planned parking facilities under different development scenarios [20] and predicting parking utilisation/behaviour influenced by several factors, such as parking price and needs (e.g., desire to park for a long time) [24]. Furthermore, full-capacity utilisation at three time periods (accounting for 7 h between 8 a.m. and 8 p.m.) can have a direct impact on increased traffic volumes at peak times, partly caused by cruising for parking. Therefore, cruising for parking to evaluate unnoticed demand should be incorporated as part of future research studies to make parking assessment more realistic and consider all sustainability aspects. It is evident from previous research studies that cruising for parking contributes to other adverse outcomes, such as traffic congestion, increased air pollution, and accidents [24].

Overall, this research study provides evidence-based information on the current state of demand for kerbside parking, highlighting varying utilisation of demand in a range of capacities. This emphasises the need for more information campaigns on parking priorities and alternative options for parking (e.g., off-street parking around the CBD) as an integral part of planning for future infrastructure projects in Parramatta CBD.

Although the current kerbside capacity meets the overall captured demand, excluding potential demand above the full capacity and any demand increase due to the construction workforce, further analysis is required to determine whether utilisation is within the allocated time limits by testing the significance of the difference between the actual average length of stay and allocated time using an appropriate statistical test(s) (e.g., ANOVA or t-test). It is expected that a statistical test of the difference between the recorded average length of stay and allocated times of respective short-term parking will show a significant difference between these two sets of data. Similarly, differences between the actual length of stay and allocated times of other short-term parking zones can also be carried out using statistical testing as required.

In this research study, we identified potential improvements to the methodology, including improving candidate selection process (for manual data collection) using appropriate selection criteria, training of selected candidates by incorporating on-site training/pilot data collection, and monitoring of the data collection process with additional resources and quality assurance methods.

Future research directions include an extension of the baseline simulation model with future impact scenarios, taking into consideration future expected changes to parking infrastructure and demand growth, particularly demand impacted by the planned construction activities in Parramatta CBD and exacerbated parking demands due to people preferring to drive a private car rather than use public transport post COVID. Furthermore, parking assessment needs to consider not only the parking demand and supply but also the necessity to "plan for car-free cities", making space for people and vehicles that really need to be there (e.g., freight activity). However, achieving economically sound solutions to minimise the gap between demand and supply would be challenging while aiming for sustainable mobility principles with social benefits. Furthermore, in the examination of FAS, it can be determined that the substitution factor is lower and hence demand is less realistic. The current investigation is limited to lower substitution factors; therefore, future studies should consider different elasticities and substitutable moves, particularly when considering sustainable mobility principles. In this context, it is critical that modelling/simulation of impact scenarios incorporate all the key indicators, including economic factors and social conditions, and evaluate their interdependencies as a basis for development of travel demand management strategies and guidelines for broader stakeholders. However, the potential impact on parking demand due to COVID-19 interruption, particularly people preferring to drive rather than use public transport, is not incorporated into the current impact scenarios. Therefore, future work needs to consider these factors when modelling impact scenarios. Modelling and simulation of future impact scenarios can be used to provide guidelines for the development of travel demand management strategies to manage the rapidly changing and transforming infrastructure and demand landscape in Parramatta CBD, as influenced by several factors. Furthermore, simulation modelling of impact scenarios can be used to guide policymakers with the required information to address environmental and economic impacts due to vehicles searching for parking [38].

Author Contributions: Conceptualization, P.S. and U.G.; methodology, P.S., U.G. and M.S.; validation, U.G.; formal analysis, P.S. and U.G.; investigation, P.S., U.G. and M.S.; resources, P.S., U.G. and M.S.; data curation, P.S. and U.G.; writing—original draft preparation, P.S.; writing—review and editing, P.S., U.G. and M.S.; visualization, U.G.; project administration, P.S.; funding acquisition, P.S. and U.G. All authors have read and agreed to the published version of the manuscript.

Funding: This research was funded by Transport for New South Wales (NSW), Australia. Funding body: Transport for NSW—20551-57889. The funding Reference: P00024581.

Institutional Review Board Statement: Not applicable.

Informed Consent Statement: Not applicable.

Data Availability Statement: Data used for the analyses are available upon request.

Conflicts of Interest: The authors declare no conflict of interest.

Appendix A

Table A1. Kerbside usage during the week (16–20 October 2017) and usage of a typical day (a sample of parking events in 39 parking spaces (31 STP and 8 LZ)).

Hour of Day	STP							LZ						
	16th	17th	18th	19th	20th	Typical Day	Max Capacity	16th	17th	18th	19th	20th	Typical Day	Max Capacity
0	1	2	2	4	1	4	31	0	1	0	0	0	1	8
1	0	2	2	3	1	3	31	0	0	0	0	0	0	8
2	0	3	2	1	0	3	31	0	0	0	0	0	0	8
3	1	2	3	0	1	3	31	0	1	1	0	2	2	8
4	2	3	2	2	3	3	31	0	1	0	0	0	1	8
5	13	19	14	14	12	19	31	2	6	3	3	4	6	8
6	22	22	15	15	21	22	31	3	4	1	3	4	4	8
7	14	17	16	12	15	17	31	2	1	3	3	1	3	8
8	20	23	24	17	22	24	31	4	4	2	3	0	4	8
9	25	30	29	20	29	30	31	6	3	5	4	4	6	8
10	22	22	30	22	30	30	31	6	6	5	5	5	6	8
11	25	28	27	21	25	28	31	6	3	3	6	6	6	8
12	21	31	27	22	31	31	31	5	4	4	4	4	5	8
13	24	28	28	17	29	29	31	7	2	3	2	6	7	8
14	31	27	26	19	26	31	31	6	6	4	5	5	6	8
15	25	23	22	20	28	28	31	4	4	2	0	5	5	8
16	22	18	18	18	16	22	31	6	2	3	5	4	6	8
17	22	24	11	14	18	24	31	5	4	3	3	3	5	8
18	29	22	25	10	25	29	31	7	4	4	3	5	7	8
19	28	29	25	11	27	29	31	7	5	4	3	4	7	8
20	27	20	22	9	28	28	31	5	5	2	3	3	5	8
21	18	14	13	7	24	24	31	3	2	3	1	1	3	8
22	10	8	5	3	10	10	31	1	3	2	2	1	3	8
23	6	0	6	1	8	8	31	1	0	0	0	0	1	8

References

1. Wang, R.; Yuan, Q. Parking Practices and Policies under Rapid Motorization: The Case of China. *Transp. Policy* **2013**, *30*, 109–116. [CrossRef]
2. Chester, M.; Fraser, A.; Matute, J.; Flower, C.; Pendyala, R. Parking Infrastructure: A Constraint on or Opportunity for Urban Redevelopment? A Study of Los Angeles County Parking Supply and Growth. *J. Am. Plann. Assoc.* **2015**, *81*, 268–286. [CrossRef]
3. Arnott, R.; Inci, E.; Rowse, J. Downtown Curbside Parking Capacity. *J. Urban Econ.* **2015**, *86*, 83–97. [CrossRef]
4. Samaranayake, P. *Kerbside Assessment: Freight and Servicing Activity in Parramatta CBD*; Western Sydney University (Internal): Penrith, Australia, 2018.
5. Piccioni, C.; Valtorta, M.; Musso, A. Investigating Effectiveness of On-Street Parking Pricing Schemes in Urban Areas: An Empirical Study in Rome. *Transp. Policy* **2019**, *80*, 136–147. [CrossRef]
6. Su, W.; Chow, M.-Y. Computational Intelligence-Based Energy Management for a Large-Scale PHEV/PEV Enabled Municipal Parking Deck. *Appl. Energy* **2012**, *96*, 171–182. [CrossRef]
7. Powell, F.; Bowie, C.; Halsted, L.; Beetham, J.; Baker, L. *The Costs and Benefits of Inner City Parking Vis-à-Vis Network Optimisation*; NZ Transport Agency: Wellington, New Zealand, 2015; ISBN 978-0-478-44528-2.
8. Kobus, M.B.W.; Gutiérrez-i-Puigarnau, E.; Rietveld, P.; Van Ommeren, J.N. The On-Street Parking Premium and Car Drivers' Choice between Street and Garage Parking. *Reg. Sci. Urban Econ.* **2013**, *43*, 395–403. [CrossRef]

9. Gragera, A.; Albalate, D. The Impact of Curbside Parking Regulation on Garage Demand. *Transp. Policy* **2016**, *47*, 160–168. [CrossRef]
10. de Vos, D.; van Ommeren, J. Parking Occupancy and External Walking Costs in Residential Parking Areas. *J. Transp. Econ. Policy JTEP* **2018**, *52*, 221–238.
11. Lehner, S.; Peer, S. The Price Elasticity of Parking: A Meta-Analysis. *Transp. Res. Part Policy Pract.* **2019**, *121*, 177–191. [CrossRef]
12. Stokoe, M. Space for Freight—Managing Capacity for Freight in Sydney—A CBD Undergoing Transformation. *Transp. Res. Procedia* **2019**, *39*, 488–501. [CrossRef]
13. van Ommeren, J.; McIvor, M. The Marginal External Costs of Street Parking, Optimal Pricing and Supply: Evidence from Melbourne. In Proceedings of the ITEA Annual Conference, Hong Kong, China, 25–29 June 2018; p. 28.
14. Hampshire, R.C.; Shoup, D. What Share of Traffic Is Cruising for Parking? *J. Transp. Econ. Policy JTEP* **2018**, *52*, 184–201.
15. De Aquino, J.T.; De Souza, J.V.; Silva, V.D.C.L.D.; Jerônimo, T.; De Melo, F.J.C. Factors That Influence the Quality of Services Provided by the Bus Rapid Transit System: A Look for User's Perception. *Benchmarking Int. J.* **2018**, *25*, 4035–4057. [CrossRef]
16. Arnott, R.; Rowse, J. Curbside Parking Time Limits. *Transp. Res. Part Policy Pract.* **2013**, *55*, 89–110. [CrossRef]
17. Lee, R.; Sener, I.N. The Effect of Light Rail Transit on Land-Use Development in a City without Zoning. *J. Transp. Land Use* **2017**, *10*, 541–556. [CrossRef]
18. Transport for NSW. *Kerbside Assessment—Freight and Servicing Activity in Parramatta CBD*; Western Sydney University: Penrith, Australia, 2018.
19. Young, W.; Miles, C.F. A Spatial Study of Parking Policy and Usage in Melbourne, Australia. *Case Stud. Transp. Policy* **2015**, *3*, 23–32. [CrossRef]
20. Levy, N.; Render, M.; Benenson, I. Spatially Explicit Modeling of Parking Search as a Tool for Urban Parking Facilities and Policy Assessment. *Transp. Policy* **2015**, *39*, 9–20. [CrossRef]
21. Mora, C.; Manzini, R.; Gamberi, M.; Cascini, A. Environmental and Economic Assessment for the Optimal Configuration of a Sustainable Solid Waste Collection System: A 'Kerbside' Case Study. *Prod. Plan. Control* **2014**, *25*, 737–761. [CrossRef]
22. Shoup, D.C. The Trouble with Minimum Parking Requirements. *Transp. Res. Part Policy Pract.* **1999**, *33*, 549–574. [CrossRef]
23. Wang, J.J.; Liu, Q. Understanding the Parking Supply Mechanism in China: A Case Study of Shenzhen. *J. Transp. Geogr.* **2014**, *40*, 77–88. [CrossRef]
24. Shoup, D.C. Cruising for Parking. *Transp. Policy* **2006**, *13*, 479–486. [CrossRef]
25. Shoup, D. *The High Cost of Free Parking*; Routledge: Abingdon, UK, 2011.
26. Weinberger, R. Death by a Thousand Curb-Cuts: Evidence on the Effect of Minimum Parking Requirements on the Choice to Drive. *Transp. Policy* **2012**, *20*, 93–102. [CrossRef]
27. Weinberger, R.; Kaehny, J.; Rufo, M. *U.S. Parking Policies: An Overview of Management Strategies*; Institute for Transportation and Development Policy: New York, NY, USA, 2010; p. 79.
28. Barter, P.A. Off-Street Parking Policy without Parking Requirements: A Need for Market Fostering and Regulation. *Transp. Rev.* **2010**, *30*, 571–588. [CrossRef]
29. van Ommeren, J.; Wentink, D.; Dekkers, J. The Real Price of Parking Policy. *J. Urban Econ.* **2011**, *70*, 25–31. [CrossRef]
30. Barter, P.A. Off-Street Parking Policy Surprises in Asian Cities. *Cities* **2012**, *29*, 23–31. [CrossRef]
31. van Ommeren, J.; Russo, G. Time-Varying Parking Prices. *Econ. Transp.* **2014**, *3*, 166–174. [CrossRef]
32. Simićević, J.; Milosavljević, N.; Maletić, G.; Kaplanović, S. Defining Parking Price Based on Users' Attitudes. *Transp. Policy* **2012**, *23*, 70–78. [CrossRef]
33. Cats, O.; Zhang, C.; Nissan, A. Survey Methodology for Measuring Parking Occupancy: Impacts of an on-Street Parking Pricing Scheme in an Urban Center. *Transp. Policy* **2016**, *47*, 55–63. [CrossRef]
34. Nurul Habib, K.M.; Morency, C.; Trépanier, M. Integrating Parking Behaviour in Activity-Based Travel Demand Modelling: Investigation of the Relationship between Parking Type Choice and Activity Scheduling Process. *Transp. Res. Part Policy Pract.* **2012**, *46*, 154–166. [CrossRef]
35. Jakle, J.A.; Sculle, K.A. *Lots of Parking: Land Use in a Car Culture*; University of Virginia Press: Charlottesville, VA, USA, 2004.
36. Smart City Expo Atlanta From Public Transit to Traffic: A Glimpse at Post-Pandemic Transportation. Available online: https://smartcityexpoatlanta.com/blog/from-public-transit-to-traffic-a-glimpse-at-post-pandemic-transportation/ (accessed on 1 September 2022).
37. Zou, B.; Kafle, N.; Wolfson, O.; Lin, J.J. A Mechanism Design Based Approach to Solving Parking Slot Assignment in the Information Era. *Transp. Res. Part B Methodol.* **2015**, *81*, 631–653. [CrossRef]
38. Brooke, S.L.; Ison, S.; Quddus, M. On-Street Parking Search: A UK Local Authority Perspective. *J. Transp. Land Use* **2017**, *10*, 13–26. [CrossRef]
39. Meng, F.; Du, Y.; Li, Y.C.; Wong, S.C. Modeling Heterogeneous Parking Choice Behavior on University Campuses. *Transp. Plan. Technol.* **2018**, *41*, 154–169. [CrossRef]
40. Cavadas, J.; Antunes, A.P. Optimization-Based Study of the Location of Park-and-Ride Facilities. *Transp. Plan. Technol.* **2019**, *42*, 201–226. [CrossRef]
41. Chen, X.; Liu, Z.; Currie, G. Optimizing Location and Capacity of Rail-Based Park-and-Ride Sites to Increase Public Transport Usage. *Transp. Plan. Technol.* **2016**, *39*, 507–526. [CrossRef]

42. Ottosson, D.B.; Chen, C.; Wang, T.; Lin, H. The Sensitivity of On-Street Parking Demand in Response to Price Changes: A Case Study in Seattle, WA. *Transp. Policy* **2013**, *25*, 222–232. [CrossRef]
43. Millard-Ball, A.; Weinberger, R.R.; Hampshire, R.C. Is the Curb 80% Full or 20% Empty? Assessing the Impacts of San Francisco's Parking Pricing Experiment. *Transp. Res. Part Policy Pract.* **2014**, *63*, 76–92. [CrossRef]
44. Straker, I.; Ison, S.; Humphreys, I.; Francis, G. A Case Study of Functional Benchmarking as a Source of Knowledge for Car Parking Strategies. *Benchmarking Int. J.* **2009**, *16*, 30–46. [CrossRef]
45. Rodrigue, J.-P.; Dablanc, L.; Giuliano, G. The Freight Landscape: Convergence and Divergence in Urban Freight Distribution. *J. Transp. Land Use* **2017**, *10*, 557–572. [CrossRef]
46. Rodier, C.J.; Shaheen, S.A.; Eaken, A.M. Transit-Based Smart Parking in the San Francisco Bay Area, California: Assessment of User Demand and Behavioral Effects. *Transp. Res. Rec.* **2005**, *1927*, 167–173. [CrossRef]
47. Greater Cities Commission Growing a Stronger and More Competitive Greater Parramatta. Available online: https://greatercities.au/central-city-district-plan/productivity/jobs-and-skills-city/growing-stronger-and-more-competitive (accessed on 29 August 2022).
48. NSW Department of Planning and Environment A Metropolis of Three Cities. Available online: http://www.planning.nsw.gov.au/Plans-for-your-area/A-Metropolis-of-Three-Cities/A-Metropolis-of-Three-Cities (accessed on 1 September 2022).
49. NSW Government Parramatta Light Rail. Available online: https://www.parramattalightrail.nsw.gov.au/ (accessed on 22 July 2022).
50. Samaranayake, P.; Gunawardana, U. Parking Assessment in the Context of Growing Construction Activity and Infrastructure Changes: Simulation of Impact Scenarios. *Sustainability* **2022**, *14*, 5098. [CrossRef]
51. Deloitte. Parramatta Crane Survey 2017. 2017. Available online: https://www2.deloitte.com/tl/en/pages/finance/articles/parramattacrane-survey-2017.html (accessed on 19 April 2021).
52. FlexSim.Com. *FlexSim 3D Simulation Modeling Software*; FlexSim Sofware Products, Inc.: Orem, UT, USA, 2021.
53. *MATLAB (R2021A)*; The Mathworks Inc.: Natick, MA, USA, 2021.

Article

The Assessment of Greyfields in Relation to Urban Resilience within the Context of Transect Theory: Exemplar of Kyrenia–Arapkoy

Vedia Akansu [1,*] and Aykut Karaman [2]

[1] Faculty of Architecture, Near East University, 99138 Mersin, Turkey
[2] Department of Architecture, Faculty of Engineering and Architecture, Altinbas University, 34218 Istanbul, Turkey
* Correspondence: vedia.akansu@neu.edu.tr

Abstract: Greyfields are construction sites that emerge from the expansion of cities towards rural settings. They are unused structures in settlement areas that negatively impact the habitats and lead to ecologically, economically, and socially problematic zones. This study aims to examine the Greyfield problem, which emerges as one of the outcomes of urban sprawl, within the context of Transect Theory and urban resilience. We analyze the Greyfield problem in the Arapkoy rural settlement, which is located along the north coastline of Kyrenia, Cyprus. This study presents the impact of Greyfield sites on environmental, social, and economic values within the framework of Transect Theory. Thus, a road map for the redevelopment of Greyfields into public use is put forward to be used for future planning activities, which is a necessity in enabling urban resilience.

Keywords: Greyfield; urban transect; urban resilience; rural areas; ecology

Citation: Akansu, V.; Karaman, A. The Assessment of Greyfields in Relation to Urban Resilience within the Context of Transect Theory: Exemplar of Kyrenia–Arapkoy. *Sustainability* **2023**, *15*, 1181. https://doi.org/10.3390/su15021181

Academic Editors: Qingsong He, Jiayu Wu, Chen Zeng and Linzi Zheng

Received: 22 November 2022
Revised: 4 January 2023
Accepted: 5 January 2023
Published: 8 January 2023

1. Introduction

Resilience is the capacity of a system to maintain its original organizational structure in terms of unity and identity over time. It is about absorbing external shocks due to its self-organizational capacity. In other words, the concept of resilience refers to the capacity of a city to recover from a wide range of shocks and risks [1–3]. Due to its wide potential applicability in local practice, urban resilience provides a popular paradigm for urban planning and policy making at different spatial scales, including cities, regions, and neighborhoods [4,5]. Some common elements that can be identified in ecological, economic, and social resilience are the notion of memory, conservation, stability, correlation, and feedback [6]. A resilient city is more talented in adapting to an increasingly complex economic environment. It will successfully cope with sudden natural disasters and social emergencies, better protecting the health of the eco-system, and meeting information requirements [7–10]. With the acceleration of urban agglomeration, urbanization has become a complex system with multiple, closely linked factors that impact themselves and the outside world [11–15]. The concept of resilience is a holistic planning approach in the sense that it addresses the areas to be protected together with the dynamics associated with development areas. It assesses ecological, economic, and social data together and intervenes from urban to rural areas with dynamic tools [16]. The effect of irregularity and destabilization in urban and rural areas increases the probability of problems. Irregular structures, particularly those built in geologically unsuitable areas, and inadequacy of planning are fundamental factors in the occurrence of issues [17]. Transect Theory aims to organize all of the elements in the environment and includes urban and rural settings and centers. It is also a system that puts forwards different settlement types and concepts in varied urban densities [18]. Therefore, besides transportation axes, areas with a high density of buildings and their uses, design features of facades, masses, public spaces, junctions, car parks, pavements, street silhouettes, lighting elements, green areas, and landscape elements are identified.

Furthermore, attention has been paid to ensuring that all urban elements in the system are in the appropriate place and continuous [19,20]. In this regard, the urban–rural scales classification scheme offers a settlement proposal for city centers and natural open spaces in rural areas. Thus, suburbanization is crucial in improving cities that insensibly expand with decentralization.

Greyfield sites are found in built-up urban areas and are one of the best choices for redevelopment because they utilize sites that are no longer profitable, are not significantly contaminated, and reduce financial risks. They are usually found in the form of old retail malls, office parks, abandoned municipal properties, and parking lots [21]. Small-scale, piecemeal, fragmented, suboptimal small-lot subdivision is spreading like a virus through Greyfield suburbs with high redevelopment potential, removing up to 50% of private green space and blocking prospects for better designed, regenerative, precinct-scale, medium-density "missing middle" redevelopment [22,23]. Greyfields typically emerge as a result of urban expansion. These are fields that are left unused, which are architecturally old-fashioned, have insufficient or unusual infrastructure, lack redevelopment capital, and create a negative impression [24]. In other words, "Greyfield" is a term used to describe the extensive band of aging, occupied, residential tracts of inner and middle suburbs that are physically, technologically, and environmentally obsolescent, and which represent economically outdated, failing, or under-capitalized real-estate assets [25]. These problematic areas cause ecological, economic, and social problems that are generally observed in suburban settlements [26]. They emerge as problem areas that pose a threat to the resilience characteristics needed for livable cities. Although they are unused areas, Greyfields cause the loss of rural features (i.e., agriculture and animal husbandry) in lands that are opened for settlement, deterioration of topography, and destruction of ecology. They also change the morphological structure of urban and rural areas.

Abundant literature exists highlighting the benefits derived from Greyfields in the USA and Australia as residential and commercial exemplars [25–41]. However, this literature has not dealt with the externality of Greyfield redevelopment, and there are a lack of case studies about the redevelopment of Greyfields in European and Asian countries [42]. In the Australian and American contexts, Greyfields are described as retail malls or commercial buildings that failed to succeed and which are usually located in suburban areas characterized as having potential for redevelopment. Although heavily described as "derelict buildings" in the literature, Raamos (2011) broadly defines Greyfields as unproductive urban spaces that can yield economic, social, and environmental benefits [43]. Within the context of North Cyprus, although the structures that are discussed in this paper are mainly derelict buildings, the majority of them were constructed in a large lot with the purpose of built-to-sell, thus can be considered commercial buildings. Therefore, for this study the "Greyfields" term has been adopted. These buildings are left uncompleted at different stages of construction, or they are abandoned due to the political or financial difficulties that construction companies experience. Thus, this unique characteristic of Greyfields in North Cyprus contributes to the relevant literature.

Another theme that is important for the study is the Transect Theory. This theory is a cut or path through part of the environment, showing a range of different habitats. Biologists and ecologists use transects to study the many symbiotic elements that contribute to habitats, where certain plants and animals thrive [44]. The concept of the transect in the urban environment is embedded in traditional cities, as a tool used informally by humanity [45]. Andres Duany and other urbanists applied this concept formally to human settlements and used it in planning beginning in the late 1980s [44,46]. The idea has evolved into an organizing theory, permeating since about 2000 with the consideration of new urbanism [44,47]. In this regard, Transect Theory has been adopted as the theoretical framework for study as it is a system that engulfs all of the elements of urban, rural, and suburban areas and it supports the planning approaches that protect ecological, economic, and social values. Therefore, these characteristics make Transect Theory a more useful tool when it is used along with urban resilience. Within this context, this paper aims to examine the Greyfield problem, which emerges as one of the outcomes of urban sprawl, within the context of Transect Theory, and to analyze the Greyfield problem in the Arapkoy rural settlement in relation to urban resilience. The Arapkoy rural settlement is located along the north coastline of Kyrenia, Cyprus.

This article, following this brief introduction, is structured into the following sections: the methodology description, the presentation of the case studies, and the discussion of results, with some final remarks and future perspectives on the research. The main aim of this paper is to help contribute to the research on the Greyfield problem, which emerges as one of the outcomes of urban sprawl, within the context of Transect Theory and to analyze the Greyfield problem in relation to urban resilience, since the Greyfield is still a quite recent topic and in need of further development.

2. Materials and Methods

2.1. General Methodology

The framework of this study was developed based on the existing relevant literature, and data were collected through secondary acquisition of reports and a questionnaire. Due to the building dynamics and density, and existing Greyfield sites in both green and built areas, the Central and East Kyrenia settlements were chosen as the foci of this study. Secondary data such as reports provided numerical facts about the studied region, whereas questionnaires were applied both for residents and experts to minimize the disadvantage that comes from using secondary data. This way, the statistical data provided an understanding of the status in the regions examined, and questionnaire results depicted the current state and existing problems. The questionnaire was conducted over 6 months from the last quarter of 2016 until the first quarter of 2017. Ecological, economic, and social values of problematic Greyfield sites for urban resilience were evaluated within the scope of data collected by questionnaire with the stakeholders, including residents, non-governmental organizations, and relevant official regional institutions. In the conducted questionnaire, Arapkoy resident participants were referred to as "residents" and individuals such as architects, engineers, and city planners with active duties in public organizations and non-governmental bodies that are related to the region were referred to as 'experts'. Questionnaires were completed by 67 individuals consisting of two groups: one group including 31 experts, and the other group including 36 residents. The questionnaire was conducted over 6 months. It started during the last quarter of 2016 and was completed during the first quarter of 2017. The questionnaire was designed in two main sections: the first section was designed to put forward the five most important problems in light of the residents' and expert participants' views, and the second section was designed to put forward the ecological, economic, and social status describing city resilience in percentages. In the first section, 20 factors were determined. These factors included: transportation, roads, security, increasing crime rates, street lighting, car parks, water mains, sewer system, air pollution, garbage and solid waste, unused and unnecessary stock of housing, unplanned construction, fields for activity, health services, unemployment and lack of employment, lack of infrastructure services, a rapid increase in population, lack of legal regulations, visual pollution, noise, and limited green fields in urban areas. The second section consisted of 10 questions with a 5-point Likert scale, where participants were asked to rate their agreement levels with the statements. Statements were as follows: Greyfield sites contribute to the future of the regional economy; ensure protection and transfer of the region to future generations; contribute to the development of tourism, create pressure on the natural environment; prevent protection of environmental and ecological values; the existence of housing above the carrying capacity of the area leads to environmental pollution; the area is not guided with planning; Greyfield sites diversify the locals' income sources; Greyfield sites limit the agricultural activities, which are a primary source of income; they create regional risk areas in terms of urban resilience (i.e., natural disasters such as flood and earthquake, technological risks; human originated risks and obstacles to the development of urban and rural areas.

The residents were chosen randomly among Arapkoy residential users. On the other hand, the experts were chosen randomly among those who work in Arapkoy and know the region. As Arapkoy has a population of 542 according to the 2011 census [48]. it was suggested to be a sufficient sampling size for the research. For example, a study aiming at measuring the environmental attitudes among local people via a user survey in Guzelyurt, North Cyprus (having a population of 18,946) used a random sample of 60 residents [49]. In addition, an article cited in our study involving a survey with a sample of 367 focused on Helsinki with a population of almost 642,000 residents.

Data on dwelling count and census results were derived from the State Planning Organization. Because the last official census was in 2011 and the results were comparatively old, statistical data from the Cyprus Turkish Building Contractors Association (CTBCA) were used, covering the 2011–2016 period. The authors acknowledge the relatively aged nature of the secondary data and the lack of other comprehensive reports.

Based on the sketch studies made for the relevant zones, a suitable transect schema (Figure 1) was created by presenting the building density, usage status, and purpose of construction.

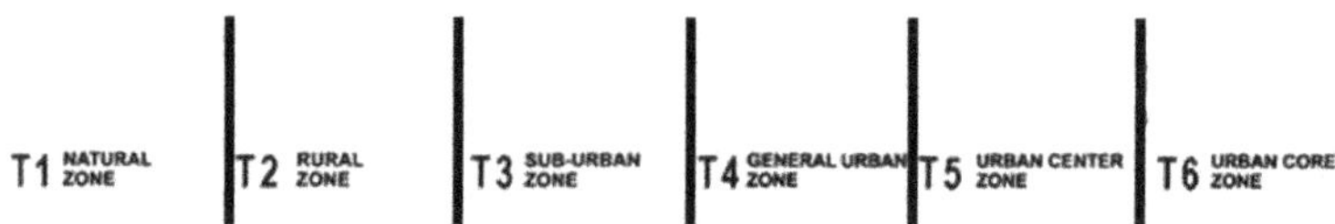

Figure 1. Transect schema structure.

The transect approach is an analytical method and a planning strategy. It can be formally described as a system that seeks to organize the elements of urbanism (street, land use building, lot, and all of the other physical elements of the human habitat) in ways that conserve the integrity of the different types of urban/rural environments. These environments can be viewed as variations along a continuum that range from rural to urban. Along this continuum, human environments vary in their level of urban intensity. Cleaving to this system of organization, urban open spaces are saved in their urban state, while rural settlements are preserved in their rural state, and the mixing of elements—a rural element in an urban open space and vice versa—is avoided [50].

The transect schema is divided into six zones: natural (T1), rural (T2), suburban (T3), general urban (T4), center (T5), and urban core zone (T6) [19,51]. According to Emily Talen, each transect zone is an immersive environment, a place where all of the elements reinforce each other to produce and consolidate the character of a specific place [50].

The significance of this approach comes with its part as a taxonomic machine. It is like a new zoning system, compared to the conventional separated-use zoning systems. The new system provides the base for real neighborhood structure, which requires walkable streets, mixed-use, transportation options, and dwelling diversity. The T- zones vary by the rate and position of the intensity of their natural, built, and social factors, in addition to their part in the creation of an environment that could be perceived as immersive [52]. Also, the creation of a more environmentally conscious sprawl, the preservation of natural and agricultural land, the creation of flexible communities, balancing every aspect of the city, enabling diversity, and the management of urban development raise the quality of the erected environment in general and produce better places to live [47,51].

Following this, a comparative analysis was used to assess the urban resilience of the suburban and rural regions where the heavy impact of construction was observed. Arguments have also been presented on how the comparative analysis of urban resilience can add value to future planning activities.

2.2. Study Area

Cyprus (Lat 34.33 and 35.41 N; Long 32.23 and 34.55 E) has a typical Mediterranean topographical and architectural building style. The topographic structure is influential in the formation of ecological and social regions. This topographical impact indicates that individuals belong to social groups, and they live in particular urban or rural residential areas, which are insulated according to their environment [53]. The transportation axes are also shaped as a result of the topographical structure [54].

However, Cyprus Island has been divided into north and south since 1974 due to ethnopolitical reasons. The effects of construction density have not been evident until the early 2000s in North Cyprus, the context of this study, and it is observed that the social, cultural, and ecological values have been preserved.

The "Annan Plan" prepared by the United Nations (UN) in 2001 was discussed to provide a comprehensive solution to the Cyprus conflict. Because it was proposed as a resolution plan by the UN, it found credibility in the international arena. Thus in 2007, this plan put North Cyprus into a "rapid structuring" process, especially with the assurance of eliminating some of the uncertainties about property issues [55]. Parallel with these developments and changes, progress was observed in construction, particularly in the Kyrenia region in 2003–2004.

A demand mainly from the UK, North Europe, and other countries was observed in the northern part of Cyprus, especially in the Kyrenia region [56]. Thus, the construction sector and urban architectural environments started to develop and spread toward the rural areas. Intensive housing was observed primarily on the northern coastline of Kyrenia. The different building styles such as villas, apartments, and duplex houses alongside hotel buildings, which showed rapid increase, brought diversity to the architectural environment in Kyrenia and other regions [57]. The increasing demand in the construction sector positively impacted the economy, and people started to leave their existing jobs and enter the construction business. Houses with a similar style were built with the aim of economic gain, without considering customer demands and needs. However, legal regulations related to the construction sector "Streets and Buildings Chapter 96 of the Laws" were insufficient. Construction permits were given to every plot of land with road access, including pasture, agricultural land, and qualified land [55]. However, the construction sector collapsed due to the surplus supply of houses, and the construction companies experienced economic problems alongside title deed problems, which were rooted in the political conflict on the island.

Today, some of these buildings are left half-completed, and some of them are completed but unsold. They create ecologically, economically, and socially problematic areas and cause problems in ensuring urban–rural resilience.

3. CTBCA Report

To resolve the problems experienced, the CTBCA assessed the situation with statistical studies aimed at determining the new housing densities of the northern Cyprus settlements based on KADEM data, which is a private research company. The household data of the census results of 2007 and 2011 were examined. In the CTBCA report, Cyprus's northern coastline was investigated in four different regions.

- Kyrenia–Eastern Region (Kyrenia Karaoglanoglu in the East, Sadrazamkoy in the West, Besparmak Mountains in the South)
- Kyrenia–West Region (Kyrenia Karakum in the West, Buyukkonuk Junction in the east of Kaplica)
- Famagusta–Bogaz Region (Tuzla in the South, Bogaz in the North)
- Bogaz–Karpaz Region (Bogaztepe in the West, Dipkarpaz in the East)

Construction was investigated in terms of building categories such as houses, flats, and shops based on the regions. Statistical information on their development density is given below (Figure 2):

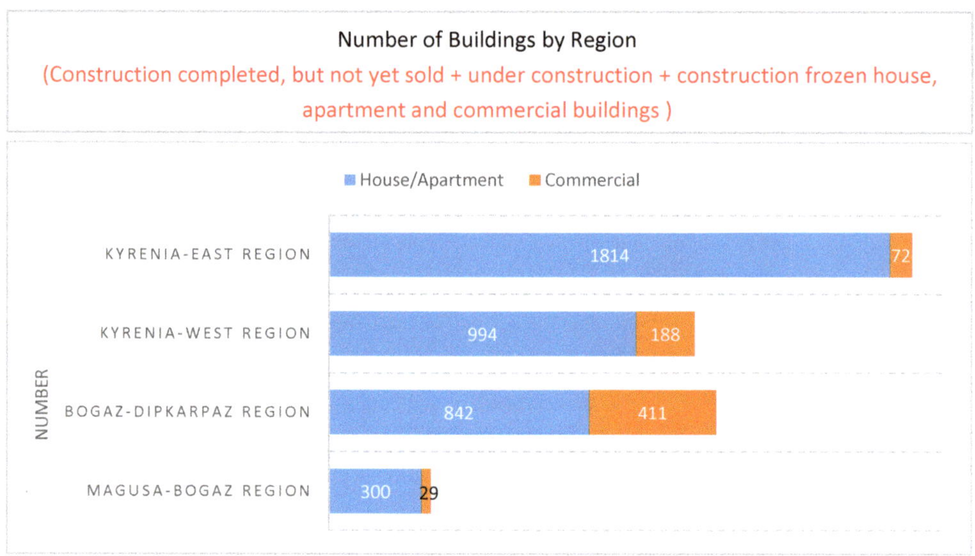

Figure 2. The number of buildings on the North Cyprus coastline [58].

Study on the construction count—phase 1 was compiled by CTBCA in 2011. The data indicated that the eastern coastline of Kyrenia was the most intense settlement with 1886 buildings; of those, 72 were commercial buildings. Furthermore, 86.9% of the 1886 buildings were built by construction companies. The significant majority of these buildings (90.9%) were constructed in the east settlements of Kyrenia (Figure 3).

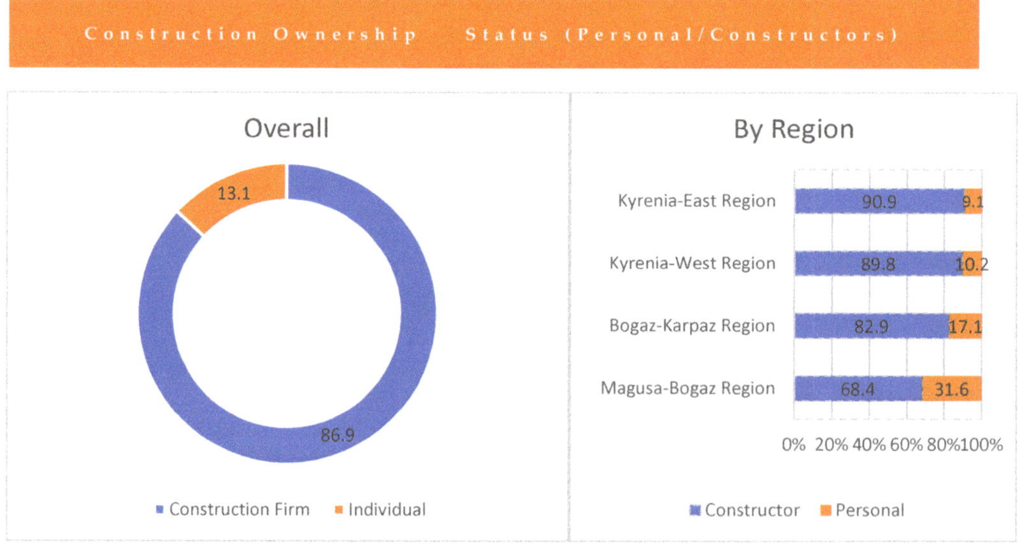

Figure 3. The assessment of the construction firms or individuals who built residences on the Cyprus coastline [58].

The inadequacy of the planning not only caused the irregular construction on the coastline, but it also revealed that there was an increase in the number of derelict houses.

Therefore, the location of the construction was assessed according to the regions. A total of 7.9% of the building constructions were halted, whereas 52.9% were still under construction, and 39.2% of building construction were completed but not yet sold (Table 1).

Table 1. Evaluation of the construction location according to the designated regions in North Cyprus [58].

REGION	THE STATUS OF CONSTRUCTION BY REGION AS OF 2011							
	Construction Completed		Under Construction		Construction Halted		Total	
Magusa–Bogaz Region	111	33.7%	191	58.1%	27	8.2%	329	100%
Bogaz–Dipkarpaz Region	418	33.4%	710	56.7%	125	9.9%	1253	100%
Kyrenia–West Region	486	41.1%	649	54.9%	47	4.0%	1182	100%
Kyrenia–East Region	809	42.9%	910	48.2%	167	8.9%	1886	100%
Total	1824	39.2%	2460	52.9%	366	7.9%	4650	100%

As mentioned above, while many residences were waiting to be completed in the construction phase, many of them did not have buyers and remain derelict. It was observed that the Kyrenia–East region, followed by the Bogaz–Dipkarpaz regions were affected by this situation the most. The longitudinal assessment presented in Table 2 covers the years between 2007 and 2011.

Table 2. Evaluation of the construction location according to the designated regions in North Cyprus [58].

REGION	NUMBER OF CONSTRUCTIONS (Construction completed, but not yet sold + under construction + construction frozen house, apartment, and commercial buildings)											
	2007						2011					
	Under construction+ frozen		Construction completed, but not yet sold		Total		Under construction+ frozen		Construction completed, but not yet sold		Total	
Bogaz–Dipkarpaz Region	424	52.54%	383	47.46%	807	100%	418	33.36%	835	66.64%	1253	100%
Kyrenia–East Region	2812	81.81%	626	18.21%	3437	100%	1077	57.10%	809	42.90%	1886	100%

Table 2 presents statistical data on Greyfields in the Kyrenia–East Region settlement. There was a total of 3437 Greyfields in 2007 and 1886 in 2011. Today, new construction projects continue while the Greyfields remain idle.

Thus, grey-area formation typologies, which are discussed in regional development, are found dominantly in newly constructed parts of rural areas, which exist on city borders.

Greyfield Statistics for Rural Areas

The difficulties experienced in urban planning cause problems in rural areas as well. Planning studies that cover all urban and rural areas are needed for the preservation of rural characteristics and their transfer to future generations.

The new dwelling increase in settlements located east of the northern coastline of Kyrenia, where most housing is located, is as follows (Figure 4).

Figure 4 shows the number of new houses built between 2004 and 2007. New residences were built in Besparmak (59), Karakum (73), Alagadi (102), Karaagac (120), Esentepe (170), Tatlısı (202), Esentepe-Tatlısu (206), Arapkoy (300), Bahceli (548), Alagadi-Esentepe (874), Kucuk Erenkoy (983), and Catalkoy (1063). An increase in the number of dwellings among rural settlements between 2004 and 2007 was mostly observed in Arapkoy due to its proximity to the city of Kyrenia. Furthermore, it was determined that a total of 4700 new houses were built between 2004 and 2007 between the Karakum and Tatlisu areas in the East–West direction and between the north coast and the Besparmak mountain in the North–South direction [59]. According to CTBCA data in 2007, the Kyrenia–East Region was consistent with statistical data from 2007, with 1263 (27%) of the houses built in the region having residents. Of the remaining 3437 (73%) houses, 2812 were under construction, and 626 of them were vacant although they were completed [59]. Figure 4 shows the status of vacant and occupied dwellings according to the regions. When vacant and occupied percentages were examined for 2011, it was evident that the majority of the constructions were uncompleted, and for those which were completed (1886), 73.1% were vacant (Figure 5).

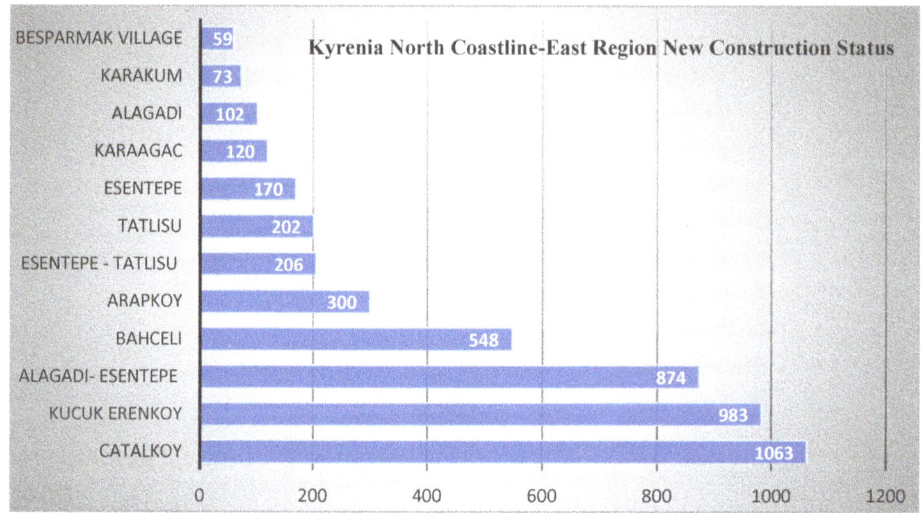

Figure 4. Kyrenia–East Region housing increase from 2004 to 2007.

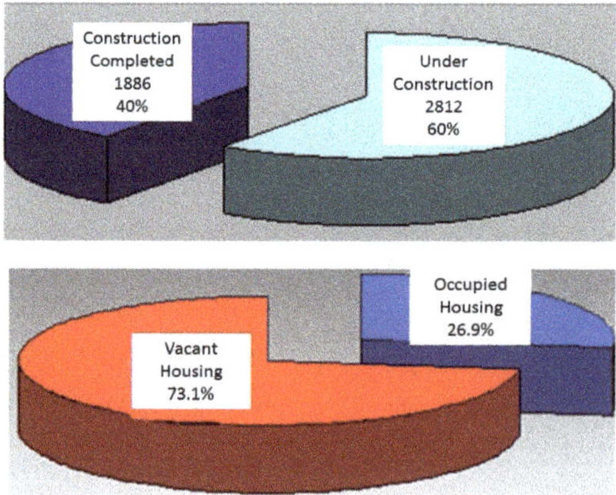

Figure 5. Vacant and occupied dwelling percentages in the Kyrenia–East Region as of 2007.

The inventory study of houses built after 2007 in the Kyrenia region was compiled according to the household count in the 2011 census results by the Cyprus Turkish Building Contractors Association [58].

The overall dwelling occupancy rate, which was 26.9%, reached the highest rate in Karakum with 60.35% and 50% in the region between Esentepe-Tatlisu. The percentage of vacant houses in Besparmak village was 100%, 91.8% in Kucuk Erenkoy, and 85.3%

in the Alagadi regions. However, a high number of abandoned houses were observed in Bahceli and Arapkoy in the east settlements of Kyrenia. The Arapkoy settlement was designated as a study area because it was close to Kyrenia city and was considered a grey area. Furthermore, it was determined that there were intensive residential construction and land subdivisions in the area. It was concluded that the investments in industrialization, tourism, and commercial or multi-purpose construction were at a very low level [58] (Figure 6).

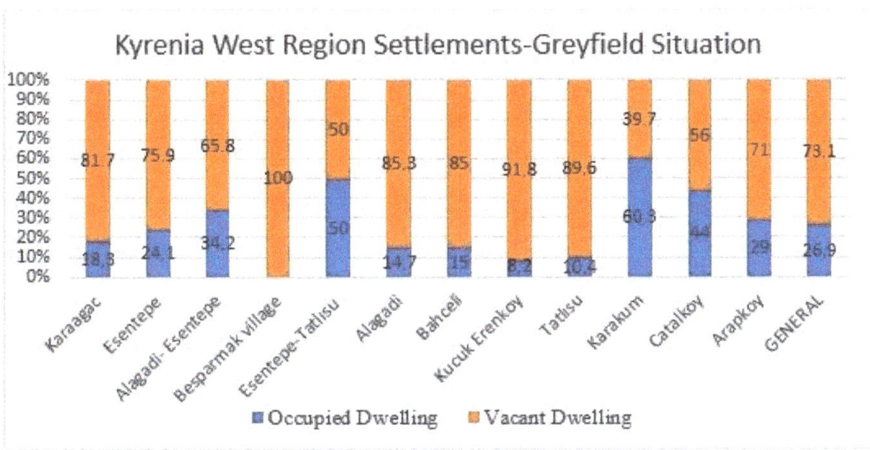

Figure 6. Percentages of occupied and vacant house settlements in the east of Kyrenia [58].

The inventory of houses built after 2004 in the Kyrenia region was compiled within the scope of the 2011 census results by the Cyprus Turkish Building Contractors Association [58].

The overall percentage of occupied dwellings was 26.9%. However, this percentage reached up to 60.35% in Karakum and 50% in Esentepe-Tatlisu. On the other hand, the rate of vacant dwellings was 100% in Besparmak village, 91.8% in Kucuk Erenkoy, and 85.3% in the Alagadi regions. It was observed that there were completed houses in the Bahceli and Arapkoy settlements (Figure 5), but they remained abandoned. In terms of the types of construction, it was determined that predominantly housing and plot parceling constructions had taken place in the region. It was concluded that the investments in terms of industrialization, tourism, and commercial and multi-purpose building were limited [60].

The construction information database booklet (2016) of the CTBCA stated that there was an increase in construction permit applications by 8% in Nicosia, 32% in Famagusta, 37% in Kyrenia, and 20% in the Guzelyurt district according to the 2015 data. On the other hand, there was a 38% decrease in the Iskele district. These figures suggested that the Kyrenia district experienced the highest increase in construction permit applications [60]. In the construction files recorded in 2015 across North Cyprus, the Kyrenia district ranked first with 716 files. Of the 716 files, 123 of the applications were completed, and 593 of them were ongoing as of 2016 [60].

4. Results and Discussion

4.1. The Rural Planning Issues and Resilience in Greyfield Sites

Urban resilience not only focuses on the resilience of large cities but additionally includes towns and smaller living areas [61]. Rural areas play an essential role in ensuring urban resilience. An evaluation index system of urban resilience was made using systematic, scientific, comprehensive, and availability principles, combined with the particular development of the cities. This index system was composed of four sub-resilience systems: ecological, economic, social, and infrastructure resilience [62].

Unplanned construction and distorted urban settlement in Kyrenia led Greyfields to be frequently observed in rural areas. The Arapkoy settlement is crucial because it accommodated exemplars of the aforementioned problem.

Within the context of this research, resident and expert opinion was sought to determine the impact of disorderly construction on urban–suburban and rural areas on the east coast of Kyrenia. In this section, 20 factors were determined: transformation, roads, security, increasing crime rates, street lighting, car parks, water main, sewer system, air pollution, garbage and solid waste, unused and unnecessary stock of housing, unplanned construction, fields for activity, health services, unemployment and lack of employment, lack of infrastructure services, and rapid urbanization in urban areas. Of these, participants were asked to rank the five most important factors. The results are presented in Table 3.

Table 3. The five most important problems determined by resident and expert participants on the Kyrenia coastline.

The Five Most Important Problems on the Kyrenia Coastline			
	RESIDENTS		*EXPERTS*
1	Unplanned Construction	1	Unplanned Construction
2	Sewer System	2	Roads (Asphalt Paving and Pavements)
3	Air Pollution and Transportation (Insufficient Public Transport)	3	Transportation (Insufficient Public Transport)
4	Visual Pollution (Greyfield Sites)	4	Insufficient Infrastructure Services
5	Limited Green Fields in Urban Areas, Roads (Asphalt Paving and Pavements), Garbage and Solid Waste	5	Car Park

Both residents and experts agreed that unplanned construction and insufficient public transport, asphalt paving, and pavements were the most important problems on the Kyrenia coastline.

The experts also identified inadequate infrastructure services and car parks, while the residents reported limited green spaces.

Further to this, a comparative analysis was provided to examine the opinions of both the residents and experts on the impact of Greyfield sites based on their responses to 10 questions on a 5-point Likert scale (strongly disagree to strongly agree). The questions determined during the survey were selected for the ecological, economic, and social maintenance of the Greyfield regions, including how Greyfield sites contribute to the future of the regional economy; ensure the protection and transfer of the region to future generations; contribute to the development of tourism; create pressure on the natural environment; prevent protection of environmental and ecological values; how the existence of housing above the carrying capacity of the area leads to environmental pollution; how the area is not guided by planning; how Greyfield sites diversify the locals' income sources; how Greyfield sites limit the agricultural activities, which are a primary source of income; how they create regional risk areas in terms of urban resilience (i.e., natural disasters such as flood and earthquake, technological risks; human originated risks and obstacle to the development of urban and rural areas). The results are presented in Table 4.

Table 4. Arapkoy Rural Area—Greyfield Site Assessments.

Greyfield Sites-Related Questionnaire Questions	Arapkoy Rural Area Greyfield Sites Impact Assessment									
	Strongly Agree		Agree		Neither Agree Nor Disagree		Disagree		Strongly Disagree	
	Resident	Expert	Resident	Expert	Resident	Expert	Resident	Expert	Resident	Expert
Greyfield sites contribute to the future of the regional economy	11%	3%	22%	13%	11%	13%	34%	32%	22%	39%
Ensure protection and transfer of the region to future generations	6%	3%	19%	0%	17%	3%	36%	29%	22%	65%
Contribute to the development of tourism	12%	3%	28%	3%	17%	3%	20%	42%	23%	49%
Create pressure on the natural environment. Prevent protection of environmental and ecological values	11%	49%	39%	39%	14%	3%	31%	3%	5%	6%
The existence of housing above the area's carrying capacity leads to environmental pollution	17%	62%	67%	29%	5%	0%	3%	3%	8%	6%
The area is not managed with a plan	25%	65%	47%	29%	19%	0%	3%	3%	6%	3%
Greyfield sites diversify the locals' income sources	8%	10%	31%	10%	28%	26%	22%	22%	11%	32%
Greyfield sites limit agricultural activities, which is a primary source of income	25%	29%	31%	29%	14%	23%	16%	12%	14%	7%
They create regional risk areas in terms of urban resilience (i.e., natural disasters such as floods and earthquakes, and technological and human-originated risks).	25%	45%	42%	39%	8%	10%	19%	6%	6%	0%
Greyfield sites are obstacles to the development of urban and/or rural areas	17%	26%	46%	58%	17%	6%	17%	10%	3%	0%

As it is stated in Table 4, there were items that the residents and experts did not agree with. Regarding economic issues, 56% of the residents and 71% of the experts disagreed or strongly disagreed that Greyfield sites *"contribute to the future of the regional economy"*. A total of 94% of the experts and 58% of the residents stated their disagreement that Greyfields *"ensure protection and transfer of the region to future generations"*. When participants were asked if they think Greyfield sites *"contribute to the development of tourism"*, 28% of the residents agreed and 49% of the experts disagreed with that statement.

The questions that both experts and residents agreed on included *"existence of housing above the area's carrying capacity leads to environmental pollution"*(experts (62%) and residents (67%)). A total of 65% of the experts and 47% of the residents agreed that the Greyfield sites *"lack planned management"*.

When the participants were asked whether *"Greyfield sites create regional risk areas in terms of urban resilience"*, both experts (45%) and residents (42%) agreed with this statement, and 58% of the experts and 46% of the residents also agreed that *"Greyfield sites are obstacles to the development of urban and rural areas"*. Also, regarding the statement *"Greyfield sites creating pressure on the natural environment and preventing the protection of environmental and ecological values"*, 49% of the experts and 39% of the residents agreed with this statement.

The Arapkoy rural area Greyfield sites create pressure on the natural environment and prevent ecological values from being preserved. The participants also suggested that the Greyfield sites were not managed with a plan. This was partially due to the existence of housing above the area's carrying capacity, which also caused environmental pollution. As a result of the questionnaire, it was concluded that the Greyfield sites limited the residents' agricultural and farming activities, which was their source of income, created regional risk areas in terms of urban resilience, and caused obstacles to spatial development. These conclusions suggested that the Greyfield sites should be taken into consideration in planning activities that focus on enabling resilience, as Greyfield sites affect the environmental, economic, and sociological factors of a region. It would not be possible to consider resilience and habitable settlements without taking into consideration the aforementioned problems while conducting planning activities.

4.2. The Urban–Rural Transect in Greyfield Sites

The transect scheme aims to organize the city with all of its details in the planning stage. It is also an effective method in bringing the Greyfields into use. By considering the grey areas together with the transect scheme, the assessment of the existing situation is introduced to transform grey areas into habitable areas. However, research has indicated that this problem arises as a result of urban expansion. As Transect Theory suggests, Greyfields are divided into regions and considered as a whole (Figure 7).

Urban–Rural Scale Classification Scheme (Transect Diagram)

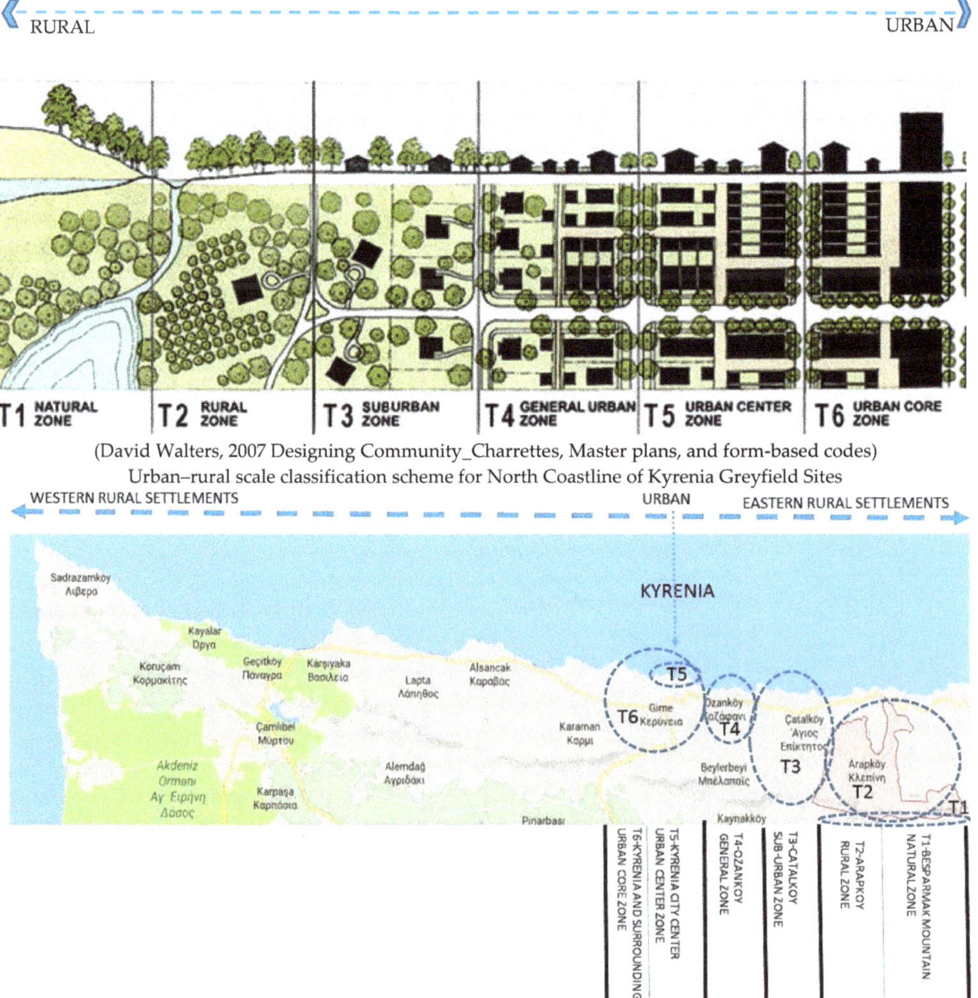

Figure 7. Urban–rural scale classification scheme: Kyrenia North Coastline. (Adapted from David Walters, 2007 Designing Community Charrettes, Master plans and form-based codes).

A Greyfield transect of 27.9 km in length was defined from the Kyrenia urban area to the eastern rural area and formed by six sections. The first section (T6) was in the city center close to Kyrenia and the last section (T1) was in the natural zone of the Besparmak Mountain area. The model transect zone diagram was based on the example of Kyrenia and its eastern settlements (Figure 8).

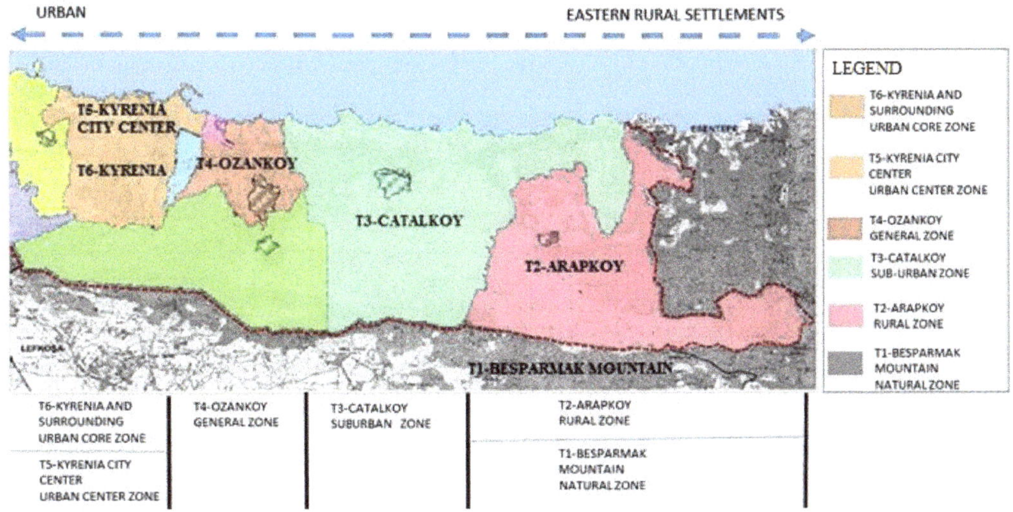

Figure 8. Urban–Rural Transect Theory. City of Kyrenia north coastal settlements.

The construction pattern, which was followed as a result of the urban–rural expansion, is presented. This enabled us to examine the state of the planning activities for the Greyfield sites (Table 5).

Table 5. Urban–rural scales classification of City of Kyrenia and north coastal settlements.

Evaluation	**Transect Theory Regions**					
	T6- Urban Core Zone Kyrenia and surrounding	T5- Urban Centre Zone Kyrenia City Centre	T4- General Urban Zone Ozankoy	T3- Suburban Zone Catalkoy	T2- Rural Zone Arapkoy	T1- Natural Zone Besparmak Mountain
		Regions between T1 and T5 settlement zones are developing similarly as a result of urban expansion.				
Density of Buildings	Dense building masses co-exist with historical and traditional textures.	Dense, multi-story buildings and single-story buildings co-exist.	Collective housing is observed intensely.	Collective housing is observed intensely.	There are constructions of mixed, traditional, and collective housing.	It is not open to settlement.
			They are developing and intertwined with each other.			
Status of Use	There are no Greyfield sites.	Although observed occasionally, Greyfield sites are not intense.	Due to its proximity to the city, Greyfield sites are not commonly observed.	It has Greyfield sites.	Greyfield sites are dense in new residential areas.	In some cases, a decision on housing construction can be made autonomously.
				71% of Greyfield sites are in rural areas, decreasing to 56% in the suburban zone.		
Aim of Construction	Generally, the purpose of construction is oriented toward a built-to-sell architecture with the aim of profit-making. Collective housing areas gradually increase as the zone changes from the general urban zone, to the suburban zone to the rural zone, respectively. There is an increase in commercial areas as the zone changes to an urban core zone and an urban center zone.					

As shown Table 5, the northern coastline of Kyrenia was divided into six zones based on the Transect Theory. The study is followed by a comparative analysis of the northern coastline of Kyrenia and east settlements based on the studies of [19,63].

T1—Natural Zone (Five Finger Mountain Region): This zone includes areas that are not habitable for settlement due to its topography, hydrology, and vegetation but it covers areas for wildlife and its habitat. However, in the north coastline settlements of Kyrenia, these zones are being opened for construction due to necessity, and their topography, hydrology, and biological diversity are being destroyed. These areas, which were determined as the borders of forest areas, are gradually becoming included in residential borders, thus causing natural areas to lose their ecological characteristics. Furthermore, since greenfield sites are being built in rural areas, at present, it is observed that these rural areas are beginning to lose their ecological characteristics.

T2-Rural Zone (Arapkoy): This zone covers open fields, cultivated or sparsely settled areas. These are forests, agricultural lands, pastures, and irrigation areas. Several rural settlement areas in the north coastline of Kyrenia including fields, agricultural areas, and forests are opened for construction. They have become a part of the built-to-sell architecture. As is presented in Table 5, in natural zone areas where dense ecological characteristics can be observed, five-story architecture has been permitted. This creates the grounds for both regions to lose their features. There are high numbers of Greyfield sites in these areas; however, there are no legal regulations so far.

T3-Sub Urban Zone (Catalkoy): Although this zone is similar to low-density settlement suburban areas, it differs from those because it hosts many occupational fields. Its vegetation is still natural. The blocks are very wide, and the roads are not compatible with the natural structure of the settlement. Suburban settlements in the north coastline of Kyrenia are residential areas that develop as a continuation of the high-density city and in particular have collective housing and apartments. Its vegetation is disappearing gradually, and there are settlements where infrastructure works that are incompatible with the natural structure are observed.

T4-General Urban Zone (Ozankoy): This zone is denser compared to the zones mentioned above and primarily includes housing settlements. It has various residential architectural orders with mixed-use, including detached and terrace houses. The landscape is varied and medium-sized block islands define the streets. The traditional urban areas of Kyrenia were generally composed of detached or twin houses. However, today this has been combined with a high density of apartments and commercial buildings. The streets are similar throughout the city.

T5-Urban Centre Zone (Kyrenia City Centre): Retail shops, offices, terrace houses, and apartments that are connected to the main roads constitute the center of this zone. It has narrow streets with wide pavements and trees. A high density of commercial buildings is observed in the Kyrenia city center, and apartments are the main living spaces; to a much lesser density there are also houses. It is not possible to describe the use of the mixed area with residential and commercial spaces; however, it is observed that the commercial sites are in zone density. Apart from the center of this zone, the pavements are narrow, and some streets do not have any pavements. Attempts to make the city center green were insufficient.

T6-Urban Core Zone (Kyrenia and surrounding): This is the core of the city. The tallest buildings, and unique and various types of public buildings exist in this zone with very little or no natural vegetation. Detached houses and a high density of tall apartments with ground floors designed as commercial spaces co-exist in certain parts of this zone.

Roads are very crucial in the formation of the city center, its periphery, and suburban and rural areas. In the northern coastline of Kyrenia, the density of buildings, their use, facades, masses; junctions; public spaces; car parking lots; pavements; street silhouettes; lighting; characteristics, and the design of landscape elements and grey areas for each row evidence complication between them. The zones (T1–T6) were created based on the Urban–Rural Scaling Scheme, and starting from Kyrenia city, the characteristics of these zones have been observed to develop irregularly. When the Kyrenia city center is examined because construction permits are given based on the street width, the height of the buildings is irregular. Towards the suburban settlements, the effects of dense construction still can be observed. No regulation has been put in place yet to redevelop the Greyfield sites for public use. It is of utmost importance that the future planning studies consider urban, suburban, and rural area planning as a whole, as in the Transect Theory. Moreover, planning should focus on the re-integration of ecological, economic, and social areas for public use, which is necessary for urban resilience.

Assessing statistical data for construction from 2004 to 2011, and when assessing the questionnaire findings, it can be argued that the problem of Greyfields is an unresolved problem that adversely has been affecting the region economically, ecologically, and socially. Being a touristic and coastal city, Kyrenia attracts many people as well as investors. There are privately owned universities, five-star hotels, and private schools located in this city, making it attractive for those who would like to have a built-to-sell business. Due to the unplanned developments and an increasing number of Greyfields, in 2016 a written decree was released regulating, and compared to before, limiting the conditions of building work. However, it did not provide a solution for the existing problem and caused adversity for those who would request to use their land for their personal use. The existing vacant houses and uncompleted construction problems have not been resolved. This is partly because these buildings are private investments; any intervention from the government in the form of regulation would cause a significant debate among stakeholders. The island of Cyprus was divided into two different societies and administrations due to political division. In this situation, two different currencies (Turkish lira and Euro) are used on the island, and the Turkish lira currency is used in the Kyrenia settlement. Another aspect is that due to international law and current political conflict in Cyprus, the property title deeds that are provided by the government are not recognized internationally. This has been reflecting the on-demand and sellability of these properties to the international market. Although there is a deadlock in terms of international recognition of the title deeds, there are still some interested international buyers, and that keeps the property prices in British pounds. In contrast, as of October 2010, the currency rate was more than 1 to 10. The Turkish lira is calculated over the current exchange rate. This situation makes properties out of reach for most of the local buyers, and no intervention from the government has lead to a sustained state of the existing problem.

Within the scope of the research, no data was found about the Subprime crisis (2008) of this situation.

5. Conclusions

Greyfield sites frequently emerge in suburban and rural areas. Within this context, Greyfield sites that exist in suburban and rural areas should be redeveloped for public use to ensure urban resilience. Transect Theory supports a planning strategy that starts from the urban areas towards rural; thus, it has a crucial role in guaranteeing the usability of Greyfield sites and urban resilience. Therefore, for future planning activities:

- Urban, suburban, and rural settlements are necessary to be considered as a whole.
- It is crucial to acknowledge that urban resilience in rural areas is vital in enabling urban areas as habitable.
- Greyfield sites are ecologically, economically, and socially problematic areas, and it is important to acknowledge the necessity of re-integration of Greyfield sites for public use.

Planning activities that would be carried out within this scope are very crucial in providing a method for bringing urban resiliency to the desired level. This would be enabled through intervention by regular tools that would create an opportunity to redevelop Greyfield sites, which emerge mainly in suburban and rural settlement areas.

Planning activities that would be carried out within this scope provide an opportunity to redevelop Greyfield. These planning activities are crucial in terms of providing a method that would bring urban resilience to the desired level from urban to rural, through regular intervention with tools. Within this context, it is expected that this study would contribute to the decision-making processes related to city planning, city design, and ecological planning.

As a result of the research, the Greyfield situation regarding the city of Kyrenia is as follows in Table 6.

Table 6. Kyrenia Greyfield Area ecological, economical, and social factors.

ECOLOGICAL	• Create pressure on natural environment. Prevent protection of environmental and ecological values • Existence of housing above the area's carrying capacity leads to environmental pollution • They create regional risk areas in term of urban resilience (i.e., natural disasters such as flood and earthquake, technological, and human-originated risks) • Greyfield sites are obstacles to the development of urban and/or rural areas • Greyfield sites limit the agricultural activities, which is a primary source of income • The area is not managed with a plan	RESIDENT AND EXPERT(AGREE/STRONGLY AGREE)
	• Ensure protection and transfer of the region to future generations	RESIDENT AND EXPERT(DISAGREE/STRONGLY DISAGREE)
ECONOMICAL	• Greyfield sites are obstacles to the development of urban and/or rural areas • Greyfield sites limit the agricultural activities, which is a primary source of income • The area is not managed with a plan	RESIDENT AND EXPERT(AGREE/STRONGLY AGREE)
	• Greyfield sites contribute to the future of the regional economy	RESIDENT AND EXPERT(DISAGREE/STRONGLY DISAGREE)
	• Greyfield sites diversity the locals' income sources • Contribute to the development of tourism	RESIDENT (AGREE)/EXPERT (STRONGLY DISAGREE)
SOCIAL	• Greyfield sites are obstacles to the development of urban and/or rural areas • The area is not managed with a plan	RESIDENT AND EXPERT(AGREE/STRONGLY AGREE)
	• Ensure protection and transfer of the region to future generations	RESIDENT AND EXPERT(DISAGREE/STRONGLY DISAGREE)
	• Contribute to the development of tourism	RESIDENT (AGREE)/EXPERT (STRONGLY DISAGREE)

Regarding the ecological issues, the residents and experts disagreed or strongly disagreed about *"ensuring protection and transfer of the region to future generations"*. However, they agreed or strongly agreed on the following aspects: *"creating pressure on the natural environment"*, *"preventing protection of environmental and ecological values"*, *"existence of housing above the area's carrying capacity leads to environmental pollution"*, *"they create regional risk areas in terms of urban resilience (i.e., natural disasters such as flood and earthquake, technological and human-originated risks)"*, *"Greyfield sites are obstacles to the development of urban and/or rural areas"*, *"Greyfield sites limit the agricultural activities which is a primary source of income"*, and *"the area is not managed with a plan"*.

Regarding the economic issues, the residents and experts disagreed or strongly disagreed that *"Greyfield sites contribute to the future of the regional economy"*. However, they agreed or strongly agreed that *"Greyfield sites are obstacles to the development of urban and/or rural areas"*, *"Greyfield sites limit the agricultural activities, which is a primary source of income"*, and *"the area is not managed with a plan'*. In addition, *"Greyfield sites diversify the locals' income sources"* and *"contribute to the development of tourism"* was agreed upon by the residents, while the experts disagreed on these points.

Regarding social issues, the residents and experts disagreed or strongly disagreed that the Greyfields *"ensure protection and transfer of the region to future generations."* However, they agreed or strongly agreed that *"Greyfield sites are obstacles to the development of urban and/or rural areas"* and *"the area is not managed with a plan"*. In addition, the residents agreed that the Greyfields, *"contribute to the development of tourism"*, while the experts disagreed.

As stated above by the residents and experts, in the evaluation of the ecological, economic, and social situation, it was revealed that Greyfield areas are problematic areas for urban resilience in the region. This study revealed the necessity of considering and evaluating the region as a whole, together with the Transect Theory, to ensure the durability of the urban–rural areas for future studies in the region. This study presents the effect of Greyfield areas on the environmental, social, and economic values within the framework of Transect Theory. Thus, a roadmap for reopening gray areas for public use has been proposed for use in future planning studies, which is a requirement for ensuring urban resilience. It is thought that this study will benefit city planners, architects, engineers, municipalities, government organizations, and researchers who will work in this direction.

Author Contributions: Conceptualization, V.A. and A.K.; methodology, V.A. and A.K.; software, V.A.; validation, V.A. and A.K.; formal analysis, V.A.; investigation, V.A.; resources, V.A.; data curation, V.A.; writing—original draft preparation, V.A.; visualization, V.A.; writing—review and editing, A.K.; supervision, A.K. All authors have read and agreed to the published version of the manuscript.

Funding: This research received no external funding.

Institutional Review Board Statement: Not applicable.

Informed Consent Statement: Not applicable.

Data Availability Statement: All data are available publicly as explained in the full article.

Conflicts of Interest: The authors declare no conflict of interest.

References

1. Leichenko, R. Climate change and urban resilience. *Curr. Opin. Environ. Sustain.* **2011**, *3*, 164–168. [CrossRef]
2. Meerow, S.; Newell, J.P.; Stults, M. Defining urban resilience: A review. *Landsc. Urban Plan.* **2016**, *147*, 38–49. [CrossRef]
3. Buyukozkan, G.; Ilıcak, O.; Feyzioglu, O. A review of urban resilience literature. *Sustain. Cities Soc.* **2022**, *77*, 103579. [CrossRef]
4. Elmqvist, T.; Andersson, E.; Frantzeskaki, N.; McPhearson, T.; Olsson, P.; Gaffney, O.; Folke, C. Sustainability and resilience for transformation in the urban century. *Nat. Sustain.* **2019**, *2*, 267–273. [CrossRef]
5. Wardekker, A.; Wilk, B.; Brown, V.; Uittenbroek, C.; Mees, H.; Driessen, P.; .Runhaar, H. A diagnostic tool for supporting policymaking on urban resilience. *Cities* **2020**, *101*, 102691. [CrossRef]
6. Girard, L.F. Creativity and the Human Sustainable City: Principles and Approaches for Nurturing City Resilience. In *Sustainable City and Creativity: Promoting Creative Urban Initiatives*; Routledge: London, UK, 2016; p. 55.
7. Zhai, G.F. The significance and realization path of building a modern metropolitan system. *People's Forum* **2019**, *19*, 58–59. [CrossRef]
8. Yang, X.D.; Li, Z.W.; Zhang, J.Y.; Li, H.L. Spatiotemporal evaluation of urban resilience from the perspective of sustainable development. *Urban Issues* **2021**, *03*, 29–37. [CrossRef]
9. Li, C.J.; Lu, T.; Fu, B.J.; Wang, S.; Holden, J. Sustainable city development challenged by extreme weather in a warming world. *Geogr. Sustain.* **2022**, *3*, 114–118. [CrossRef]
10. Núria, B.P.; Benayas, J.; Jorge, M.R.; Marta, S.; Elías, S.C. The role of urban resilience in research and its contribution to sustainability. *Cities* **2022**, *126*, 103715. [CrossRef]
11. Yang, Y.; Qi, W.; Ma, L.; Liu, Y. Spatial optimization strategies of population function in China's world-class urban agglomerations during 14th five-year plan period. *Proc. Chin. Acad. Sci.* **2020**, *35*, 835–843. [CrossRef]
12. Mallick, S.K.; Das, P.; Maity, B.; Rudra, S.; Pramanik, M.; Pradhan, B.; Sahana, M. Understanding future urban growth, urban resilience, and sustainable development of small cities using prediction-adaptation-resilience (PAR) approach. *Sustain. Cities Soc.* **2021**, *74*, 103196. [CrossRef]
13. Amirzadeh, M.; Sobhaninia, S.; Sharifi, A. Urban resilience: A vague or an evolutionary concept? *Sustain. Cities Soc.* **2022**, *81*, 103853. [CrossRef]
14. Lu, H.; Lu, X.; Jiao, L.D.; Zhang, Y. Evaluating urban agglomeration resilience to disaster in the Yangtze Delta city group in China. *Sustain. Cities Soc.* **2022**, *76*, 103464. [CrossRef]
15. Lu, H.; Zhang, C.; Jiao, L.D.; Wei, Y.; Zhang, Y. Analysis on the spatialtemporal evolution of urban agglomeration resilience: A case study in Chengdu Chongqing Urban Agglomeration, China. *Int. J. Disaster Risk Reduct.* **2022**, *79*, 103167. [CrossRef]
16. Keck, M.; Sakdapolrak, P. What is Resilience? Lessons Learned and Ways Forward. *Erdkunde* **2013**, *67*, 5–19. [CrossRef]

17. Turkoglu, H. Afete Direncli Sehir Planlama ve Yapılasma, İSMEP Rehber Kitaplar Beyaz Gemi Sosyal Proje Ajansı. 2014. Available online: http://www.guvenliyasam.org (accessed on 12 January 2018).
18. Bressi, T.W. (Ed.) *The Seaside Debates: A Critique of the New Urbanism*; Rizzoli: New York, NY, USA, 2002.
19. Walters, D. *Designing Community Charrettes, Masterplans and Form-Based Codes*; Elsevier: Oxford, UK, 2007.
20. Steuteville, R.; Platen, P.; Roberts, R.; Brown, S. *New Urbanism: Comprehensive Reports & Best Practice Guide*; New Urban Publications Inc.: Ithaca, NY, USA, 2001.
21. Greenfield, J. Greyfield Redevelopment in the Greater Toronto Area: Strategies to Overcome Barriers. Unpublished Master' Thesis, Ryerson University, Toronto, ON, Canada, 2013.
22. Newton, P.; Glackin, S.; Witheridge, J.; Garner, L. Beyond small lot subdivision: Pathways for municipality-initiated and resident-supported precinct-scale medium-density residential infill regeneration in greyfield suburbs. *Urban Policy Res.* **2020**, *38*, 338–356. [CrossRef]
23. Newman, P.; Kenworthy, J. *The End of Automobile Dependence: How Cities Are Moving beyond Car-Based Planning*; Island Press: Washington, DC, USA, 2015.
24. Shannon, K. Greyfield Site Development and Environmental Assessments. 2013. Available online: http://info.firstcarbonsolutions.com/blog/bid/317438/Greyfield-Site-Development-and-Environmental (accessed on 5 May 2016).
25. Newton, P.W.; Newman, P.W.; Glackin, S.; Thomson, G. The Global Greyfields Transition: Why Urban Redevelopment in Low-Density, Car-Based Middle Suburbs Needs a New Model. In *Greening the Greyfields*; Palgrave Macmillan: Singapore, 2022; pp. 1–48.
26. Pavlou, K. Out of the Greyzone: Exploring Greyfield Design and Redevelopment. Ph.D. Thesis, The University of Guelph, Guelph, ON, Canada, 2013.
27. Feronti, S.M. Greyfield Redevelopment for Community Revitalization: An Exploration of Applications. Ph.D. Thesis, University of Florida, Gainesville, FL, USA, 2003.
28. Chilton, K. *Greyfields: The New Horizon for Infill and Higher Density Regeneration*; Center for Environmental Policy and Management in University of Louisville: Louisville, KY, USA, 2004.
29. Merritt, A.W. Redeveloping Greyfields: Definitions, Opportunities and Barriers. Master's Thesis, Massachusetts Institute of Technology, Cambridge, MA, USA, 2006.
30. Gilder-Busatti, A. Quantifying the Use of New Urbanism and Smart Growth Principles in the Redevelopment of Greyfield Sites in Central Maryland. Master's Thesis, Morgan State University, Baltimore, MD, USA, 2011.
31. McKay, D. Redeveloping Greyfields in Greater Toronto Area. Master's Thesis, University of Toronto, Toronto, ON, Canada, 2007.
32. Randolph, B.; Freestone, R. *Problems and Prospects for Suburban Renewal: An Australian Perspective*; Research Paper; City Future Research Centre, University of New South Wales: Sydney, Australia, 2008; p. 11.
33. Newton, P.W. Beyond Greenfields and Brownfields: The Challenge of Regenerating Australia's Greyfield Suburbs. *Built Environ.* **2010**, *36*, 81–104. [CrossRef]
34. Newton, P.W.; Tucker, S.N. Hybrid Buildings: Towards Zero Carbon Housing. *Archit. Sci. Rev.* **2010**, *53*, 95–106. [CrossRef]
35. Newton, P.W.; Murray, S.; Wakefield, R.; Murphy, C.; Khor, L.; Morgan, T. *Towards a New Development Model for Housing Regeneration in Greyfield Residential Sites*; AHURI Final Report No. 171; Australian Housing and Urban Research Institute, Swinburne-Monash Research Centre: Melbourne, Australia, 2011.
36. Newton, P.W.; Murray, S.; Wakefield, R.; Murphy, C.; Khor, L.; Morgan, T. How Do We Regenerate Middle Suburban 'Greyfield' Areas? *AHURI Res. Policy Bull.* **2012**, *150*.
37. Atkinson, J.P. Greyfield Development in Vallejo, California: Opportunities, Constraints, and Alternatives. Master's Thesis, Faculty of California Polytechnic State University, San Luis Obispo, CA, USA, 2013.
38. Kichline, M.; Cozzone, K.; Farrell, T. *Commercial Landscape: Transforming Greyfields into Dynamic Destinations*; Chester County Board of Commissioners: Chester County, PA, USA, 2017.
39. Newton, P.; Frantzeskaki, N. Creating a national urban research and development platform for advancing urban experimentation. *Sustainability* **2021**, *13*, 530. [CrossRef]
40. Newton, P.W.; Newman, P.W.; Glackin, S.; Thomson, G. Distributed Green Technologies for Regenerating Greyfields. In *Greening the Greyfields*; Palgrave Macmillan: Singapore, 2022; pp. 71–87.
41. Newton, P.W.; Newman, P.W.; Glackin, S.; Thomson, G. Changing Attitudes to Housing and Residential Location in Cities: The Cultural Clash and the Greyfield Solution. In *Greening the Greyfields*; Palgrave Macmillan: Singapore, 2022; pp. 121–133.
42. Seo, J.W.; Lee, J.L. Characteristics and Retrofit Constraints of Grayfield in Korean Cities. *Sustainability* **2019**, *11*, 2381. [CrossRef]
43. Ramos, R.P. Analysis of Mixed-Use Schemes in Regeneration Areas. Ph.D. Thesis, Ulster University, Coleraine, UK, 2011.
44. Duany, A.; Falk, B. *Transect Urbanism: Readings in Human Ecology*, 1st ed.; Oro Editions: San Francisco, CA, USA, 2020.
45. Nizam, F. Transect urbanism and form-based codes. *Urban. Arhit. Constr.* **2021**, *2021*, 5–9.
46. Duany, A.; Talen, E. Transect planning. *J. Am. Plan. Assoc.* **2002**, *68*, 245–266. [CrossRef]
47. A CNU Journal. Your Guide to a Unifying Urban Theory. 2020. Available online: https://www.cnu.org/publicsquare/2020/12/10/your-guideunifying-urban-theory (accessed on 21 November 2022).
48. Available online: http://devplan.org (accessed on 7 June 2018).
49. Yıldırım, S.; Asilsoy, B.; Özden, Ö. Urban Resident Views about Open Green Spaces: A Study in Guzelyurt (Morphou), Cyprus. *Eur. J. Sustain. Dev.* **2020**, *9*, 441–450. [CrossRef]

50. Talen, E. Help for urban planning: The transect strategy. *J. Urban Des.* **2002**, *7*, 293–312. [CrossRef]
51. A CNU Journal. Great Idea: The Rural-Tourban Transect. 2017. Available online: https://www.cnu.org/publicsquare/2017/04/13/great-idearural-urban-transect (accessed on 12 January 2019).
52. CATS. Center for Applied Transect Studies. 2021. Available online: http://transect.org/index.html (accessed on 21 November 2022).
53. Erkal, T.; Tas, B. *Jeomorfoloji ve İnsan (Uygulamalı Jeomorfoloji)*; Yeditepe Yayınevi: Istanbul, Turkey, 2013.
54. Ozsahin, E. *Kent Planlaması ve Morfolojisi', Kent Calısmaları II*; Karakuyu, M., Keceli, A., Celikoglu, S., Eds.; Pagem Academy: Ankara, Turkey, 2015; pp. 214–231, Chapter: 10.
55. Hoskara, S.; Hoskara, E. Annan Planı Sonrasında Kuzey Kıbrıs'ta İnsaat Sektorune, Mimarlık ve Planlamaya Elestirel bir Bakıs-A Critical Approach to Construction, Architecture and Planning in Northern Cyprus after Annan Plan. *Mimar. Derg.* **2007**, *334*, 53–61.
56. Oktay, M.; Orcunoglu, H. Evaluation of Traditional and Recent Residential Environments from Users' Point of View: The Case of Ozankoy, North Cyprus. In Proceedings of the ENHR 2007 Conference on Sustainable Urban Areas, Rotterdam, The Netherlands, 25–28 June 2007.
57. Akansu, V. Comperative Analysis of House (Ing) Developments in Lapta Town and Kayalar Village Kyrenia District. Master's Thesis, Near East University, Nicosia, Cyprus, 2011.
58. CTBCA. *Kuzey Kıbrıs Insaat Sayım Calısması-1: Asama*; Cyprus Turkish Building Constructors Association: Nicosia, Cyprus, 2011.
59. CTBCA. *Ev Envanter Calısması 2004 Yılı ve Sonrasında İnsa Edilen Evler Rapor 1-Girne Bolgesi*; Cyprus Turkish Building Constructors Association: Nicosia, Cyprus, 2007.
60. CTBCA. *2015 Yılı İnsaat Bilgileri VeriBankası Kitapcıgı*; Cyprus Turkish Building Constructors Association: Nicosia, Cyprus, 2016.
61. Meyer, N.; Auriacombe, C. Good urban governance and city resilience: An afrocentric approach to sustainable development. *Sustainability* **2019**, *11*, 5514. [CrossRef]
62. Zhou, Q.; Zhu, M.; Qiao, Y.; Zhang, X.; Chen, J. Achieving resilience through smart cities? Evidence from China. *Habitat Int.* **2021**, *111*, 102348. [CrossRef]
63. Duany, A.; Robert, P.; Talen, E. A General Theory of Urbanism: Towards a System of Assessment Based upon Garden City Principles (Duany Plater-Zeyberk & Company, 2014). 2018. Available online: http://www.dpz.com/uploads/Books/DRAFT20140624-A_General_Theory_of_Urbanism.pdf (accessed on 30 March 2018).

MDPI

Article

A Low-Carbon Land Use Management Framework Based on Urban Carbon Metabolism: A Case of a Typical Coal Resource-Based City in China

Lingwei Li [1], Yongping Bai [1,*], Xuedi Yang [2], Zuqiao Gao [2], Fuwei Qiao [3], Jianshe Liang [1] and Chunyue Zhang [1]

1 College of Geography and Environmental Science, Northwest Normal University, Lanzhou 730070, China
2 College of Earth and Environmental Sciences, Lanzhou University, Lanzhou 730000, China
3 College of Economics, Northwest Normal University, Lanzhou 730070, China
* Correspondence: baiyp@nwnu.edu.cn

Abstract: It is of great significance to study urban carbon metabolism and explore the low-carbon land use management framework from the perspective of "ecological-production-living" space, an important means for the government to strengthen spatial regulation. In the study, first of all, a carbon metabolism network model was established based on the evolution of the "ecological-production-living" space. Secondly, an ecological network analysis (ENA) method was used to identify the ecological relationships between land use types under the effect of carbon metabolism. In addition, ArcGIS software was used to visualize the spatial distribution of carbon flow and ecological relationships. Finally, a low-carbon oriented land use management framework was proposed based on the above research. Yulin, a typical coal resource-based city in China, was taken as a case study for verification. The results showed that Yulin had net carbon emissions from 2010 to 2020, indicating that the evolution of "ecological-production-living" space had a negative impact on the carbon metabolism. Industrial, mining and transportation land dominated carbon emissions, while forestland played an important role in carbon sequestration. Under the effect of carbon metabolism, a controlling and exploitative relationship was the main ecological relationship, and a mutualism relationship accounted for the smallest proportion, indicating that the urban ecological conflict was obvious in the evolution of the "ecological-production-living" space. Based on the above research, a land use management framework was proposed, which divided urban space into six types of control units. In conclusion, the results provided experience for other coal resource-based cities to promote low-carbon and sustainable land use.

Keywords: carbon metabolism; land use management framework; "ecological-production-living" space; low-carbon

Citation: Li, L.; Bai, Y.; Yang, X.; Gao, Z.; Qiao, F.; Liang, J.; Zhang, C. A Low-Carbon Land Use Management Framework Based on Urban Carbon Metabolism: A Case of a Typical Coal Resource-Based City in China. *Sustainability* **2022**, *14*, 13854. https://doi.org/10.3390/su142113854

Academic Editors: Qingsong He, Jiayu Wu, Chen Zeng and Linzi Zheng

Received: 3 October 2022
Accepted: 20 October 2022
Published: 25 October 2022

1. Introduction

Global climate change caused by increasing carbon emissions has become the focus of attention. Studies have shown that urban areas account for 75% of global carbon dioxide emissions (the Intergovernmental Panel on Climate Change, IPCC, 2006). The increase in urban carbon emissions is mainly due to the expansion of economic scale [1]. Large-scale urbanization caused by economic growth further stimulates energy consumption and land use and land cover change (LUCC) [2,3], which profoundly affects the urban carbon metabolism system. LUCC caused by urbanization is usually characterized by the transition of carbon sink land (e.g., forestland and grassland) to carbon source land (e.g., industrial, mining, and transportation land) [4,5]. Studies have shown that carbon emissions generated by such LUCC account for one third of urban carbon emissions [6]. Therefore, countries worldwide have attempted to optimize the urban carbon metabolism system through adjustments of land use management measures, which has become a strategy to promote low-carbon development and realize sustainable urbanization [7,8].

"Ecological-production-living" space is the general term for an ecological space, production space and living space. Ecological space refers to the geographical space that has an ecological protection function and can provide ecological products and ecological services [9]. Production space refers to a specific functional area where people engage in production activities [10]. Living space refers to the territorial space for people to live, consume, relax and entertain [11]. The 18th Party Congress of the Communist Party of China [12], held in November 2012, is a very important congress held at the decisive stage of building a moderately prosperous society in all respects. The theme of the report is to unswervingly advance along the path of socialism with Chinese characteristics and strive to finish building a moderately prosperous society in all respects. The report made the first summary of the development essentials of "ecological-production-living" space from a strategic perspective, namely "intensive and efficient production space, livable and comfortable living space, and beautiful ecological space". The method of spatial division coincides with the "three pillars" concept of sustainable development: environment, society and economy widely recognized at home and abroad [13]. The sustainable development of the city is based on the sustainable development of "ecological-production-living" space. As a development goal, "ecological-production-living" space has important guiding significance to the urban land layout. Therefore, the analysis of urban carbon metabolism and the discussion of the land use management framework from the perspective of "ecological-production-living" space are helpful for urban low-carbon and sustainable development.

Urban carbon metabolism refers to the flow of carbon between the biosphere, atmosphere and artificial components of the urban system, including carbon emissions from the biosphere to the atmosphere, carbon sequestration from the atmosphere to the biosphere, and carbon transition through LUCC [14]. The balance between these processes determines whether cities function as carbon sources or sinks.

At present, the content of the research on urban carbon metabolism can be summarized into two aspects. On the one hand, the estimation of carbon emissions and carbon sequestration is the basis of carbon metabolism research [15,16]. When estimating carbon emissions, most studies were based on 2006 IPCC Guidelines for National Greenhouse Gas inventories. For example, Liang et al. [17] calculated energy-based CO_2 emissions by the IPCC carbon emissions coefficient method, covering 30 provinces in China. Cellura et al. [18] used the IPCC and the Life Cycle Assessment (LCA) approach to estimate the energy-related greenhouse gas (GHG) balance for Italian cities. However, for the case study of China, the Guidelines for Compiling Provincial GHG Inventories issued by China's National Development and Reform Commission in 2011 may be more suitable, as the emission factors provided in the guideline are obtained by measuring carbon emissions in some Chinese cities, which is more in line with China's national conditions. For example, Ma et al. [19] used the emission factors in the guideline to measure the carbon emissions of Yantai, China. When estimating carbon sinks, researchers mainly used the carbon sink coefficient method to calculate the carbon sinks of carbon sink land (such as woodland, grassland, water, etc.). For example, Kong et al. [20] estimated carbon emissions and sequestration of urban turfgrass systems in Hong Kong. Escobedo et al. [21] analyzed the efficacy of subtropical urban forests in offsetting carbon emissions from cities. On the other hand, analyzing and simulating the carbon metabolism process from the perspective of "network" has gradually become an emerging research direction. In the early stage, some scholars constructed a carbon metabolic network model with socioeconomic sectors as metabolic compartments, aiming to provide suggestions for carbon emission reduction in various industries of the city [22]. With the deepening influence of LUCC on urban carbon emissions, the research on the carbon metabolism network based on land conversion with land use types as metabolic compartments has been gradually carried out [23]. Ecological network analysis (ENA) method was widely used in the analysis of the carbon metabolic network. The reason is that the method can not only quantitatively identify the key nodes and paths that affect the urban metabolic system, but also quantitatively analyze the ecological relationships among components, which can provide information for urban low-carbon construction [24]. For example, Chen et al. [25] took Dongguan as a case study, established a carbon metabolism network model including 42 socio-economic sectors based on the input-output table, and combined with the ENA method to reveal the main contributors and action paths of carbon emissions in all social sectors. The results showed that the total direct and embodied carbon flows were mainly concentrated in the manufacturing industry. Zhang et al. [26] took Beijing as an example, constructed a carbon metabolism model based on LUCC, and used the ENA method to analyze the structure and function of the network, as well as the ecological relationship between components. The results showed that carbon throughflow of the network decreased and positive relations mostly outweighed negative relations.

The research objects of urban carbon metabolism were mainly cities that do not rely on resources and less coal resource-based cities [27,28]. In 2013, the Chinese State Council issued the Sustainable Development Plan for National Resource-based Cities (2013–2020), which clearly points out that resource-based cities are cities with mining and processing of natural resources such as minerals and forests as the leading industries. A coal resource-based city refers to the city with mining and processing of local coal resources as the leading industry. Coal resource-based cities are a kind of special city in the world, mainly distributed in the United States, Russia, Australia, China and other countries with rich coal resources. There are 58 coal resource-based cities in China, accounting for 61% of the number of mining cities. Coal resource-based cities play an important role in China's economic development, but most of them have an extensive economic development mode, large energy consumption and arduous low-carbon construction task. Therefore, they are the main undertakers to achieve the goal of "carbon peaking and carbon neutrality".

In general, the research on urban carbon metabolism has achieved rich results, which have an important reference for this paper, but there are still the following shortcomings: (1) The carbon emission accounting list was not comprehensive enough, especially for farmland and industrial land; (2) As a new target of land use layout at the present stage, few studies studied carbon metabolism from the perspective of "ecological-production-living" space, which cannot fully meet the needs of decision makers for land use management practices; (3) The spatial characteristics of the carbon metabolism network were not sufficiently described; (4) It provided policy suggestions at the macro level, but did not provide a land use management framework from the practical level; (5) Lack of attention to coal resource-based cities, an important carrier of carbon emission reduction.

Therefore, from the perspective of "ecological-production-living" space, the study constructed a carbon metabolism network based on a comprehensive carbon emissions and carbon sinks accounting system and LUCC. Then, the ENA method was used to identify the ecological relationships between components under the effect of carbon metabolism. Additionally, ArcGIS software was used to analyze the spatial characteristics of carbon flow and ecological relationships. Finally, a low-carbon land use management framework based on carbon metabolism was proposed. Figure 1 shows the technical route of the paper. Taking Yulin, a typical coal resource-based city, as an example, the above technical thinking was verified to form a complete and propagable technical system. The following aspects are solved: (1) What is the comprehensive impact of LUCC on urban carbon metabolism? What is the direction and spatial distribution of carbon transfer induced by LUCC? What are the key nodes and paths associated with net carbon emissions? (2) What is the ecological relationship between land use types under the effect of carbon metabolism and its distribution characteristics? (3) How do we construct the land use management framework based on carbon metabolism? The purpose of the study is to provide a scientific basis for the technical path of sustainable land use management in coal resource-based cities.

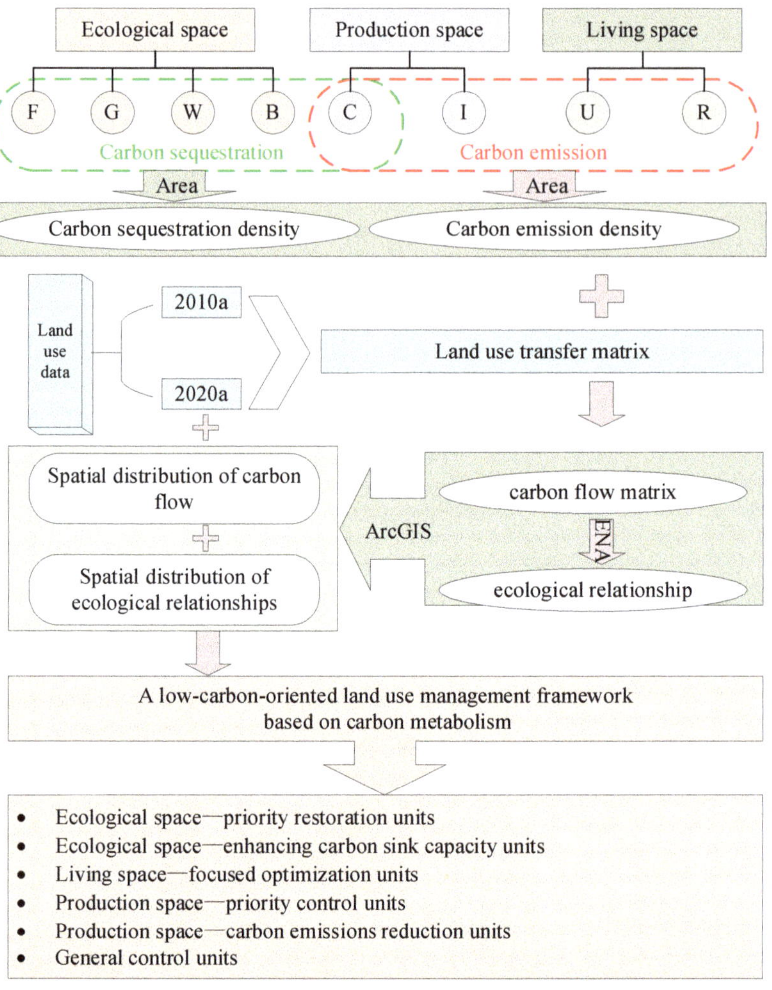

Figure 1. Technical diagram to construct a low-carbon land use management framework. Note: F represents woodland; G represents grassland; W represents water; B represents unused land; C represents cultivated land; I represents industrial, mining and transportation land; U represents urban land; R represents rural residential land.

2. Materials and Methods

2.1. Study Area

Yulin is located in the northernmost part of Shaanxi Province in China (36°57′ N~39°35′ N, 107°28′ E~111°15′ E), at the boundary of the Loess Plateau and the Mu Us Sandy Land. It has jurisdiction over one county-level city, two districts and nine counties, with an area of 4.29×10^4 km^2 (Figure 2). As a city with growing resources, Yulin is rich in energy resources, known as China's "Kuwait", and is China's emerging energy and chemical industry base. At present, 48 kinds of mineral resources in 8 categories have been discovered in Yulin. In the north, there is the Jurassic coal field, one of the seven largest coal fields in the world. In the west, there is the largest integrated gas field in China's interior. In the south, there is the largest rock salt deposit in China. Among them, the most advantageous resource is the coal resource. Yulin has the largest reserves of high quality coal in China. About 54% of the land in Yulin City contains coal; its predicted coal reserves are 2.720×10^{11} t, and its proven reserves are 1.490×10^{11} t, accounting for 12% of China's proven reserves. Yulin relies on coal resources to develop and grow, and the coal chemical industry is the pillar industry of its economic and social development. In the Sustainable Development Plan for National Resource-based Cities (2013–2020) issued by the Chinese State Council, Yulin was defined as a coal resource-based city and the supply and reserve base of China's energy resources. In recent years, Yulin's urbanization and industrialization process accelerated. The process of urban development and expansion relied on coal resources mining, processing and other related industries, causing resource development to rise. While driving regional economic development and social progress, the inertial dependence on carbon-based energy such as coal also brought challenges to the low-carbon development of Yulin. As a typical representative of the growing coal resource-based cities, Yulin is faced with a more complex and urgent task of low-carbon construction under the "carbon peak and carbon neutrality" goal. Therefore, it is typical and universal to select Yulin as the research area to study urban carbon metabolism.

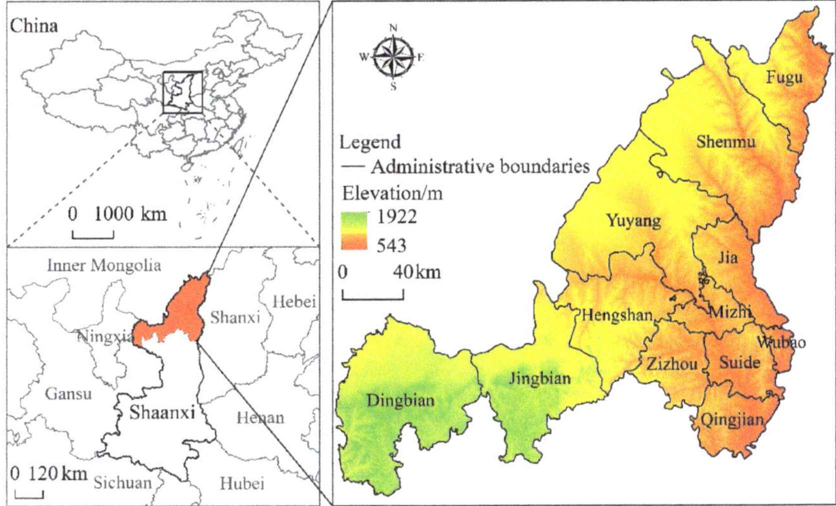

Figure 2. Study area.

2.2. Data Sources

The data used in this paper includes DEM data, land use data and statistical data. DEM data were obtained from Geospatial Data Cloud (http://www.gscloud.cn, accessed on 5 January 2022). The land use data were obtained from Resource and Environment Science and Data Center (http://www.resdc.cn/Default.aspx, accessed on 5 January 2022), with a resolution of 1 km × 1 km. Statistical data were obtained from *Yulin Statistical Yearbook*, *Shaanxi Statistical Yearbook* and *Yulin Statistical Bulletin of National Economic and Social Development*.

2.3. Research Methods

2.3.1. Land Classification of "Ecological-Production-Living" Space

Referring to relevant research [29–32], according to the land use classification system of the Chinese Academy of Sciences, the "ecological-production-living" space was divided on the basis of the dominant function of land use (Table 1).

Table 1. The "ecological-production-living" space classification.

"Ecological-Production-Living" Space	Land Use Type and Code
Ecological space	Woodland (F) Grassland (G) Water (W) Unused land (B)
Production space	Cultivated land (C) Industrial, mining and transportation land (I)
Living space	Urban land (U) [1] Rural residential land (R) [2]

[1] Urban land refers to the land within the scope of planned city proper. [2] Rural residential land refers to the rural settlements independent of the town. Information source is the remote sensing monitoring data of land use types in China from Resource and Environment Science and Data Center of the Chinese Academy of Sciences (http://www.resdc.cn/Datalist1.aspx?FieldTyepID=1,3, accessed on 5 January 2022).

2.3.2. Carbon Metabolism in "Ecological-Production-Living" Space

The carbon metabolism of this paper fully considers the carbon emissions from the biosphere to the atmosphere, the carbon sequestration from the atmosphere to the biosphere, and the potential carbon transfer among land use types in "ecological-production-living" space.

(1) Carbon emission from the biosphere to the atmosphere

Urban carbon emission sources include all land use types of production space and living space. Carbon emissions for these land use types are calculated as follows.

The carbon emission sources of cultivated land mainly include agricultural production activities, animal enteric fermentation, animal manure management and rice fields. Carbon emissions from agricultural production activities are calculated as:

$$C_A = B \times k_B + D \times k_D + E \times k_E + F \times k_F + G \times k_G + H \times k_H + W \times k_W \quad (1)$$

where C_A is the carbon emissions of agricultural production activities; B, D, E and F are the usage of chemical fertilizer, pesticide, agricultural film and agricultural diesel oil, respectively; G and H are the irrigation and planting area of agricultural land, respectively; W is the total power of agricultural machinery; k is the carbon emission coefficient, which refers to Tian and Zhang [33], as shown in the Supplementary Table S3.

The CH_4 emissions from animal enteric fermentation are calculated by multiplying the year-end number by the corresponding emission factors. The CH_4 emissions from animal manure management are calculated by multiplying the number of animals by the corresponding emission factors. The CH_4 emissions from rice field are calculated by multiplying the sown area by its emission factors. The above emission factors refer to Guidelines for Compiling Provincial Greenhouse Gas Inventories, as shown in the Supplementary Tables S3 and S4.

Carbon emissions from industrial, mining and transportation land mainly come from transportation, energy activities and industrial production processes. In the paper, the carbon emission calculation method of transportation in the study refers to Du et al. [14]. Additionally, the greenhouse gas calculation method recommended by the IPCC is adopted to calculate the carbon emissions of energy activities:

$$C_e = \sum_{i=1}^{n} E_i \times N_i \times \delta_i \times O_i \times 44/12 \quad (2)$$

where C_e is the total amount of carbon emissions from energy activities, n is energy types; E_i, N_i, δ_i and O_i are the consumption, average low calorific value, carbon content and carbon oxidation rate of i energy, respectively. The molecular weight ratio of CO_2 and C is 44/12. The relevant index values refer to Wang et al. [34].

For Yulin, this paper considers carbon emissions in the production of cement and calcium carbide. It is obtained by multiplying the output of cement or calcium carbide by the corresponding emission factors, which refer to Guidelines for Compiling Provincial Greenhouse Gas Inventories, as shown in the Supplementary Table S3.

The carbon emissions of urban land mainly come from residents' respiration and living energy consumption, but the carbon emission source of rural land is the burning of straw besides two parts. First of all, the carbon emissions of residents' breathing are calculated by multiplying the urban or rural permanent population by the carbon emission coefficient of human breathing. The carbon emission coefficient of human breathing is 79 kg/per./a. [14]. Secondly, the carbon emissions of living energy consumption are calculated according to Equation (2). Finally, the carbon emissions from straw combustion are calculated as Equation (3):

$$C_s = \sum_{i=1}^{m} P_i \times S_i \times \theta_i \times a \times b \tag{3}$$

where C_s is carbon emissions from straw burning; m is the number of crop species; P_i is the production of *i* crop; S_i and θ_i are the ratio of grain to straw and the carbon emission coefficient of straw burning of *i* crop, respectively; *a* is the open burning ratio of straw; *b* is the burning efficiency of straw. The values of above indicators refer to Du et al. [14].

(2) Carbon sequestration from the atmosphere to the biosphere

The components that play the role of carbon sink are all land use types of ecological space and cultivated land. Their carbon sinks are calculated as follows.

The formula for calculating the carbon sink of woodland, grassland, water and unused land is:

$$C_i = S_i \times K_i \tag{4}$$

where C_i, S_i and K_i are carbon sink, area and carbon sink coefficients of *i* component, respectively. The carbon sink coefficients refer to existing research [35–37] and are shown in the Supplementary Table S1.

The carbon sequestration of cultivated land is calculated by Equation (5):

$$C_c = \sum_{i=1}^{n} P_i / H_i \times (1 - r_i) \times f_i \tag{5}$$

where C_c is the carbon sequestration of cultivated land, *n* is the number of crop species, P_i, H_i, r_i, f_i are the economic output, economic coefficient, moisture content and carbon absorption rate of the *i* crop, respectively. The values of relevant indicators refer to Tian and Zhang [33], as shown in the Supplementary Table S2.

(3) Carbon flow caused by LUCC

A carbon flow from component *j* to component *i* was represented as f_{ij} [38].

$$f_{ij} = \Delta W \times \Delta S \tag{6}$$

$$\Delta W = W_j - W_i = V_j / S_j - V_i / S_i \tag{7}$$

where ΔW is the difference in carbon metabolism density between component *j* and *i*; ΔS is the area difference between *j* and *i*; W_j and W_i are the net carbon metabolism density of *j* and *i*, respectively; V_j and V_i are the net carbon metabolism of component *j* and *i*, respectively; S_j and S_i are area of component *j* and *i*, respectively.

2.3.3. Ecological Network Utility Analysis

In order to characterize the effective direct utility amount among the components, the direct utility matrix D is defined, as shown in Equation (8) [39]. Based on this, we calculate the dimensionless total utility matrix U, as shown in Equation (9) [39]:

$$d_{ij} = (f_{ij} - f_{ji})/T_i \tag{8}$$

$$U = (u_{ij}) = D^0 + D^1 + \cdots D^n = (I - D)^{-1} \tag{9}$$

where d_{ij} is the element of matrix D; f_{ij} is the direct flow from j to i; f_{ji} is the direct flow from i to j; T_i is the carbon flux of i; u_{ij} is the element of matrix U; n is the number of components; I is identity matrix.

In the matrix U, positive elements represent positive utility, and negative elements represent negative utility. According to the matrix U, the ecological relationship between the two components can be judged. The possible types of ecological relationships are shown in the Supplementary Table S5. Among them, four common ecological relationships are competition, control, exploitation, and mutualism. Control and exploitation relationships indicate that one component benefits from the relationship and the other is harmed. Competitive relationships indicate that two compartments compete with each other and both lose their utility. Mutualism relationships indicate that both compartments increase their utility in the process of interaction.

The utility of the ecological relationship to the system is evaluated using the mutualism index M, as shown in Equation (10) [39]:

$$M = N_+/N_- \tag{10}$$

where N_+ and N_- are the numbers of positive and negative elements in matrix U, respectively. $M > 1$ indicates that the evolution of "ecological-production-living" space has a positive effect on the urban carbon metabolism balance, and the larger the value, the stronger the positive effect; when $M < 1$, the effect is opposite.

3. Results

3.1. Carbon Metabolism in "Ecological-Production-Living" Space

Table 2 shows the carbon emissions and carbon sinks of Yulin City's "ecological-production-living" space in 2010 and 2020. In terms of carbon emissions, Yulin's total carbon emissions in 2020 were 4.46 times that of 2010. The dominance of carbon emissions from industrial, mining and transportation land during the study period was prominent, accounting for about 99% of the total carbon emissions on average, indicating that the carbon emissions of industrial, mining and transportation land determined the total carbon emissions of Yulin City. Therefore, the carbon emission reduction work of Yulin City needs to start from the carbon emission reduction in the industrial and transportation industry. In 2020, compared with 2010, the carbon emissions of rural land decreased by 14.78%, and the carbon emissions of cultivated land, urban land, industrial and transportation land increased by 32.03%, 28.37% and 348.33%, respectively. Figure S1 shows the distribution of "ecological-production-living" space in Yulin City in 2010 and 2020. Combined with Figure S1, the reasons are analyzed. It is found that the area of rural land increased from 2010 to 2020, but its carbon emissions decreased. This indicates that the carbon emission density of rural land decreased from 2010 to 2020. Possible causes are as follows. On the one hand, with the implementation of the rural revitalization strategy, the farmers' production and living conditions have improved, the level of rural electrification has enhanced, and rural activities such as cooking and heating have been gradually replaced by clean energy, so the carbon emissions from rural coal fired have reduced. On the other hand, in recent years, with the implementation of a ban on straw burning in Yulin city, carbon emissions from straw burning in rural areas have decreased. However, with the continuous improvement of the rural mechanization level, the widespread use of chemical fertilizers, pesticides and agricultural machinery has led to an increase in carbon emissions from cultivated land. In addition, Yulin City is in the stage of rapid industrialization and urbanization, so that energy consumption is rapidly increasing. As a result, carbon emissions from urban land as well as industrial and transportation land during the study period increased.

Table 2. Carbon emissions and carbon sinks of "ecological-production-living" land in Yulin ($\times 10^4$ t).

Land Use Type	2010		2020	
	Carbon Emissions	Carbon Sinks	Carbon Emissions	Carbon Sinks
C	26.54	119.23	35.04	192.26
F		1349.03		1411.92
G		3.93		4.00
W		1.37		1.18
U	12.55		16.11	
R	15.16		12.92	
I	8524.81		38,219.30	
B		0.02		0.02
Sum	8579.07	1473.59	38,283.36	1609.38

In terms of carbon sinks, the total carbon sinks of Yulin in 2020 were 1.10 times that of 2010. Forest land was the most important components of carbon sink in 2010 and 2020, accounting for 89.64% of the total carbon sinks on average. In addition, the carbon sink capacity of cultivated land could not be ignored, whose carbon sinks accounted for about 10.02% of the total carbon sinks on average. Compared with 2010, the carbon sinks in ecological space increased by 4.63% in 2020. Among them, the carbon sinks of forest land and grassland increased by 4.66% and 1.78%, respectively, the carbon sinks of waters decreased by 13.87%, and the carbon sinks of unused land remained unchanged. It can be seen from Figure S1 that the area of the ecological space decreased from 2010 to 2020, but here we see an increase in its carbon sink. The reason is that the area of forest land and grassland in ecological space increased, indicating that the carbon sink role of forest land and grassland in the ecological space is more obvious. In addition, from 2010 to 2020, the carbon sinks of cultivated land increased by 61.25%, but according to Figure S1, the area of cultivated land decreased. This indicates that the carbon sink capacity of cultivated land was enhanced from 2010 to 2020. Overall, the carbon emissions of Yulin City in 2010 and 2020 were greater than the carbon sinks, and the carbon emissions were 5.82 and 27.79 times the carbon sinks, respectively, indicating that the carbon metabolism imbalance in Yulin City intensified during the study period.

Based on the land use transition matrix (Table S6), the "carbon flow" matrix of Yulin's "ecological-production-living" space was obtained (Table S7). Figure 3 shows the carbon flow network among components. The results showed that the net carbon flow in Yulin from 2010 to 2020 was -2.76×10^8 t, which indicated that the evolution of Yulin's "ecological-production-living" space had led to an increase in carbon emissions. Among them, the main path for negative carbon flow was the transfer of ecological space and cultivated land into industrial, mining and transportation land, accounting for 81.24% and 16.24% of the total negative carbon flow, respectively. The positive carbon flow was only 26.07% of the negative carbon flow; its main path was the transfer of industrial, mining and transportation land into various lands of ecological space and cultivated land, accounting for 25.66% and 12.07% of the total positive carbon flow, respectively. This indicated that industrial, mining and transportation land was the key component of carbon metabolism in Yulin city. In addition, the transfer of other land use types to forest land was also an important path for positive carbon flow, and the positive carbon flow on the path accounted for 10.37% of the total positive carbon flow. Among them, the positive carbon flow along the path from cultivated land to forest land accounted for 42.69%, indicating that the implementation of the project of returning farmland to forestland had a certain promoting effect on the balance of urban carbon metabolism.

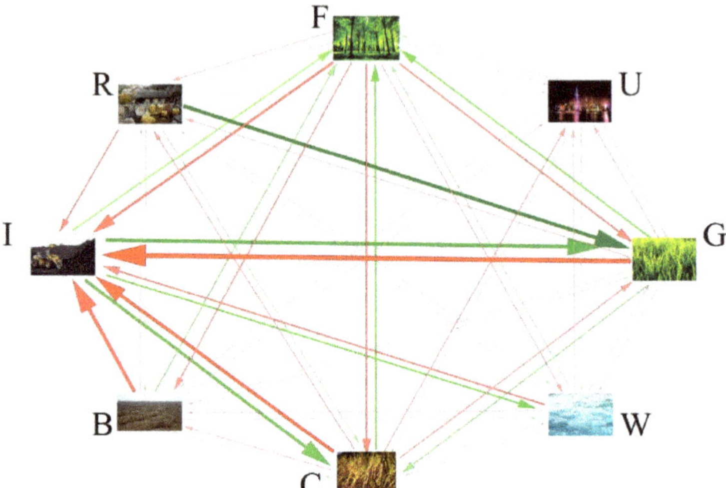

Figure 3. Carbon flow network. Note: Green represents positive carbon flow, red represents negative carbon flow, and the darker the color, the greater the flow.

In order to more intuitively display the spatial distribution of net carbon flow in the "ecological-production-living" space of Yulin city from 2010 to 2020, the study finally divided both positive and negative carbon flow into four grades, which achieved the best grading effect, as shown in Figure 4. From 2010 to 2020, the patches without carbon transfer in the horizontal direction of Yulin City accounted for 55.76% of the total area, and the patches with negative carbon flow and positive carbon flow accounted for 23.24% and 21.00%, respectively. Although the area of the plaques with positive and negative carbon flow was not much different, it could be seen from above that the overall net carbon flow was a large negative number, which indicated that the carbon emission intensity of the plaques with negative carbon flow was far greater than that with positive carbon flow. It could be seen from Figure 4a that according to the flow grade from low to high, the area proportions of the plaques with negative carbon flow accounted for 77.76%, 1.40%, 13.75% and 7.09%, respectively. The lowest grade patches with negative carbon flow accounted for the largest proportion, whose carbon flow was mainly caused by the transfer of cultivated land to grassland. In addition, although the plaques with carbon flow higher than 100×10^4 t only accounted for 7.09%, their contribution to the negative carbon flow was 54 times that of other grades of plaques. Moreover, the study found that the spatial distribution of these patches with the highest grade of negative carbon flow was consistent with that of the newly added industrial, mining and transportation land mentioned above, which further highlighted the negative effects of industrial, mining and transportation land on urban carbon metabolism. It could be seen from Figure 4b that according to the flow grade from low to high, the area proportions of the plaques with positive carbon flow accounted for 82.02%, 0.97%, 16.73% and 0.28%, respectively. The lowest grade patches with positive carbon flow accounted for the largest proportion and were widely distributed, whose carbon flow was mainly caused by the transfer of grassland to cultivated land and the transfer of unused land to grassland. In view of the small area, small number, and scattered distribution of patches with the highest grade of positive carbon flow, green circles were added around them for convenient observation. It was found that the plaques with the highest grade of positive carbon flow were mainly distributed in the coal mining areas of Shenmu, Fugu and Yuyang. This was closely related to the fact that the Yulin City government took the Shenmu-Fugu coal distribution area, Yulin-Shenmu coal distribution area and Fugu coal distribution area as key governance areas of the mining environment during the study period, and carried out vegetation restoration and land reclamation for abandoned mines.

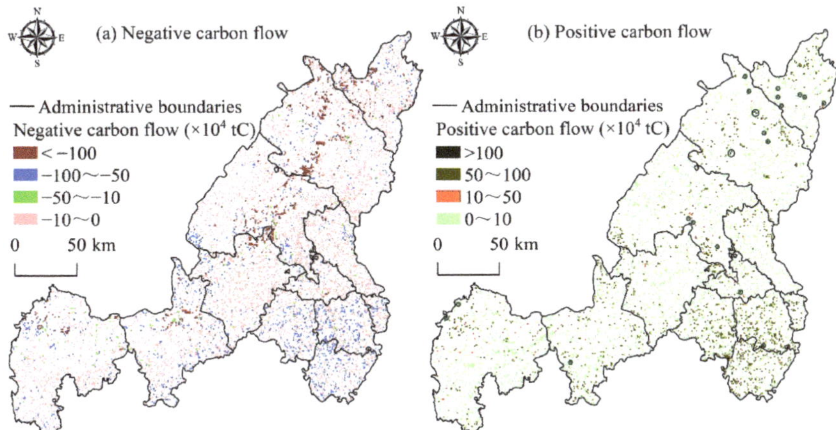

Figure 4. Spatial distribution of net carbon flow in "ecological-production-living" space of Yulin City from 2010 to 2020.

3.2. Ecological Relationships among "Ecological-Production-Living" Space Land

Figure 5 shows the ecological relationships between various land use types in the "ecological-production-living" space in Yulin City under the action of carbon metabolism, including the four relationships: competition, control, exploitation, and mutualism. Among them, control and exploitation are identical in essence although they are presented in opposite ways. Therefore, these two relationships are combined for statistics in the study. The results showed that the number of control and exploitation relationships accounted for 60.72% of the total, and they were widely distributed among the components. Among them, grassland, industrial, mining and transportation land were very important components of the exploitation relationship, both of which accounted for 23.53% of the relationship. It showed that the increase in grassland area and the expansion of industrial, mining and transportation land had important effects on urban carbon metabolism. Competition relationships accounted for 28.57% of all relationships, 75% of which existed in cultivated land and the land of ecological space. This indicated that under the action of the system, there was fierce competition for carbon storage among land use types, including cultivated land and all types of land in ecological space. The reason was that all these land use types have the function of carbon sequestration, and competition between them was inevitable under the premise of limited carbon storage in the system. Mutualism relationships accounted for only 10.71% of all relationships. There were mutualism relationships between grassland and urban land, between grassland and industrial, mining and transportation land, and between urban land and rural land. Under the premise that carbon emission was fixed and there was no obvious land transfer between these two types of land, the carbon emitted by urban land and industrial, mining and transportation land was absorbed by grassland, which promoted the expansion of grassland. At the same time, the enhanced carbon absorption brought about by the expansion of grassland also reduced the pressure on the carbon emission of the system, which was beneficial to the expansion of urban land and industrial and transportation land. For urban land and rural land, although the direct "carbon flow" will not bring about a mutualism relationship, the transfer of these two types of land to other types of land to a certain extent will reduce the carbon emissions of these two types of land relatively. In general, the carbon metabolism system of "ecological-production-living" space in Yulin was dominated by the control and exploitation relationship, and the proportion of mutualism relationships in all relationships was smaller than that of competition relationships. This showed us that the relationship among the components in the carbon metabolism system was relatively conflicted and the reciprocal level of the carbon metabolism system was not high.

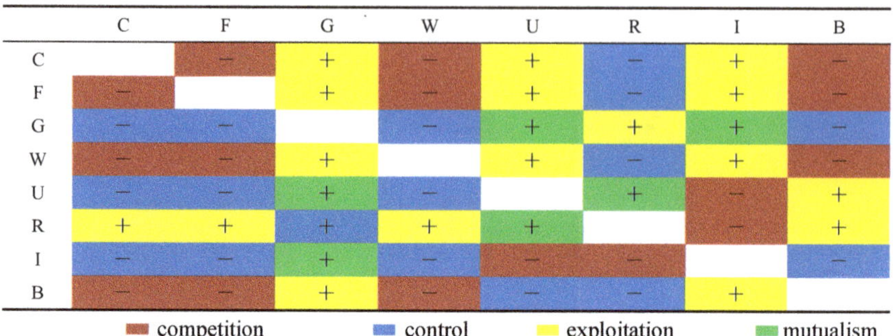

Figure 5. Ecological relationships between components of "ecological-production-living" space.

The spatial distribution of ecological relationships is shown in Figure 6. The white area in the figure is the area where the land use type has not changed. It could be seen from the figure that the control and exploitation relationships were widely distributed, mainly in Shenmu, Dingbian and Yuyang, accounting for 30.70%, 30.35% and 28.42% of the relationship, respectively. This is because the area of mutual transfer between cultivated land and grassland, grassland and rural land, forestland and grassland in these areas is relatively large. The competition relationships were mainly distributed in Yuyang, Jingbian and Suide, accounting for 16.11%, 14.15% and 12.00%, respectively. The main reason is that the area of mutual transfer between rural land and cultivated land, and cultivated land and forestland, in these areas is relatively large. In addition, there was also a large transfer area between cultivated land and forest land. The mutualism relationships were mainly concentrated in Shenmu, Fugu and Yuyang, accounting for 45.41%, 21.00% and 16.27%, respectively. The reason for this is that the transfer area between grassland and industrial, and mining and transportation land, in these areas was relatively large. Generally speaking, the distribution of control and exploitation relationships were relatively broad, the distribution of competition relationships were relatively scattered, and the distribution of mutualism relationships were more concentrated.

The mutualism index M considers both the ecological relationship of direct and indirect carbon flow, which reflects the impact of the comprehensive effect of carbon flow on urban carbon metabolism, and can more comprehensively evaluate the carbon metabolism of urban "ecological-production-living" space. The value of M obtained by the calculation is $0.94 < 1$, indicating that the positive effect of carbon flow is less than the negative effect in the carbon metabolism system, and the evolution of the "ecological-production-living" space exacerbates the disturbance of urban carbon metabolism, which is consistent with the calculation results of net carbon flow.

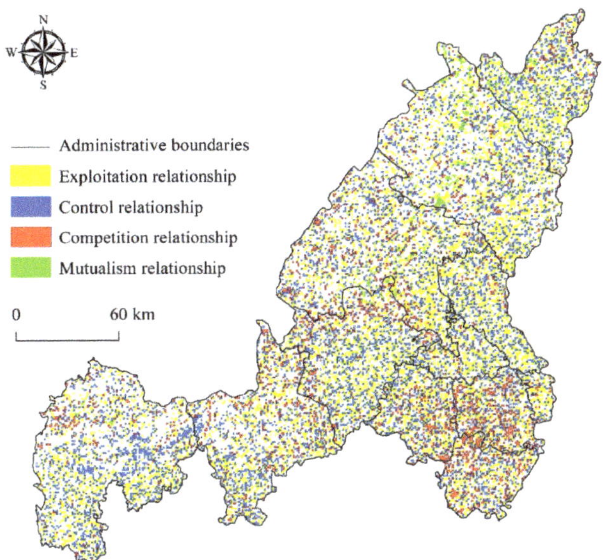

Figure 6. Spatial distribution of ecological relationships.

3.3. Low-Carbon Land Use Management Framework in"Ecological-Production-Living" Space

According to the dominant factors such as the distribution pattern of carbon flow and its ecological relationships, the status quo of "ecological-production-living" space and so on, a low-carbon land use management framework was proposed. In the paper, an overlay analysis was carried out on the spatial distribution map of Yulin's "ecological-production-living" space in 2020, and the spatial distribution map of carbon flow and its ecological relationships from 2010 to 2020, and the whole area of Yulin was divided into six types of management and control units (Figure 7).

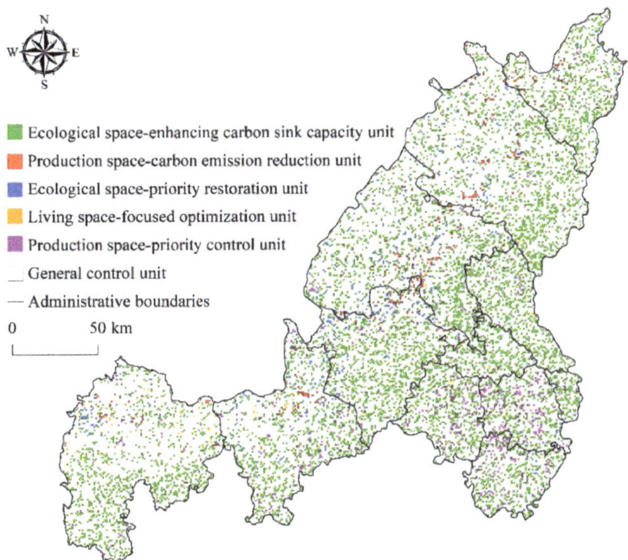

Figure 7. Management and control units of "ecological-production-living" space in Yulin.

(1) Ecological space—priority restoration units

These units refer to the area with negative carbon flow and a competitive ecological relationship in the ecological space at the end of the study period. The main reason is that water and bare land with a lower carbon sink density compete with cultivated land and forest land with a higher carbon sink density, resulting in a decline in the carbon sink capacity of the system and a negative effect on urban carbon metabolism. These units accounted for 1.27% of the whole area, and were mainly distributed in Yuyang, Hengshan, Shenmu, Jingbian and Dingbian, accounting for 26.67%, 18.70%, 17.04%, 16.67% and 10.56%, respectively. The changes in land use types during the study period of these units indicated that there were phenomena of human destruction of the ecological environment, such as deforestation and abandonment of land. Therefore, it was necessary to strengthen management, resolutely stop and punish all man-made destruction activities that lead to the continuous deterioration of ecological functions, and it was also necessary to strengthen the follow-up construction work such as supplementary afforestation and land reclamation.

(2) Ecological space—enhancing carbon sink capacity units

These units refer to the area with negative carbon flow and control and the exploitation relationship in ecological space at the end of the study period. Grassland can be found to be a very important component in forming an exploitation relationship. A total of 87.81% of such units were derived from the exploitation of other land by grassland. Among them, 78.11% of the units originated from the exploitation of cultivated land by grassland, which may be influenced by the policy of returning farmland to grassland. For these units, on the basis of suitability evaluation, priority should be given to restoring the cultivated land that needs to be transferred to forest land, so as to improve the carbon sequestration capacity. In addition, 7.9% of the units originated from the exploitation of forest land by grassland. However, it was the large carbon sink density difference between grassland and forestland that caused 91.31% of the negative carbon flow. These units accounted for 18.54% of the whole area, and were widely distributed but mainly in Shenmu, Yuyang, Dingbian, Hengshan and Jingbian, accounting for 15.66%, 14.11%, 12.98%, 11.08% and 10.61%, respectively. These units may be anthropogenically affected by logging, mining, etc., so the forest land is degraded and turned into grassland. Therefore, it was necessary to revegetate these units to improve the ecological restoration effect.

(3) Living space—focused optimization units

These units refer to the area with negative carbon flow and control and exploitation relationship in the living space at the end of the study period. They accounted for 0.32% of the total area, mainly distributed in Dingbian, Jingbian, Yuyang, Shenmu and Hengshan, accounting for 33.33%, 15.22%, 11.59%, 9.42% and 9.42%, respectively. They originated from the exploitation of ecological space and cultivated land by living space, resulting in an increase in carbon emission sources and a decrease in carbon sinks, which had an important impact on the balance of urban carbon metabolism. Therefore, the expansion of such units should be controlled in time to avoid the exploitation of the carbon sink land. At the same time, resource allocation should be used to attract the population to live in a concentrated area, so as to improve the efficiency of resource utilization, and then achieve the goal of reducing carbon emissions.

(4) Production space—priority control units

These units refer to the area with a negative carbon flow and competitive ecological relationship in the production space at the end of the study period. They accounted for 1.44% of the total area, mainly distributed in Suide, Qingjian, Zizhou, Jingbian and Dingbian which belong to the loess hilly region, accounting for 22.39%, 15.85%, 14.71%, 14.05% and 8.50%, respectively. They mainly included the following two types. One was the competition between industrial, mining and transportation land and rural land; such units accounted for only 0.33%. They reflected the industrialization of rural areas to a certain extent, and were conducive to promoting economic development and social progress in rural areas. Therefore, it was necessary to accelerate the exploration of low-carbon technologies and reduce carbon emissions in the production process. The other was the competition between cultivated land and forest land; such units accounted for 99.67%. The reason for this was that in order to achieve cultivated land requisition-compensation, the supplementary cultivated land mainly came from forestland. Therefore, these units should be based on the premise of farmland protection and ecological protection, and the farmland that is not suitable for farming should be transferred into forest land to enhance carbon sinks.

(5) Production space—carbon emissions reduction units

These units refer to the area with a negative carbon flow and a control and exploitation relationship in the production space at the end of the study period. They were mainly the exploitation of ecological space and cultivated land by industrial and transportation land, and their negative carbon flow exceeds 5.00×10^8 t, which seriously affected the carbon metabolism balance of Yulin. These units accounted for 0.81% of the whole area, and were mainly distributed in Yuyang, Shenmu, Hengshan, Fugu and Jingbian, accounting for 26.88%, 24.57%, 13.01%, 12.14% and 10.98%, respectively. The main reason is that these areas are the coal mines of Yulin. As a growing resource-based city, Yulin's resource exploitation is on the rise. Especially under the "dual circulation" strategy, it is bound to speed up resource exploitation and utilization. Therefore, on the one hand, these units should eliminate high energy-consuming industries and support new energy industries; on the other hand, it is necessary to optimize the layout of industrial production space, establish industrial parks to form low-carbon, efficient, and high-quality production spaces.

(6) General control units

These units refer to areas other than the above-mentioned units. They account for 77.62% of the total area and are widely distributed. They include three types, namely units with no change in land use type during the study period, units with positive carbon flow, and units with negative carbon flow but with a mutualism relationship. Such units have no negative effect on the urban carbon metabolism balance. Therefore, under the premise of prioritizing the quality of the ecological environment, management and utilization should be carried out according to the original land use type, and a low-carbon mode of land use should be sought at the same time.

4. Discussion

The study analyzed the carbon metabolism process of the "ecological-production-living" space in Yulin, a coal resource-based city, and found that the carbon emissions from industrial, mining and transportation land dominated the total carbon emissions, and forest land was the most important component of carbon sink. This is the same as the research results of Xia [24], but different from those of Du et al. [14]. Xia [24] analyzed the carbon metabolism process in Hangzhou from the perspective of land use, and found that industrial and transportation land was the main source of carbon emissions in Hangzhou from 1995 to 2015, whose carbon emissions accounted for about 86.89% of the total carbon emissions on average, and forest land is the most important component for carbon sink, whose carbon sinks accounted for 70% of the total carbon sink on average. Du et al. [14] studied the carbon metabolism of the "ecological-production-living" space in Zhaotong City, a coal resource-based city. He took the industrial land and road land as two independent components and compared its carbon emissions with those of the cultivated land, urban land, and rural land. It was found that cultivated land had the largest carbon emissions in 2010, and industrial land had the largest carbon emissions in 2018. However, by summing up the carbon emissions of industrial land and road land, it was found that in 2010 and 2018, the carbon emissions of industrial land and road land in Zhaotong City accounted for 41.96% and 79.00%, respectively, which is the most important source of carbon emissions. Compared with this study, the proportion of carbon emissions from industrial and road land in Zhaotong City is lower than that of Yulin City. The possible reason is that the industrial development level of Zhaotong City is lower than that of Yulin City. For example, in 2010, the industrial added value of Yulin City was about 7.44 times that of Zhaotong City. Therefore, the industrial carbon emission of Zhaotong City is lower than that of Yulin City. In addition, Du et al. [14] found that cultivated land was the most important land for carbon sink in Zhaotong City during the study period, which was different from the results of the study. The possible reason is that the proportion of the added value of Zhaotong City's primary industry in GDP is higher than that of Yulin. For example, in 2010, the proportion of primary industry in Zhaotong City and Yulin City was 19.5% and 5.3%, respectively. Therefore, the proportion of crop planting areas in Zhaotong City may be higher than that in Yulin City, resulting in a more prominent role of cultivated land as a carbon sink. To sum up, irrespective of whether they are resource-based cities or non-resource-based cities, industrial and transportation land are the main sources of carbon emissions in cities, and forest land and cultivated land are the main components of carbon sinks in cities.

The limitation of the study is that the resolution of land use data is coarse. This is because the land use classification of the higher resolution land use data in Yulin at this stage cannot meet the needs of the land use classification of "ecological-production-living" space in this study, so the data with a resolution of 1 km is selected. In the future, it is necessary to use higher-resolution spatial data to carry out research to accurately grasp the spatial evolution of Yulin City's "ecological-production-living" space and carbon flow changes, so as to formulate regional land use management framework more accurately.

5. Conclusions

Taking Yulin, a typical coal resource-based city in China, as an example, we proposed a low-carbon land use management framework based on the urban carbon metabolism network. The results showed that industrial, mining and transportation land had the largest carbon emissions, accounting for more than 90% of the total carbon emissions. Carbon sequestration of forestland was prominent, accounting for 89.64% of the total carbon sequestration. The main path of negative carbon flow was the transfer of ecological land and cultivated land to industrial, mining and transportation land. The positive carbon flow was mainly caused by the transfer of industrial, mining and transportation land to ecological land and cultivated land. The ecological relationships between land use types under the effect of carbon metabolism was dominated by control and exploitation relationships, and mutualism relationship accounted for the lowest proportion. The value of mutualism index M was $0.94 < 1$, indicating that the evolution of "ecological-production-living" space in Yulin had a negative effect on urban carbon metabolism. Based on the spatial analysis of carbon flow and the ecological relationship, combined with the current situation of "ecological-production-living" space, and oriented by low-carbon development, six types of control units are divided: Ecological space—priority restoration units; Ecological space—enhancing carbon sink capacity units; Living space—focused optimization units; Production space—priority control units; Production space—carbon emissions reduction units; General control units. This provides a reference framework of low-carbon land use management on a practical level for international coal resource-based cities.

Supplementary Materials: The following supporting information can be downloaded at: https://www.mdpi.com/article/10.3390/su142113854/s1, Figure S1: "ecological-production-living" space of Yulin City in 2010 and 2020; Table S1: Carbon sink coefficients of ecological land; Table S2: Economic coefficient, moisture content and carbon absorption rate of main crops in Yulin; Table S3: Carbon emission coefficients of production space; Table S4: Carbon emission coefficients of major animals; Table S5: Ecological relationships classification in ecological network analysis; Table S6:"Ecological-production-living" land transfer matrix of Yulin from 2010 to 2020 (10^4 hm^2); Table S7: Carbon flow matrix of "ecological-production-living" space in Yulin from 2010 to 2020 (10^4 t).

Author Contributions: Conceptualization, L.L. and F.Q.; methodology, L.L. and X.Y.; software, L.L. and Z.G.; validation, L.L. and J.L.; formal analysis, L.L. and J.L.; investigation, L.L. and C.Z.; resources, L.L. and Y.B.; data curation, L.L. and X.Y.; writing—original draft preparation, L.L.; writing—review and editing, Y.B., X.Y., Z.G., F.Q., J.L. and C.Z.; visualization, L.L. and C.Z.; supervision, Y.B.; project administration, Y.B.; funding acquisition, Y.B. All authors have read and agreed to the published version of the manuscript.

Funding: This research was funded by "the National Natural Science Foundation of China, grant number 40771054"; "the Key R&D Program of Gansu Province, grant number 18YF1FA052"; "the Research Fund for the Doctoral Program of Higher Education of China, grant number 20106203110002".

Institutional Review Board Statement: Not applicable.

Informed Consent Statement: Not applicable.

Data Availability Statement: Not applicable.

Conflicts of Interest: The authors declare no conflict of interest.

References

1. Zhang, W.; Duan, X. The research progress in the relationship among economic growth, industrial structure, and carbon emissions. *Prog. Geogr.* **2012**, *31*, 442–450.
2. He, W.J.; Zhang, B. A comparative analysis of Chinese provincial carbon dioxide emissions allowances allocation schemes in 2030: An egalitarian perspective. *Sci. Total Environ.* **2021**, *765*, 142705. [CrossRef] [PubMed]
3. Liu, X.P.; Wang, S.J.; Wu, P.J. Impacts of urban expansion on terrestrial carbon storage in China. *Environ. Sci. Technol.* **2019**, *53*, 6834–6844. [CrossRef] [PubMed]
4. Abbasi, M.A.; Parveen, S.; Khan, S.; Kamal, M.A. Urbanization and energy consumption effects on carbon dioxide emissions: Evidence from Asian-8 countries using panel data analysis. *Environ. Sci. Pollut. Res.* **2020**, *27*, 18029–18043. [CrossRef] [PubMed]
5. Hong, C.; Burney, J.A.; Pongratz, J.; Nabel, J.; Mueller, N.D.; Jackson, R.B.; Davis, S.J. Global and regional drivers of land-use emissions in 1961–2017. *Nature* **2021**, *589*, 554–561. [CrossRef] [PubMed]
6. Friedlingstein, P.; Houghton, R.A.; Marland, G.; Hackler, J.; Boden, T.A.; Conway, T.J.; Canadell, J.G.; Raupach, M.R.; Ciais, P.; Le Quéré, C. Update on CO_2 emissions. *Nat. Geosci.* **2010**, *3*, 811–812. [CrossRef]
7. Pacheco-Torres, R.; Roldán, J.; Gago, E.J.; Ordóñez, J. Assessing the relationship between urban planning options and carbon emissions at the use stage of new urbanized areas: A case study in a warm climate location. *Energy Build.* **2017**, *136*, 73–85. [CrossRef]
8. Zhang, R.; Matsushima, K.; Kobayashi, K. Can land use planning help mitigate transport-related carbon emissions? A case of Changzhou. *Land Use Policy* **2018**, *74*, 32–40. [CrossRef]
9. Xu, W. The connotation, relationship and optimization path of "ecological-production-living" space. *Dongyue Trib.* **2022**, *43*, 126–134.
10. Li, J.; Sun, W.; Yu, J. Change and regional differences of production-living-ecological space in the Yel-low River Basin: Based on comparative analysis of resource-based and non-resource-based cities. *Resour. Sci.* **2020**, *42*, 2285–2299.
11. Dou, Y.; Ye, W.; Li, B.; Liu, P. Tourism adaptability of traditional village based on living-production-ecological spaces: A case study of Zhangguying Village. *Econ. Geogr.* **2022**, *42*, 215–224.
12. Hu Jintao's Report at the Eighteenth National Congress of the Communist Party of China. Available online: http://www.gov.cn/ldhd/2012-11/17/content_2268826.htm (accessed on 29 September 2022).
13. Li, G.; Fang, C. Quantitative function identification and analysis of urban ecological-production-living spaces. *Acta Geogr. Sin.* **2016**, *71*, 49–65.
14. Du, J.; Fu, J.; Hao, M. Analyzing the carbon metabolism of "Production-Living-Ecological" space based on ecological network utility in Zhaotong. *J. Nat. Resour.* **2021**, *36*, 1208–1223. [CrossRef]
15. Eduardo, B.S.L.; Luís, F.M.N.; José, A.P.O. Carbon accounting approaches and reporting gaps in urban emissions: An analysis of the Greenhouse Gas inventories and climate action plans in Brazilian cities. *J. Nat. Resour.* **2020**, *245*, 118930. [CrossRef]
16. Mark, S. Updated potential soil carbon sequestration rates on U.S. agricultural land based on the 2019 IPCC guidelines. *Soil Tillage Res.* **2020**, *204*, 104719. [CrossRef]
17. Liang, X.; Fan, M.; Xiao, Y.; Yao, J. Temporal-spatial characteristics of energy-based carbon dioxide emissions and driving factors during 2004–2019, China. *Energy* **2022**, *261*, 124965. [CrossRef]
18. Maurizio, C.; Maria, A.C.; Sonia, L. Energy-related GHG emissions balances: IPCC versus LCA. *Sci. Total Environ.* **2018**, *628*, 1328–1339.
19. Ma, Y.; Shang, Y.; Fen, Y. Analysis on influence of spatial factors on carbon emissions of communities in Northern China: Taking Yantai as an example. *Urban Dev. Stud.* **2022**, *29*, 118–124.
20. Kong, L.; Shi, Z.; Chu, L.M. Carbon emission and sequestration of urban turfgrass systems in Hong Kong. *Sci. Total Environ.* **2014**, *473*, 132–138. [CrossRef]

21. Francisco, E.; Sebastian, V.; Wayne, Z. Analyzing the efficacy of subtropical urban forests in offsetting carbon emissions from cities. *Environ. Sci. Policy* **2010**, *13*, 362–372.
22. Li, J.; Huang, G.; Liu, L. Ecological network analysis for urban metabolism and carbon emissions based on input-output tables: A case study of Guangdong province. *Ecol. Model.* **2018**, *383*, 118–126. [CrossRef]
23. Breetz, H.L. Regulating carbon emissions from indirect land use change (ILUC): U.S. and California case studies. *Environ. Sci. Policy* **2017**, *77*, 25–31. [CrossRef]
24. Xia, C. Multi-Scale Studies on Urban Carbon Metabolism from the Perspective of Land Use and Scenario Analysis of Emission Reduction. Ph.D. Thesis, Zhejiang University, Hangzhou, China, 2019.
25. Xia, L.; Zhang, Y.; Yu, X.; Li, Y.; Zhang, X. Hierarchical structure analysis of urban carbon metabolism: A case study of Beijing, China. *Ecol. Indic.* **2019**, *107*, 105602. [CrossRef]
26. Zhang, Y.; Xia, L.; Fath, B.D.; Yang, Z.; Yin, X.; Su, M.; Liu, G.; Li, Y. Development of a spatially explicit network model of urban metabolism and analysis of the distribution of ecological relationships: Case study of Beijing, China. *J. Clean. Prod.* **2016**, *112*, 4304–4317. [CrossRef]
27. Zou, K.; Shu, Y.; Li, G.; Yan, Q.; Bao, Y.; Zhang, S.; Zhang, D. Urban Land-carbon Framework Construction Based on Ecological Network Analysis and Its Space-time Evolution Research. *J. Ecol. Rural Environ.* **2022**, *38*, 972–982.
28. Lin, G.; Jiang, D.; Fu, J.; Cao, C.; Li, X.; Li, Z.; Huang, S. Carbon flow analysis for production-living-ecological space: A case study of Tangshan, China. *Sci. Technol. Rev.* **2020**, *38*, 107–114.
29. Zhang, H.; Xu, E.; Zhu, H. An ecological-living-industrial land classification system and its spatial distribution in China. *Resour. Sci.* **2015**, *37*, 1332–1338.
30. Kong, D.; Chen, H.; Wu, K. The evolution of "Production-Living-Ecological" space, eco-environmental effects and its influencing factors in China. *J. Nat. Resour.* **2021**, *36*, 1116–1135. [CrossRef]
31. Hu, W.; Wang, L.; Shu, M. Reflections on delimiting the three basic spaces in the compilation of urban and rural plans. *City Plan. Rev.* **2016**, *40*, 21–26.
32. Jiang, D.; Lin, G.; Fu, J. Discussion on scientific foundation and approach for the overall optimization of "Production-Living-Ecological" space. *J. Nat. Resour.* **2021**, *36*, 1085–1101. [CrossRef]
33. Tian, Y.; Zhang, J. Regional differentiation research on net carbon effect of agricultural production in China. *J. Nat. Resour.* **2013**, *28*, 1298–1309.
34. Wang, S.; Tian, S.; Cai, Q.; Wu, Q. Driving factors and carbon transfer of industrial carbon emissions in Guangdong province under the background of industrial transfer. *Geogr. Res.* **2021**, *40*, 2606–2622.
35. Fang, J.; Guo, Z.; Pu, S.; Chen, A. Estimation of carbon sinks in China's terrestrial vegetation from 1981 to 2000. *Sci. China Ser. D* **2007**, *37*, 804–812.
36. Lai, L. Research on Carbon Emission Effects of Land Use in China. Ph.D. Thesis, Nanjing University, Nanjing, China, 2010.
37. Duan, X.; Wang, X.; Lu, F.; Ouyang, Z. Carbon sequestration and its potential by wetland ecosystems in China. *Acta Ecol. Sin.* **2008**, *28*, 463–469.
38. Wei, J.; Xia, L.; Chen, L.; Zhang, Y.; Yang, Z. A network-based framework for characterizing urban carbon metabolism associated with land use changes: A case of Beijing city, China. *J. Nat. Resour.* **2022**, *371*, 133695. [CrossRef]
39. Xia, C.; Li, Y.; Ye, Y.; Shi, Z.; Liu, J.; Li, X. Analyzing urban carbon metabolism based on ecological network utility: A case study of Hangzhou City. *Acta Ecol. Sin.* **2018**, *38*, 73–85.

MDPI
St. Alban-Anlage 66
4052 Basel
Switzerland
Tel. +41 61 683 77 34
Fax +41 61 302 89 18
www.mdpi.com

Sustainability Editorial Office
E-mail: sustainability@mdpi.com
www.mdpi.com/journal/sustainability

www.ingramcontent.com/pod-product-compliance
Lightning Source LLC
LaVergne TN
LVHW070202170726
843515LV00007B/1980

* 9 7 8 3 0 3 6 5 8 0 4 7 0 *